The Cambridge Encyclopedia of the Sun

The Sun is our powerhouse, sustaining life on Earth. It energizes our planet and fuels the engine of life. Its warmth drives our weather, lifting water from the seas and producing winds that drive clouds over the continents.

The Cambridge Encyclopedia of the Sun is a fundamental, up-to-date reference source of information about the Sun, from basic material to detailed concepts. It is written in a concise, light and uniform style, without being unnecessarily weighted down with specialized materials or the variable writing of multiple authors. It is filled with vital facts and information for astronomers of all types and for anyone with a scientific interest in the Sun. The language, styles, ideas and profuse illustrations will attract the general reader as well as professionals. Equations are kept to a minimum, and when employed are placed within set-aside focus elements. These focus boxes enhance and amplify the discussions with interesting details, fundamental physics and important related topics. They will be read by the especially curious person or serious student, but do not interfere with the general flow of the text and can be bypassed by the general educated reader who wants to follow the main ideas. Numerous tables of fundamental data also complement the text, including dates of total solar eclipses, basic physical data for the Sun, and a listing of major solar observatories. A unique and comprehensive Glossary defines all terms and acronyms that deal with the Sun, its interaction with the Earth, and the telescopes, satellites and instruments that observe it.

The many full-color figures and photographs throughout the book help to make all the information highly accessible. The text is organized in a thematic way, with chapters on the properties of the Sun as a star and its place in the Galaxy and the Universe. There are chapters that deal with the science of the Sun's interior and its visible disk, and what makes it shine. The work also covers solar flares and the solar wind, and their impact on the Earth.

This important, one-stop, single-author encyclopedia will be a significant reference for anyone needing information on any aspect of solar astronomy, solar astrophysics, or solar–terrestrial relations. It will become the definitive reference work on our home star for professionals, graduate students, advanced undergraduates, keen amateur astronomers and the interested general reader.

KENNETH R. LANG, Professor of Astronomy at Tufts University, uses the Very Large Array radio telescope and modern spacecraft, such as SOHO and Yohkoh, to investigate activity on the Sun. He is also a well-known writer, having published nine books translated into ten languages as well as more than 150 professional articles. Lang's books include Astrophysical Formulae, published in a third enlarged edition in 1999, the popular Sun, Earth and Sky, and the prize-winning Wanderers in Space – Prix du livre de l'Astronomie in 1994. Professor Lang has been a Visiting Senior Scientist in Solar Physics at NASA Headquarters and a Fulbright Scholar in Italy. He is a member of the International Astronomical Union, the Royal Astronomical Society and the American Astronomical Society. Professor Lang lives with his wife, Marcella, and his three children, Julia, David and Marina, in Arlington, Massachusetts.

THE CAMBRIDGE
ENCYCLOPEDIA OF THE
SUN

Kenneth R. Lang

Tufts University, Medford, Massachusetts, USA

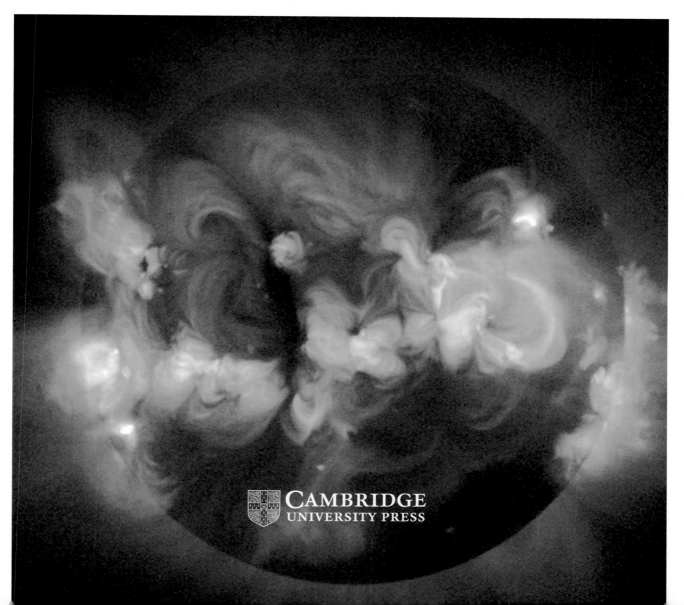

CAMBRIDGE
UNIVERSITY PRESS

PUBLISHED BY THE PRESS SYNDICATE OF THE UNIVERSITY OF CAMBRIDGE
The Pitt Building, Trumpington Street, Cambridge, United Kingdom

CAMBRIDGE UNIVERSITY PRESS
The Edinburgh Building, Cambridge CB2 2RU, UK
40 West 20th Street, New York, NY 10011-4211
10 Stamford Road, Oakleigh, VIC 3166, Australia
Ruiz de Alarcón 13, 28014 Madrid, Spain
Dock House, The Waterfront, Cape Town 8001, South Africa

http://www.cambridge.org

© Cambridge University Press 2001

First published 2001

Printed in the United Kingdom at the University Press, Cambridge

Typeface Joanna 10.25/12.5pt.   System QuarkXpress

A catalogue record for this book is available from the British Library

Library of Congress cataloguing in Publication data

Lang, Kenneth R.
The Cambridge Encyclopedia of the Sun/Kenneth R. Lang
    p.   cm.
Includes bibliographical references and index.
ISBN 0 521 78093 4
1. Sun–Encyclopedias. 1. Title.
QB521.L24 2001
523.7´03–dc21 00-049365

ISBN 0 521 78093 4 hardback

Contents

Preface

The Cambridge Encyclopedia of the Sun is a fundamental, up-to-date reference source for information about the Sun, from basic data to detailed concepts. It is written in a concise, light and uniform style, without being unnecessarily weighted down with narrow, specialized knowledge or the variable writing of multiple authors. The Encyclopedia is filled with vital facts and information for astronomers of all types and for anyone with a scientific interest in the Sun. The language, style, ideas and profuse illustrations will appeal to the general reader as well as professionals.

Equations are kept to a minimum, and when employed are placed within set-aside focus elements. These focus boxes enhance and amplify the discussion with interesting details, fundamental physics and important related topics. They will be read by the especially curious person or serious student, but do not interfere with the general flow of the text and can be bypassed by the general educated reader who wants to follow the main ideas. Numerous tables of fundamental data also complement the text, including dates of total solar eclipses, basic physical data for the Sun, and major solar observatories. A unique and comprehensive Glossary defines all terms and acronyms that deal with the Sun, its interaction with the Earth, and the telescopes, satellites and instruments that observe it.

The Encyclopedia is organized on thematic principles using chapters. After two introductory chapters that describe the Sun's physical characteristics and deal with the Sun as a star within our Galaxy and Universe, the chapters progress from the center of the Sun through its atmosphere to the Earth, and then to the telescopes and other instruments that are used to observe the Sun from the ground and space.

The first chapter deals with fundamental physical properties of the Sun and its retinue of planets, together with basic scientific principles concerning radiation, gravity, atomic structure and spectroscopy, and the age and origin of the solar system.

Chapter 2 places the Sun within the context of the other stars, the galaxies and the expanding Universe. It also contains basic scientific principles, including the distance, mass, luminosity and evolution of stars, as well as the origin and fate of the Universe.

The third chapter describes how the Sun shines by thermonuclear reactions within its hot, dense core, and provides an up-to-date treatment of the solar neutrino problem. The fusion reactions that generate the Sun's energy also create countless neutrinos that pass effortlessly through both the Sun and Earth. Massive subterranean neutrino detectors catch just a few of the trillions of ghostly neutrinos that move right through these detectors every second. Their neutrino count always comes up short, detecting only one-third to one-half of the number of neutrinos predicted by theory. Recent results suggest that the

neutrinos have an identity crisis, transforming into a currently undetectable form on their way out of the Sun. Ongoing neutrino observations may show that this metamorphosis is happening, which would solve the solar neutrino problem.

In less than a decade, three modern spacecraft, Yohkoh, Ulysses and the SOlar and Heliospheric Observatory, abbreviated SOHO, have provided more important new information about the Sun than perhaps the entire century of previous observations of the star. Instruments aboard these satellites have extended our gaze from the visible solar disk to down within the hidden solar interior and out in all directions through the Sun's tenuous atmosphere and solar wind. Their discoveries are highlighted in Chapters 4 through 7, and placed there within the context of supporting ground-based observations at visible and radio wavelengths. These chapters continue our thematic treatment, considering the solar interior, the Sun's magnetic atmosphere, solar explosions and the Sun's winds in separate chapters.

The Cambridge Encyclopedia of the Sun does not record the historical development of our understanding of the Sun, since the book's purpose is to present our current understanding as it actually is, rather than describing how that position has been reached. A detailed historical treatment of the topics given in these middle chapters of our Encyclopedia can be found in my book The Sun From Space, published by Springer-Verlag in 2000; it contains hundreds of references to the professional literature and detailed time lines of historical matters. Nevertheless, key discoveries are woven into the text of The Cambridge Encyclopedia of the Sun, with dates and important persons.

Chapter 4 discusses the internal structure of the Sun, and the way in which widespread throbbing motions are used to "see" inside it. The oscillations detected at the visible solar disk are caused by sound waves that can be used to measure the temperature, rotation and other gaseous flows within the Sun. They have taken the Sun's temperature all the way down to its energy-generating core, showing that the temperature agrees with model predictions and apparently ruling out any astrophysical solution to the solar neutrino problem.

Regions near the Sun's poles rotate at exceptionally slow speeds, while the equatorial regions spin rapidly. This differential rotation persists to about a third of the way inside the Sun, where the rotation becomes uniform from pole to pole. The Sun's magnetism is probably generated by dynamo action at the interface between the deep interior, that rotates at one speed, and the overlying gas that spins faster in the equatorial middle. Recent studies show that the rotation rates near this boundary vary periodically over the years, and that the periodic variations at different depths are out of phase with each other. They provide important clues to how the solar dynamo works.

The Sun's internal gases circulate with its rotation, but in deeply rooted zones with relatively fast or slow speed. They migrate toward the solar equator during the ending years of the Sun's 11-year activity cycle. Other internal currents of gas flow between the equator and poles.

We discover in Chapter 5 that the sharp visible edge of the Sun is an illusion. A tenuous, hot, million-degree gas, called the corona, envelops the Sun and overlies the cooler, visible solar disk. This tenuous atmosphere is molded and constrained by magnetic fields generated down inside the Sun and looping through sunspots in the photosphere into the corona. The numbers and positions of these sunspots, and the intense magnetic fields that they contain, vary over an 11-year cycle of solar magnetic activity.

The hottest, densest material in the low corona is concentrated within thin, long magnetized loops that are in a state of continual agitation. The closed magnetic loops are often coming together, releasing magnetic energy when they make contact. This provides a plausible explanation for heating the million-degree corona. Dark coronal holes contain relatively little hot coronal material; it is flowing out along the open magnetic fields in these regions.

In Chapter 6 we describe sudden, brief, and intense outbursts, called solar flares, which release magnetic energy equivalent to billions of terrestrial nuclear bombs. The Sun's flares are easily detected at invisible X-ray and radio wavelengths, and are synchronized with the sunspot cycle of magnetic activity. The most energetic flaring X-ray radiation is emitted by beams of electrons accelerated just above coronal loops and beamed down into the Sun. When they strike the dense lower atmosphere, flare-accelerated protons generate nuclear reactions that emit gamma rays. The radio emission during solar flares provides evidence for high-speed electrons hurled down into the Sun, as well as for electron beams and shock waves that are ejected out into space.

Instruments aboard solar satellites routinely detect magnetic bubbles of awesome proportion, called Coronal Mass Ejections, and abbreviated CMEs. The CMEs expand as they propagate outward from the Sun to rapidly rival it in size, carrying up to ten billion tons, ten million million kilograms, of coronal material into interplanetary space. Their associated shock waves generate intense magnetic fields in interplanetary space, and accelerate and propel vast quantities of high-speed particles ahead of them.

The sixth chapter also describes how explosive solar activity can occur when magnetic fields come together and reconnect in the low solar corona. Stored magnetic energy is released rapidly at the place where the magnetic fields touch.

Chapter 7 provides an up-to-date account of the Sun's ceaseless winds that expand out in all directions, filling the solar system with an endless flow of electrons, protons and other ions, and magnetic fields. There are two kinds of solar wind, a fast one moving at about 800 thousand meters per second, and a slow one with about half that speed. Much of the steady, high-speed wind is coming out of honeycomb-shaped magnetic fields at the base of polar coronal holes. Oxygen ions move faster than protons in the polar coronal holes, apparently absorbing more power from magnetic waves that preferentially accelerate the heavier ions. A capricious, gusty, slow wind emanates from the Sun's equatorial regions near the minimum in the 11-year solar activity cycle. The high-speed wind is accelerated very close to the Sun, within just a few solar radii, and the slow component obtains full speed further out.

In the eighth chapter, our account turns toward our home planet, Earth, where the Sun's light and heat permit life to flourish. The Earth's magnetic field shields us from the eternal solar gale, but gusts in the wind buffet our magnetic domain and sometimes penetrate within it. Forceful CMEs can create intense magnetic storms on Earth, trigger intense auroras in the skies, damage or destroy Earth-orbiting satellites, and induce destructive power surges in long-distance transmission lines on Earth. Energetic charged particles, hurled out during solar flares, endanger astronauts and can destroy satellites. Intense radiation from these flares, which occur more frequently during the maximum period of solar activity, increases the ionization of our atmosphere, disrupting long-distance radio communications and disturbing satellite orbits.

The eighth chapter also examines the Sun's role in warming and cooling the Earth. Varying magnetic activity changes the total solar brightness over the 11-year solar activity cycle. The Earth's surface temperature varies in step with the length of this cycle, and the amount of solar activity. This could be due to changing cloud cover related to the Sun's modulation of the amount of cosmic rays reaching Earth. Radioactive isotopes found in tree rings and ice cores indicate that the Sun's activity has fallen to unusually low levels at least three times during the past one thousand years, causing long cold spells lasting about a century. Further back in time, during the past one million years, our climate has been dominated by the recurrent ice ages, each about 100 thousand years in duration. They are explained by three overlapping astronomical cycles, which combine to alter the distribution and amount of sunlight falling on Earth.

The concluding chapter, Chapter 9, describes the tools of solar astronomy. They include ground-based optical telescopes that study visible sunlight, and gigantic arrays of radio telescopes that also observe the Sun from the ground. We next provide an account of satellites that have been specifically designed to study the Sun, together with their component telescopes or instruments and the things that they observe and measure. Spacecraft which are expected to study the Sun in the future are included, and updated to the time of book publication.

Finally, we include an annotated list of books for further reading, all published after 1989, and a list of Internet addresses for the topics discussed in the *Encyclopedia*. It ends with the exceptionally comprehensive *Glossary* and thorough indexes.

Many of the tables of numerical data in the *Encyclopedia* have been extracted from my technical reference books *Astrophysical Formulae I, II* (Kenneth R. Lang, Third Edition Springer-Verlag, 1999) and *Astrophysical Data: Planets and Stars* (Kenneth R. Lang, Springer-Verlag, 1991).

The illustrator Sue Lee has combined artistic talent with a scientist's eye for detail in producing some fantastic line drawings for this *Encyclopedia*. The text has been substantially improved by the careful attention of copy-editor Brian Watts.

Locating quality figures is perhaps the most time-consuming and frustrating aspect of producing a volume like this, so I am especially thankful for the support of the European Space Agency (ESA), the Institute of Space and Astronautical Science, Japan (ISAS) and the National Aeronautics and Space Administration (NASA) for providing them. Individuals that were especially helpful in locating and providing specific images include Loren W. Acton, David Alexander, Cary Anderson, Frances Bagenel, Richard C. Canfield, Michael Changery, David Chenette, Fred Espenak, Bernhard Fleck, Eigil Friis-Christensen, Claus Fröhlich, Bruce Goldstein, Leon Golub, Steele Hill, Gordon Holman, Beth Jacob, Imelda Joson, Therese Kucera, Judith Lean, William C. Livingston, Michael E. Mann, Richard Marsden, Michael J. Reiner, Thomas Rimmele, Kazunari Shibata, Gregory Lee Slater, Barbara Thompson, Haimin Wang, and Joachim Woch.

Kenneth R. Lang
Tufts University
1 January 2001

Principal units, solar quantities, and fundamental constants

The International System of Units (Systéme International, SI) is used for most quantities, but the reader should be warned that centimeter–gram–second (c.g.s.) units are employed in nearly all of the seminal papers referenced in this Encyclopedia. Moreover, c.g.s. units are still extensively used by many solar astronomers. Table P.1 provides unit abbreviations and conversions between units. Some other common units are the nanometer (nm) with 1 nm $= 10^{-9}$ meters, the angstrom unit of wavelength, where 1 angstrom $= 1$ Å $= 10^{-10}$ meters $= 10^{-8}$ centimeters, the nanotesla (nT) unit of magnetic flux density, where 1 nT $= 10^{-9}$ tesla $= 10^{-5}$ gauss, the electron volt (eV) unit of energy, with 1 eV $= 1.6 \times 10^{-19}$ joule, and the ton measurement of mass, where 1 ton $= 10^{3}$ kilograms $= 10^{6}$ grams. The accompanying Table P.2 provides numerical values for solar quantities and fundamental constants.

Quantity	SI unit	Conversion to c.g.s. unit
Length	meter (m)	100 centimeters (cm)
	nanometer (nm) $= 10^{-9}$ m	10^{-7} cm $= 10$ ångstroms $= 10$ Å
Mass	kilogram (kg)	1000 grams (g)
Time	second (s)	
Temperature	kelvin (K)	
Velocity	meter per second (m s^{-1})	100 centimeters per second (cm s^{-1})
Energy^a	joule (J)	10 000 000 erg ($= 10^{7}$ erg)
Frequency	hertz (Hz) $=$ cycle per second (c s^{-1})	
Power	watt (W) $=$ joule per second (J s^{-1})	10 000 000 erg s^{-1} ($= 10^{7}$ erg s^{-1})
Magnetic Flux Density	tesla (T)	10 000 gauss (G) ($= 10^{4}$ G)
Force	newton (N) ($=$ kg m s^{-2})	100 000 dyn ($= 10^{5}$ dyn)
Pressure	pascal (Pa) ($=$ N m^{-2}) ($=$ kg m^{-1} s^{-2})	10 dyn cm^{-2}
Electric charge	coulomb (C)	
Electric current	ampere (A)	
Electric capicitance	farad (F)	
Electric conductance	siemens (S)	

^{a}The energy of high-energy particles and X-ray radiation are often expressed in units of kilo-electron volts, or keV, where 1 keV $= 1.602 \times 10^{-16}$ joule, or MeV, where 1 MeV $= 1000$ keV.

Table P.1 **Principal SI units and their conversion to corresponding c.g.s. units**

Symbol	Name	Value
$L_{\odot}$	Luminosity of Sun	3.854×10^{26} J s^{-1}
$M_{\odot}$	Mass of Sun	1.989×10^{30} kg
$R_{\odot}$	Radius of Sun	6.955×10^{8} m
AU	Mean distance of Sun	1.496×10^{11} m
$f_{\odot}$	Solar constant	1366 W m^{-2}
$T_{e\odot}$	Effective temperature of photosphere	5780 K
c	Velocity of light	2.9979×10^{8} m s^{-1}
G	Gravitational constant	6.6726×10^{-11} N m^2 kg^{-2}
k	Boltzmann's constant	1.38066×10^{-23} J K^{-1}
h	Planck's constant	6.6261×10^{-34} J s
a	Radiation density constant	7.5659×10^{-16} J m^{-3} K^{-4}
m_e	Mass of electron	9.1094×10^{-31} kg
e	Charge of electron	1.6022×10^{-19} C
m_H	Mass of hydrogen atom	1.673534×10^{-27} kg
m_p	Mass of proton	1.672623×10^{-27} kg
ε_0	Permittivity of free space	$10^{-9}/(36\pi) = 8.854 \times 10^{-12}$ F m^{-1}
μ_0	Permeability of free space	$4\pi \times 10^{-7} = 12.566 \times 10^{-7}$ N A^{-2}

^{a}Adapted from Lang, K.R. *Astrophysical Data: Planets and Stars* (Springer–Verlag, 1991). One joule per second $= 1$ J s^{-1} is equal to one watt $= 1$ W. The unit symbols are J for joule, s for second, kg for kilogram, m for meter, K for degree kelvin, C for coulomb, N for newton, F for farad and A for ampere.

Table P.2 **Solar quantities and fundamental constants**^{a}

CHAPTER ONE

The Sun's domain

Sunflowers The Sun sustains all living creatures and plants on Earth. Here the eyes of sunflowers turn in unison to follow their life-sustaining star. (Courtesy of Charles E. Rodgers.)

1.1 Fire of life

The Sun is our powerhouse, sustaining life on Earth. It energizes our planet and fuels the engine of life. The Sun warms our world, keeping the temperature at a level that allows liquid water to exist and keeps the Earth teeming with life. Without the Sun's light and heat, all life would quickly vanish from the surface of our planet.

The Sun's warmth drives our weather, lifting water from the seas and producing winds that drive clouds over the continents. The water returns to the ground as rain and snow, flowing back to the oceans through the ground and in rivers. They make Earth fit for life.

The Sun is not only warmth, it is light, too. The miracle of plants is their ability to use sunlight to make them grow (Fig. 1.1), and in doing so they create another miracle. Using photosynthesis, plants use the Sun's energy to convert water and carbon dioxide into carbohydrates, which releases the oxygen animals breath. Animals also eat these plants for nourishment. The warmth in every animal's body was once sunlight. We all owe our lives to this savage Sun.

The Sun provides – directly or indirectly – most of the energy on the Earth that people use for fuel. Ancient plants used sunlight to grow, and animals consumed these plants. Both plants and animals stored the energy of sunlight in the organic material that composed them. When the plants and animals died and decayed, this organic material was buried, compressed and gradually turned into petroleum, coal and natural gas. Today we harness that solar energy with machines that burn these fossil fuels, using up reserves that have been stored over the last billion (one thousand million) years. They energize the lights in our homes and power the cars that we drive.

Our planet glides through space at just the right distance from the Sun for life to thrive on its energy, while other planets either freeze or fry. We sit in the comfort zone. Any closer and the oceans would boil away. Further out the Earth would be in a frozen wasteland. We receive just enough energy from the Sun to keep most of our water liquid. In comparison, the surface of Venus, the planet next-closest to the Sun, is hot enough to melt lead; further away from the Sun, the planet Mars is now frozen into a global ice age. Turn off the Sun's powerhouse, and in just a few months we could all be under ice.

We are more closely linked with the life-sustaining Sun than with any other celestial object. People seem to have realized the connection from earliest times, when the Sun was revered and held in awe. Even today, the voices of life celebrate sunrise, from a bird's melody to a rooster's crow. Light is associated with good, beauty, truth and wisdom, in sharp contrast with the dark forces of evil. Electric sunlight illuminates our nights, and fire symbolically lights the darkness in many of our rituals. In everyday life, most of us feel happier on bright days than on gloomy ones, so cheerful people have a "sunny" disposition while an unhappy day is a "dark" one. So, it is the awesome Sun that warms our soul, and draws astronomers out away from the Earth.

Fig. 1.1 **Sunflowers** The Sun sustains all living creatures and plants on Earth. Sunlight is absorbed by green plants where it strikes chlorophyll, giving it the energy to break water molecules apart and energize photosynthesis. Plants thereby use the Sun's energy to live and grow, giving off oxygen as a byproduct, and animals are nourished by eating these plants. (Courtesy of Charles E. Rodgers.)

1.2 Radiation from the Sun

Describing sunlight

To understand how the Sun operates we must examine its radiation. It spreads out and carries energy in all directions. Some of the sunlight is intercepted by astronomers at Earth, who use it to deduce physical properties of the Sun.

When radiation moves in space from one place to another, it behaves like trains of waves. It is called "electromagnetic" radiation because it propagates by the interplay of oscillating electrical and magnetic fields. Electromagnetic waves all travel through empty space at the same constant speed – the velocity of light. This velocity is usually denoted by the lower-case letter c, and it has a value of roughly 300 000 000, or 3×10^8, m s^{-1}. A more exact value is $2.997\ 924\ 58 \times 10^8$, m s^{-1}. No energy can be transported more swiftly than the speed of light.

The fact that light travels through space at a constant speed provides us with a convenient way of describing distance. We can specify the distance of an astronomical object in terms of the time it takes light to travel from the object to the Earth. This is called the light travel-time. Light from the Moon takes 1.5 seconds to reach the Earth, so we say the Moon is 1.5 light-seconds away from the Earth. Sunlight takes about eight minutes to cover the average distance between the Sun and the Earth. If the Sun suddenly switched off one day at 12 noon, we would notice it eight minutes past noon. The next nearest star is 4.29 light-years away. The most distant objects in the Universe are billions of light-years away, and the light we detect from them was generated billions of years ago.

Different types of electromagnetic radiation differ in their wavelength, although they propagate at the same speed. Like waves on water, the electromagnetic waves have crests and troughs. The wavelength is the distance between successive crests or successive troughs (Fig. 1.2). It is usually denoted by the lower-case Greek letter lambda, λ, and measured in meters.

Sometimes radiation is described by its frequency, denoted by the lower-case Greek letter nu or ν. Radio stations are, for example, denoted by their call letters and the frequency of their broadcasts. The frequency of a wave is the number of wave crests passing a stationary observer each second, measured in hertz, abbreviated Hz. One hertz is equivalent to one cycle per second. The frequency tells us how fast the radiation oscillates, or moves up and down. The product of wavelength and frequency equals the velocity of light, so $\lambda \times \nu = c$. When the wavelength increases, the frequency decreases, and vice versa.

Our eyes detect a narrow range of wavelengths that are collectively called white light. This band of sunlight is also termed visible radiation, to distinguish it from the invisible rays that cannot be seen with the eye. White light is all the wavelengths of visible radiation together.

The lenses and mirrors that are used to focus and collect visible radiation are described by the science of optics, so the study of visible light from objects in space is called optical astronomy. The optical telescopes and observatories that are now used to scrutinize the Sun are described in Section 9.1.

Although sunlight appears yellowish, it is actually a combination of colors that we designate white light. In the mid-17th century the English scientist Isaac Newton discovered that sunlight could be broken into its spectral colors, using a prism – a specially cut chunk of glass – to display the range of colors. White sunlight is similarly bent into its separate wavelengths when raindrops act like tiny prisms to give us a rainbow or a crystal or a single drop of dew resonate with color (Fig. 1.3). Our eyes and brain translate these wavelengths into colors.

This range of wavelengths of sunlight and the intensity at each wavelength is called the Sun's spectrum. The study of the spectra of the Sun and other objects or materials is called spectroscopy.

From long to short waves, the colors in the spectrum of sunlight correspond to red, orange, yellow, green, blue and violet. Red light has a wavelength of about 7×10^{-7} m, or 700 nm, and violet waves are about 400 nm long. A nanometer, abbreviated nm, is equal to a billionth of a meter, or 1 nm = 10^{-9} m.

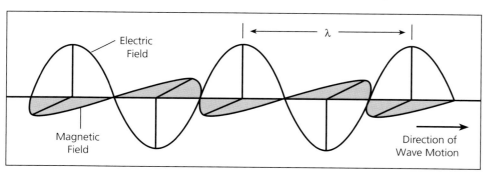

Fig. 1.2 **Electromagnetic waves**
All forms of radiation consist of electric and magnetic fields that oscillate at right angles to each other and to the direction of travel. They move through empty space at the velocity of light. The separation between adjacent wave crests is called the wavelength of the radiation and is usually designated by the lower case Greek letter lambda or λ.

Fig. 1.3 **Light painting** This picture was made by using crystals to liberate the spectral colors in visible sunlight, refracting them directly onto a photographic plate. It was obtained in the rarefied atmosphere atop Hawaii's Mauna Kea volcano, where many of the world's best optical telescopes are located. (Courtesy of Eric J. Pittman, Victoria, British Colombia.)

Why does the human eye respond just to visible light? The most intense radiation of the Sun is emitted at these wavelengths, and our atmosphere permits it to reach the ground. If our eyes were not sensitive to visible sunlight, we could not identify objects or move around on the Earth's surface. The sensitivity of our eyes is matched to the tasks of vision.

When light is absorbed or emitted by atoms, it behaves not as a wave but as a package of energy, or a particle. These packages are given the name photons. They are created whenever a material object emits electromagnetic radiation, and they are consumed when radiation is absorbed by matter. Moreover, each elemental atom can only absorb and radiate a very specific set of photon energies.

The ability of radiation to interact with matter is determined by the energy of its photons. This photon energy depends on the wavelength or frequency of the radiation. Light waves with shorter wavelengths, or higher frequencies, correspond to photons with higher energy (Focus 1.1).

Like many hot, gaseous cosmic objects, the Sun behaves like an ideal thermal radiator called a black body. Thermal radiation arises by virtue of an object's heat, and it is characterized by a single temperature. A black body absorbs all the radiation incident on it and reflects none – hence the term "black".

When the emission from a thermal radiator is arranged in order of wavelength, radiation is found at all wavelengths but with a varying intensity (Fig. 1.4). Astronomers call this a continuum spectrum since the different colors or wavelengths run together with no breaks or gaps. The wavelength, $\lambda_{\max}$, at which the thermal radiation is at a maximum is inversely proportional to temperature and is given by the Wien displacement law (see Focus 1.1). It indicates that colder objects radiate most of their energy at longer wavelengths, and hotter objects are most luminous at shorter wavelengths. In other words, as the temperature of a gas increases, most of its thermal radiation is emitted at shorter and shorter wavelengths.

Fig. 1.4 **Spectrum of thermal radiation** When the intensity of thermal radiation is spread out as a function of wavelength, the resultant spectrum is most intense in a band of wavelengths that depends on the temperature (see Fig. 1.7). This occurs at visible wavelengths when the temperature is about 6000 K. Such a hot gas emits radiation at longer wavelengths, but at a lower intensity. In contrast, non-thermal radiation is more intense at longer radio wavelengths. High-speed electrons emit non-thermal radiation in the presence of a magnetic field.

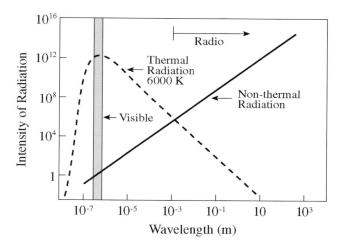

Focus 1.1 The energy of light

Electromagnetic radiation can be described by packets of energy called photons. The energy, E, transported by a particular photon is directly proportional to its frequency, ν, and inversely proportional to its wavelength, λ. The relationship is given by:

$$\text{photon energy} = E = h\nu = \frac{hc}{\lambda} = \frac{1.986 \times 10^{-25}}{\lambda} \text{ J}$$

where h is a universal constant of nature called Planck's constant with the value $h = 6.6261 \times 10^{-34}$ joule second, the frequency is given in Hz or cycles per second, and the wavelength is in meters.

From this expression we see that photons of radiation at shorter wavelengths have greater energy. This is the reason that shorter-wavelength, energetic X-rays can penetrate inside your body, while longer-wavelength, less-energetic visible light just warms your face.

The product of wavelength and frequency is equal to the velocity of light, or

$$\text{velocity of light} = c = \lambda \times \nu = 2.997\,924\,58 \times 10^{8} \text{ m s}^{-1}.$$

The amount of energy transported by a single photon is quite small. For yellow light, the wavelength $\lambda = 550$ nm, so the frequency $\nu = 5.45 \times 10^{14}$ Hz, and the photon energy, E, is only 3.61×10^{-19} J. A hundred-watt light bulb radiates 100 joules per second, so it sends out an incredible 277 billion billion, or 2.77×10^{20}, photons every second.

Astronomers very often describe energetic, short-wavelength radiation, such as X-rays or gamma rays, in terms of energy rather than its wavelength or frequency. At the atomic level, the natural unit of its energy is the electron volt, or eV. One electron volt is equivalent to 1.6022×10^{-19} J. It is the energy an electron gains when it passes across the terminals of a 1-volt battery. A photon of visible light has an energy of about two electron volts, or 2 eV. Much higher energies are associated with nuclear processes; they are often specified in units of millions of electron volts, or MeV. A somewhat lower unit of energy is a thousand electron volts, called kilo-electron volts, or keV; it is often used to describe X-ray radiation.

A hot gas will emit radiation at all wavelengths, but the peak intensity of that radiation is emitted at a maximum wavelength, $\lambda_{\max}$, that varies inversely with the temperature. This can be understood by equating the thermal energy of the gas, which increases with its temperature, T, to the photon energy of its radiation, or by

$$\text{gas thermal energy} = kT = \frac{hc}{\lambda_{\max}} = \text{photon energy},$$

where Boltzmann's constant $k = 1.38066 \times 10^{-23}$ J K^{-1}. Collecting terms and inserting the values for the constants, this expression gives

$$\text{maximum wavelength} = \lambda_{\max} = \frac{hc}{kT} \approx \frac{0.01}{T} \text{ m}$$

for a temperature, T, in degrees kelvin. The exact relationship, known as the Wien displacement law, is given by $\lambda_{\max} = 0.002\,897\,75 / T$ meters.

The Wien displacement law explains why stars have different colors. Red stars are relatively cool, near 3000 K; yellow ones like the Sun are closer to 6000 K; white are about 10 000 K; and blue stars are hotter yet, at 20 000 K or greater.

Since a thermal radiator emits radiation at all wavelengths, its total energy output is obtained by adding up, or integrating, the contributions at every wavelength. When this is done, the total power, or intrinsic luminosity, L, of a volume of gas with radius, R (in meters), and effective temperature, T_e, is obtained:

$$\text{luminosity} = L = 4\pi\sigma R^2 T_e^4 \text{ J s}^{-1}$$

where the Stefan–Boltzmann constant $\sigma = 5.670 \times 10^{-8}$ J m^{-2} K^{-4} s^{-1}. This is called the Stefan–Boltzmann law. It indicates that at a given effective temperature, bigger, giant stars have a greater luminosity (Section 2.3).

We can determine the overall amount of radiation produced by the Sun from the amount of radiation falling on the Earth and from the distance of the Sun. Satellites have been used to accurately measure the Sun's total irradiance just outside the Earth's atmosphere. They indicate that the solar constant, $f_\odot$, is

$$f_\odot = 1366 \text{ W m}^{-2},$$

with an uncertainty of ± 1.0 W m^{-2}. The solar constant is defined as the total amount of radiant solar energy per unit time per unit area reaching the top of the Earth's atmosphere at the Earth's mean distance from the Sun. Once we know the Sun's distance, we can use the solar constant to infer the Sun's luminosity, and this is done in the Section 1.3.

Fig. 1.5 **Spectral flux** The Sun's spectral flux at the Earth plotted as a function of the wavelength. At each wavelength, the amount of solar energy received at the Earth's surface is less than the amount received outside the Earth's atmosphere. The attenuation is much greater at certain wavelengths than at others owing to absorption in the Earth's atmosphere by molecules of oxygen, O_2, water, H_2O, and carbon dioxide, CO_2. There is a general reduction at shorter wavelengths, less than 320 nm, due to absorption by ozone, O_3, molecules, and to scattering by molecules, aerosols and small particles in the atmosphere. At any given location on the Earth's surface, the solar insolation, or power per square meter, can be obtained by integrating the spectral flux over all wavelengths.

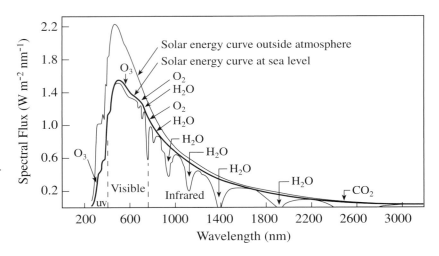

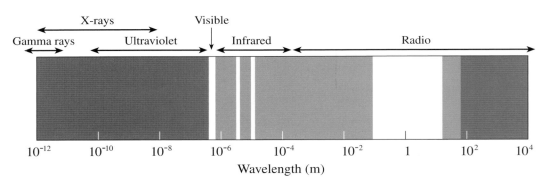

Fig. 1.6 **Electromagnetic spectrum** Radiation from the Sun and other cosmic objects is emitted at wavelengths from less than 10^{-12} m to greater than 10^4 m. The visible spectrum is a very small portion of the entire range of wavelengths. The lighter the shading, the greater the transparency of the Earth's atmosphere. Solar radiation only penetrates to the Earth's surface at visible and radio wavelengths, respectively represented by the narrow and broad white areas. Electromagnetic radiation at short X-ray and ultraviolet wavelengths, represented by the dark areas, is absorbed in our air, so the Sun is now observed in these spectral regions from above the atmosphere in Earth-orbiting satellites.

The total power received at any square meter of the Earth's surface, known as the solar insolation, is much less than the solar constant (Fig. 1.5). This is due to the absorption of sunlight in the terrestrial atmosphere, as well as the time of day. The insolation varies according to the Sun's altitude both because of the varying angle between the normal to the Earth's surface and the Sun's direction and because of the varying amount of the Earth's atmosphere that the sunlight has to shine through. When the Sun is overhead at a location, the atmosphere it shines through is least and there is therefore the lowest amount of attenuation by the atmosphere. That is essentially why the days are hottest near noontime. The total energy received at ground level is then reduced from the solar irradiance value of 1366 W m^{-2} to about 1000 W m^{-2}. Of course, the Sun is not out at night, and the insolation is zero.

The atmospheric attenuation is much greater at certain wavelengths than at others due to absorption by atmospheric molecules of ozone, oxygen, water, and carbon dioxide (Fig. 1.5). The ozone molecules in our atmosphere absorb the short-wavelength ultraviolet rays from the Sun, thereby filtering out energetic radiation that can be harmful to life on Earth.

Invisible radiation from the Sun

There is much more to the Sun than meets the eye! In addition to visible light, there is invisible radiation as well. The solar spectrum extends over an enormous range in wavelength, and different wavelength regions carry different names (Fig. 1.6). The visual or optical region contains the observed colors. The invisible domains include the infrared and radio waves, with wavelengths longer than that of red light, and the ultraviolet (UV), X-rays and gamma (γ) rays that are shorter than violet light. They are all electromagnetic waves and part of the same family, and they all move in empty space at the velocity of light, but we can't see them.

Although the most intense radiation from the Sun is emitted at visible wavelengths, it emits radiation of lower intensity across the full electromagnetic spectrum. Because of its proximity and brilliance, the Sun can be examined with precision in all of these spectral regions, from gamma rays that are smaller than an atom to radio waves that are broader than a mountain.

Astronomers now observe the Sun at most of these invisible wavelengths, broadening and sharpening our vision of it. They are aided by new telescopes, operating from the ground and in space, and by sophisticated computers that guide the telescopes, analyze their data, and create new "invisible" images of the Sun. What the astronomers see on the Sun depends on how they look at it, for each wavelength provides a different picture, somewhat like tuning in a television to different channels.

Radio waves are the only kind of invisible radiation that is not absorbed in the Earth's atmosphere, so radio astronomy provided the first new window on the Sun. The radio region of the electromagnetic spectrum covers a wide range of wavelengths between 0.001 and 1000 meters. Solar radio waves that are longer than about 10 meters are reflected by an ionized layer in the Earth's air, called the ionosphere; so these longer radio waves cannot reach the ground and must be observed from space. Nevertheless, the shorter radio radiation, with wavelengths from 0.001 to 10 meters, provides an important tool for ground-based solar astronomy. The radio telescopes and observatories that are used to study the radio Sun are described in Section 9.2.

The radio waves that reach the ground have a small photon energy compared with the photons of visible light (Focus 1.1). The low energies of the radio photons cannot easily excite the atoms of our atmosphere, so these photons easily pass through it. Radio waves can even pass through rain clouds, so the radio Sun can be observed on cloudy days and in stormy weather, just as your home radio works even when it rains or snows outside. Solar radio waves are too long to enter the eye as well as being too feeble to affect your vision.

The infrared part of the electromagnetic spectrum is located between the radio wave region and visible region (Fig. 1.6). Infrared radiation was discovered in 1840 by the astronomer William Herschel, who was already world-famous for his discovery of the planet Uranus. He put a beam of sunlight through a prism to spread it into its spectral components and he noticed that a thermometer placed beyond the red edge of the visible spectrum of the Sun became warmed by an invisible portion of sunlight. The thermometer placed in this infrared sunlight showed higher temperatures than one placed in normal visible sunlight. The warmth of a heat lamp is similarly provided by its infrared radiation.

Although our eyes are not sensitive to infrared radiation, rattlesnakes have infrared-sensitive eyes that enable them to see the heat radiated by other animals at night. Soldiers can locate the enemy at night by using infrared sensors that detect their heat, and spy satellites use infrared telescopes to detect the heat radiated by rocket exhaust and by large concentrations of troops and vehicles. Warm objects emit infrared radiation, and infrared telescopes see in the dark by detecting heat waves.

Infrared radiation from the Sun is absorbed by atmospheric molecules, such as carbon dioxide and water vapor. So the air that looks so transparent to our eyes is opaque to much of the infrared radiation coming from the Sun or other objects in outer space. The atmosphere similarly blocks the heat radiation from the Earth's surface, keeping it warmer than it would otherwise be (Section 8.5). Telescopes located above part of the atmosphere, on the tops of mountains in dry climates, can catch some of the incoming infrared radiation before it is completely absorbed.

Next in our inventory of the spectrum is visible radiation. This can slip through the Earth's atmosphere with little trouble. Its photons are too energetic to resonate with molecular vibrations and they are too feeble to excite atoms.

Just beyond violet light is the short-wavelength ultraviolet part of the electromagnetic spectrum (Fig. 1.6), where photons are sufficiently energetic to tear electrons or atoms off many of the molecular constituents of the Earth's atmosphere, particularly in the ozone layer. These energetic photons cannot reach the ground, and if they did they would cause considerable damage to our skin and eyes.

The ultraviolet radiation that reaches the ground can burn our skin, but it is easily blocked by window glass and by absorbing liquids in suntan lotion. Because there is less-absorbing atmosphere above mountains than above the sea, we burn more quickly when we go mountain climbing or skiing than when we lie at the beach by the sea.

The X-ray region of the electromagnetic spectrum extends from a wavelength of one-hundred-billionth (10^{-11}) of a meter, which is about the size of an atom, to the short-wavelength side of the ultraviolet (Fig. 1.6). X-ray radiation is so energetic that it is usually described in terms of the energy it carries. The X-ray region lies between 1 and 100 kilo-electron volts, abbreviated keV, of energy (Focus 1.1). There are soft X-rays with relatively low energies of 1 to 10 keV and modest penetrating power. Hard X-rays have higher energies, at 10 to 100 keV, and greater penetrating power. As a metaphor, one thinks of the large, pliant softballs and the compact, firm hardballs, used in the two kinds of American baseball games.

Gamma rays are even more energetic than X-rays, exceeding 100 keV in energy. They are the shortest and most energetic electromagnetic waves. Their wavelengths are as small as the nucleus of an atom, and their waves are so energetic that they can pass through an iron plate that is 0.2 meters thick. The deadly effects of gamma rays were tragically demonstrated by the nuclear bomb explosions at Hiroshima and Nagasaki, Japan.

The atmosphere effectively absorbs most of the Sun's ultraviolet radiation and all of its X-rays and gamma rays. To look at the Sun at these invisible wavelengths, we must loft telescopes above the atmosphere. This was done first by using balloons and rockets, followed by satellites that orbit the Earth above the atmosphere. Satellite-borne telescopes now view the Sun at invisible ultraviolet and X-ray wavelengths,

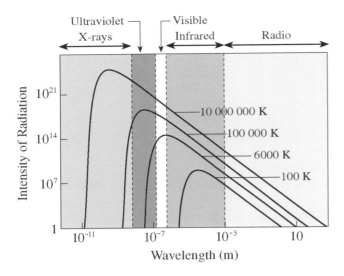

Fig. 1.7 **Thermal radiation at different temperatures** The spectral plot of thermal radiation intensity as a function of wavelength depends on the temperature of the gas emitting the radiation. At higher temperatures the most intense emission occurs at shorter wavelengths, and the thermal radiation intensity becomes greater at all wavelengths. At a temperature of 6000 K, the thermal radiation peaks in the visible, or V, band of wavelengths. A hot gas with a temperature of 10^5 K emits most of its thermal radiation at ultraviolet, or UV, wavelengths, while the emission peaks in X-rays when the temperature is 10^6 to 10^7 K.

above the Earth's absorbing atmosphere in a world where night is often brief or non-existent. One of these solar satellites is the *SOlar and Heliospheric Observatory*, or *SOHO* for short, launched on 2 December 1995, and another is the *Yohkoh*, or "sunbeam", satellite launched on 21 February 1981. They provide entirely new perspectives of the Sun, detecting the hotter parts of the solar atmosphere (Fig. 1.7) and revealing a Sun that is unlike anything we have ever seen before (Fig. 1.8). The instruments and other information about current and future solar space missions are described in Sections 9.3 and 9.4.

Fig. 1.8 **The X-ray Sun** If our eyes could see X-ray radiation, then the Sun would look something like this. Million-degree gas is constrained within ubiquitous magnetic loops, giving rise to bright X-ray emission from active regions. Relatively faint magnetic loops connect active regions to distant areas on the Sun, or emerge within quiet regions away from active ones. The extended corona rings the Sun, and dark coronal holes are found at its poles (*top* and *bottom*). This image was taken on 8 May 1992 with the Soft X-ray Telescope (SXT) onboard the *Yohkoh* mission; it has been corrected for instrumental effects and processed to enhance solar features. (*Yohkoh* is a project of international cooperation between ISAS and NASA. The SXT was prepared by the Lockheed-Martin Solar and Astrophysics Laboratory, the National Astronomical Observatory of Japan, and the University of Tokyo with the support of NASA and ISAS.)

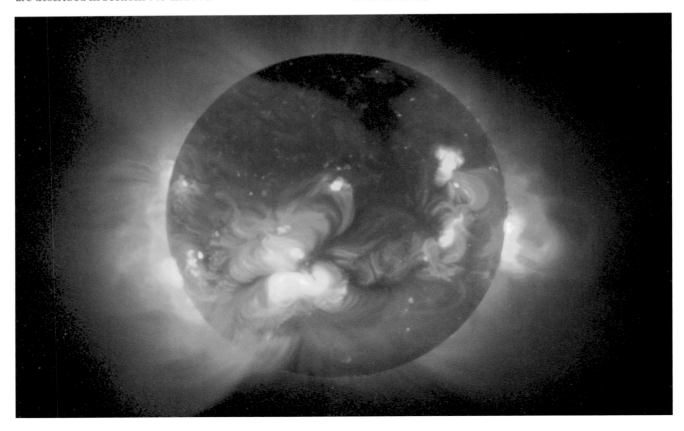

1.3 Physical characteristics of the Sun

The Sun is a perfectly ordinary star, and like the other stars it is large and massive. The immense Sun is 109 times the Earth's diameter and 333 000 times its mass. A million Earths could fit inside the Sun.

Despite its large mass, the Sun has a lower density, or mass per unit volume, than Earth. The Sun's average density is only 1409 kg m^{-3}, which is a quarter the average density of Earth.

From Earth the Sun looks small, because it is far away. Its average distance from Earth is about 1.5×10^{11}, or 150 billion, meters. The mean distance of the Sun from the Earth is a very important measurement, for it sets the scale of our solar system and enables us to infer, from other observations, the size, mass, luminosity and effective temperature of the Sun. This distance is called the astronomical unit, or AU for short. The best current value is (Focus 1.2)

Sun's mean distance = one astronomical unit =
1 AU = $1.495\,978\,7 \times 10^{11}$ m.

At that distance, light from the Sun takes about eight minutes to reach Earth. The mean light travel-time, designated τ_A, from the Sun to Earth is $\tau_A = $ AU/c = 499.004 782 seconds, where the velocity of light $c = 2.997\,924\,58 \times 10^8$ meters per second.

The Sun is much closer to the Earth than any other star is. The Sun's nearest stellar neighbor, Proxima Centauri (part of the triple-star system Alpha Centauri), is 4.29 light-years from our solar system, meaning light from Proxima Centauri takes 4.29 years to reach us. One year is equivalent to 3.1557×10^7 (31.557 million) seconds. Thus, the Sun is 0.27 million times closer than the next nearest star, so we

Focus 1.2 The Sun's distance, luminosity, size and effective temperature

The distances to the Sun and planets can be measured by triangulation from different points on the Earth. These distances are inferred from their parallax, or the angular difference in the apparent direction of an object as seen from two different locations. The solar parallax, designated $\pi_\odot$, is defined as half the angular displacement of the Sun as viewed from opposite sides of the Earth, or

$$\sin \pi_\odot = a_E/\text{AU},$$

and,

$$\text{solar parallax} = \pi_\odot = \arcsin(a_E/\text{AU}) = 8.794\,148 \text{ seconds of arc},$$

where one radian = $2.062\,648 \times 10^5$ seconds of arc, the equatorial radius of the Earth, a_E, = $6.378\,140 \times 10^6$ meters and the unit distance AU, called the astronomical unit, is the semi-major axis of the Earth's orbit about the Sun.

A rough value for the astronomical unit can be obtained by assuming that the Earth moves in a circular orbit about the Sun. The circumference of that orbit can be found by measuring the Earth's orbital speed in meters per second and multiplying by the number of seconds in a year. The speed is 29 786 m s^{-1}. Multiplying the speed by 3.1557×10^7, the number of seconds in a year, and dividing by 2π we obtain an average distance of 150 billion (1.5×10^{11}) meters, where the constant $\pi = 3.14159$.

The solar parallax has been inferred with increasingly greater accuracy during the past two centuries by measuring the parallaxes of Venus, Mars, and the nearby minor planet, or asteroid, Eros, during their closest approaches to the Earth (Fig. 1.9). This quest for accuracy in the mean distance of the Sun from the Earth has involved hundreds of trips to remote countries, tens of thousands of observations and photographs, and the lifetime work of several astronomers. It culminated in the 1960s when radio pulses, or radar (radio detection and ranging), were used to accurately determine the distance to Venus, and to thereby measure the Sun's distance with an

accuracy of about 1000 m (Fig. 1.10). The distance to Venus is obtained by multiplying half the round-trip time for the radio pulses by the speed of light. The Earth's mean distance from the Sun, the astronomical unit or AU, has been determined from the orbital distance, D_V, of Venus using Kepler's third law (Section 1.4):

$$\text{Sun's mean distance} = \text{AU} = \left(\frac{P_E}{P_V}\right)^{2/3} D_V = 1.495\,978\,7 \times 10^{11}\,\text{m},$$

where the orbital period of the Earth is $P_E = 1$ year = 3.1557×10^7 seconds, and Venus orbits the Sun at a distance of $D_V = 0.723$ AU and a period of $P_V = 0.615$ years.

We can use the solar constant, $f_\odot$, and the Sun's mean distance to determine the Sun's absolute luminosity, $L_\odot$, from:

$$\text{solar luminosity} = L_\odot = 4\pi f_\odot (\text{AU})^2 = 3.85 \times 10^{26}\,\text{W}$$

where $f_\odot = 1366$ W m^{-2}. The linear radius of the Sun, $R_\odot$, can be determined from its angular diameter, θ, using

$$\text{Sun's angular diameter} = \theta = \frac{2R_\odot}{\text{AU}} = 0.0093 \text{ radians},$$

where a full circle subtends 2π radians and 360 degrees. Since there are 60 minutes of arc in a degree, we can express the angular diameter of the Sun in minutes of arc: $\theta = 60 \times 360 \times 0.0093/(2\pi) = 31.97$ minutes of arc, where $\pi = 3.14159$.

Precise solar diameter measurements during the 1980s have resulted in a mean value for the solar near-equatorial radius, with an uncertainty of 2.6×10^4 meters:

$$\text{Solar radius} = R_\odot = \frac{\theta \times \text{AU}}{2} = 6.955 \times 10^8\,\text{m}.$$

The effective temperature, $T_{e\odot}$, of the visible solar disk can be determined using the Stefan–Boltzmann law given in the previous section

$$\text{Sun's effective temperature} = T_{e\odot} = \left[\frac{L_\odot}{4\pi\sigma R_\odot^2}\right]^{1/4} = 5780 \text{ K}.$$

Property	Value		
Mean distance, AU *(from radar measurements of the distance to Venus and Kepler's third law)*	$1.495\,978\,7 \times 10^{11}$ m		
Light travel-time from Sun to Earth	$499.004\,782$ s		
Radius, $R_\odot$ *(from distance and angular extent)*	6.955×10^8 m (109 Earth radii)		
Volume	1.412×10^{27} m³ (1.3 million Earths)		
Mass, $M_\odot$ *(from distance and Earth's orbital period using Kepler's third law)*	1.989×10^{30} kg (332 946 Earth masses)		
Escape velocity at photosphere	6.178×10^5 m s⁻¹		
Mean density	1409 kg m⁻³		
Solar constant, $f_\odot$	1366 J s⁻¹ m⁻² = 1366 W m⁻²		
Luminosity, $L_\odot$ *(from solar constant and distance)*	3.85×10^{26} J s⁻¹ = 3.85×10^{26} W		
Principal chemical constituents *(from analysis of Fraunhofer lines)*		(By number of atoms)	(By mass fraction)
hydrogen		92.1 percent	X = 70.68 percent
helium		7.8 percent	Y = 27.43 percent
all others		0.1 percent	Z = 1.89 percent
Age *(from age of oldest meteorites)*	4.566×10^9 years		
Density (center)	1.513×10^5 kg m⁻³		
Pressure (center)	2.334×10^{16} Pa		
(photosphere)	10 Pa		
Temperature (center)	1.56×10^7 K		
(photosphere)	5780 K		
(chromosphere)	$6 \times 10^3 - 2 \times 10^4$ K		
(transition region)	$2 \times 10^4 - 2 \times 10^6$ K		
(corona)	$2 \times 10^6 - 3 \times 10^6$ K		
Rotation period (equator)	26.8 days		
(30° latitude)	28.2 days		
(60° latitude)	30.8 days		
Magnetic field (sunspots) *(from Zeeman effect)*	$0.1 - 0.4$ T = $1 \times 10^3 - 4 \times 10^3$ G		
(polar)	0.001 T = 10 G		

[a] Mass density is given in kilograms per cubic meter, or kg m⁻³; the density of water is 1000 kg m⁻³. The unit of pressure is pascal, where 1.013×10^5 Pa is the pressure of the Earth's atmosphere at sea level.

Table 1.1. **The Sun's physical properties.**[a]

receive much more energy from the Sun. The Sun is so much closer to Earth than all the other stars that the intense light of the Sun keeps us from seeing other stars during the day.

The Sun produces an enormous amount of light. Its absolute, or intrinsic, luminosity is designated by the symbol $L_\odot$, where the subscript $_\odot$ denotes the Sun. We can infer the Sun's luminosity from satellite measurements of the total amount of solar energy reaching every square centimeter of the Earth every second (Focus 1.2), obtaining:

$$\text{solar luminosity} = L_\odot = 3.854 \times 10^{26} \text{ W} = 3.854 \times 10^{26} \text{ J s}^{-1}.$$

In comparison, an incandescent lamp emits 60 to 100 watts of power, so the Sun is more luminous than a million billion billion (10^{24}) light bulbs.

Because of its proximity, the Sun influences all of the objects in the solar system. The Sun's light warms our days and illuminates our world. At night, reflected sunlight makes the Moon and planets bright in the dark night sky.

The Sun's size can be inferred from its distance and angular extent (Focus 1.2). It has the value

$$\text{solar radius} = R_\odot = 6.955 \times 10^8 \text{ m}.$$

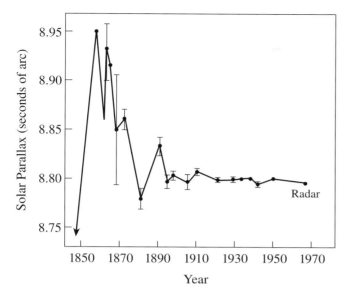

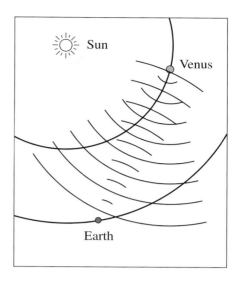

Fig. 1.9 **Distance to the Sun** Values of the solar parallax obtained from measurements of the parallax of Venus, Mars, and the asteroid Eros between 1850 and 1970. Here the error bars denote the probable errors in the determination, whereas the points for 1941, 1950 and 1965 all have errors smaller than the plotted points. In the 1960s, the newly developed radar technology enabled the determination of the Sun's distance with an accuracy of about 1000 m. The radar value of the solar parallax is 8.794 15 seconds of arc.

Fig. 1.10 **Radar-ranging to Venus** Accurate distances to the nearby planets have been determined by sending radio pulses from Earth to the planet, and timing their return a few minutes later. The figure shows the emission of a pulse toward Venus; when it bounces from Venus the radiation spreads over the sky and we receive only a small fraction of the original signal, delayed by the round-trip travel-time. If T is the round-trip time and c is the speed of light, the total distance traveled is cT and the distance to Venus is $cT/2$. For Venus, the round-trip time is 4.6 minutes when the planet is nearest Earth and increases to 28.7 minutes when it is furthest away from us. In computing the distance there is a small correction for the bending of light in the Sun's gravitational field. The greater distance and larger travel-time due to the Sun's curvature of nearby space amounts to only two-ten-millionths of the total round-trip travel-time.

That is about 109 times the radius of the Earth. If the Earth were situated at the center of the Sun, the Moon's orbit would still lie inside the Sun.

From the Sun's size and luminous output, we can infer the temperature of its visible disk, 5780 K (Focus 1.2). Another fundamental solar constant is the Sun's mass, derived in Section 1.4. It is also given, together with other physical properties of the Sun, in Table 1.1.

1.4 Gravity's center

An Earth-centered system captivated the minds of most astrologers and astronomers until the 16th century when Nicholas Copernicus (1473–1543) placed the Sun at the center of the planetary system. He laid the groundwork for the "Copernican Revolution" in his book *De revolutionibus orbium coelestium* ("On the Revolutions of the Heavenly Spheres", 1543). The planets were all supposed to orbit the Sun in roughly the same plane, now known as the ecliptic (the plane of the Earth's orbit), and in the same direction, counterclockwise as viewed from the north ecliptic pole.

Copernicus' theory had one fatal flaw; he assumed that the planets moved in circular orbits. As a result, his explanation could not be reconciled with careful observations of the changing positions of the planets in the sky, meticulously carried out by the Danish astronomer Tycho Brahe (1546–1601). As discovered by the mathematician Johannes Kepler (1571–1630), the architecture of the solar system had to be described by non-circular shapes. In 1605, after four years of computations, Kepler found that the observed planetary orbits could be described by ellipses with the Sun at one focus. This ultimately became known as Kepler's first law of planetary motion. Although the planet orbits are nearly circular, they are slightly elliptical in shape.

Also in 1605, Kepler found that as each planet moved about its elliptical orbit, it moved faster when it was closer to the Sun. In fact, he was able to state the relationship in a precise mathematical form, now known as Kepler's second law, that can be explained with the help of Figure 1.11. Imagine a line drawn from the Sun to a planet. As the planet swings about its elliptical path, the line (which will increase and decrease in length) sweeps out a surface at a constant rate. This is also known as the "law of equal areas". During the three equal time intervals shown in Figure 1.11, the planet moves through different arcs because its orbital speed changes, but the areas swept out are identical.

When the planet is closest to the Sun, it must be moving most rapidly. Its closest point is called the perihelion; and its most distant point is the aphelion, where the planet moves most slowly. The distance between the perihelion and aphelion is the major axis of the orbital ellipse. Half that distance is called the semi-major axis and designated with the symbol, a_p. The semi-major axis of the Earth's orbit, a_E, is a fundamental quantity called the astronomical unit, or AU, described in Section 1.3.

Kepler's third law took another 10 years of work to discover. It is also known as the harmonic law and it states that the squares of the planetary periods are in proportion to the cubes of their average distances from the Sun. If P_P

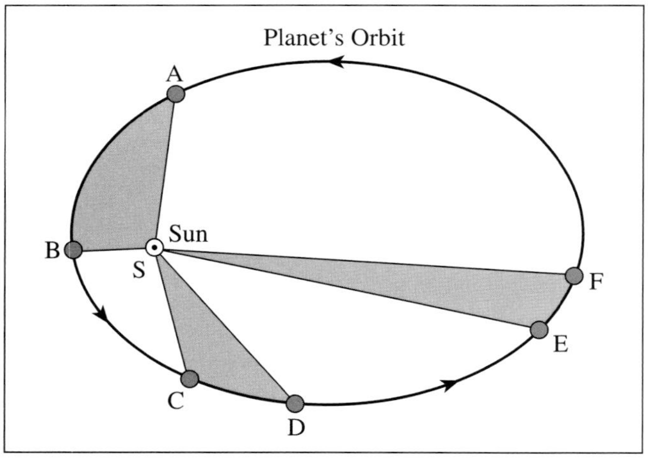

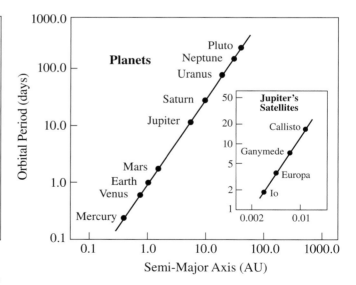

Fig. 1.11 **Kepler's first and second laws** Kepler's first law states that the orbit of a planet about the Sun is an ellipse with the Sun at one focus. According to Kepler's second law, the line joining a planet to the Sun sweeps out equal areas in equal times. This is also known as the law of equal areas. It is represented by the equality of the three shaded areas ABS, CDS and EFS. It takes as long to travel from A to B as from C to D and from E to F. A planet moves most rapidly when it is nearest the Sun (at perihelion); a planet's slowest motion occurs when it is farthest from the Sun (at aphelion).

Fig. 1.12 **Kepler's third law** The orbital periods of the planets are plotted against their semi-major axes, using a logarithmic scale. The straight line that connects the points has a slope of 3/2, thereby verifying Kepler's third law that states that the square of the orbital periods increase with the cubes of the planets' distances. This type of relation applies to any set of bodies in elliptical orbits, including Jupiter's four largest satellites shown in the inset.

Planet	Semi-major axis, a_p (AU)	Orbital period, P_p (years)[a]	Mean orbital velocity (km s^{-1})[b]
Mercury	0.387 099	0.2409	47.89
Venus	0.723 332	0.6152	35.03
Earth	1.000 000	1.0000	29.79
Mars	1.523 688	1.8809	24.13
Jupiter	5.202 834	11.8622	13.06
Saturn	9.538 762	29.4577	9.64
Uranus	19.191 391	84.0139	6.81
Neptune	30.061 069	164.793	5.43
Pluto	39.529 402	247.7	4.74

[a] One tropical year = $3.155\,692\,597\,47 \times 10^7$ seconds = 365.242 19 days
[b] 1.0 km s^{-1} = 1000 meters per second

Table 1.2 **Mean orbital parameters of the planets**

denotes the orbital period of a planet measured in years and a_p describes its semi-major axis measured in AU, then Kepler's third law states that $P_p^2 = a_p^3$. This expression is illustrated in Figure 1.12, for the major planets and for the brighter moons of Jupiter.

So, the Sun is at the hub of the solar system, with planets revolving around it with speeds that decrease at increasing distances. The more distant planets have longer periods and they move around the Sun at a slower pace. For example, Jupiter is 5.2 times as far away from the Sun as the Earth is, and it takes Jupiter 11.86 Earth years to travel once around the Sun. Parameters for the orbits of the planets are given in Table 1.2.

An interesting aspect of the planetary orbits is their regular spacing. A given planet is about twice as far away from the Sun as the next nearest planet to the Sun. This is described by the Titius–Bode law given in Focus 1.3.

Kepler's laws were explained in the late 1600s by Isaac Newton (1643–1727), the extraordinary mathematical genius who introduced the concept of universal gravitation and established the laws of motion. According to Newton's theory, mass produces a force called gravity. Every object in the Universe attracts every other object with a gravitational force that is directed along the line joining the two objects; the force is proportional to the product of their masses and inversely proportional to the square of their separation (Focus 1.4). Gravity is centered in every mass; it increases with the mass and with proximity to it. Within the solar system, the dominant mass is that of the Sun, which far surpasses the mass of any other object there.

Of all the known forces of nature, gravitation exerts its influence to the greatest distance. It is unimportant over the short distances between atoms in a crystal or a molecule, but the gravitational effect of stars and galaxies can be felt across

Focus 1.3 The Titius–Bode law

In the inner solar system each planet's orbit is about 1.5 times the distance of its inward neighbor, and this ratio increases to roughly a factor of 2.0 in the outer solar system. This is described by the Titius–Bode law, first discovered by Johann Daniel Titius (1729–96) and brought to prominence by Johann Elert Bode (1747–1826). The law states that the semi-major axes of the planets, a_n, in astronomical units can be roughly approximated by taking the sequence 0, 3, 6, 12, 24, ..., adding 4, and dividing by 10. Mathematically:

$$a_n = 0.4, \quad (n = 1)$$
and
$$a_n = 0.1\,[\,4 + 3 \times 2^{n-2}\,], (n = 2, 3, ..., 9).$$

A comparison of the semi-major axes, a_n, of the planets with this law is given in Table 1.3.

The Titius–Bode law predates the discovery of the first asteroid, at $n = 5$, by 35 years and the discovery of Uranus at $n = 8$, by 15 years. Although there is no well-accepted explanation for why this expression works so well, it probably has something to do with the origin of the solar system.

Planet	n	Value of a_p Measured (AU)	Predicted from Titius-Bode law
Mercury	1	0.387	0.4
Venus	2	0.723	0.7
Earth	3	1.000	1.0
Mars	4	1.524	1.6
Ceres (asteroid)	5	2.767	2.8
Jupiter	6	5.203	5.2
Saturn	7	9.537	10.0
Uranus	8	19.19	19.6
Neptune	9	30.07	38.8

Table 1.3 **Comparison of measured planetary distances with those predicted from the Titius–Bode Law**

the Universe. This behavior can be traced to two causes. In the first place, according to the inverse-square law, gravitational force decreases relatively slowly with distance, and this gives gravitation a much greater range than the forces that hold together the nuclei of atoms. In the second place, gravitation has no positive and negative polarity. This is in contrast to electricity, in which the repulsive and attractive forces among like and unlike charges can cancel. The effect of this cancellation is to shield distant atoms from each other's electrical forces. There is no gravitational repulsion between like masses, so there is no gravitational shielding, and every atom in the Universe feels the attraction of every other atom. These are the reasons why gravity plays

Focus 1.4 The distant reach of gravity

According to Newton, force produces an acceleration, or a change in velocity, where velocity is defined as a body's speed and direction. Moreover, any mass, M_1, produces a gravitational force, $F_{gravity}$, on another mass, M_2, given by the expression:

$$\text{Gravitational force} = F_{gravity} = \frac{GM_1M_2}{R^2},$$

where G is the universal gravitational constant $G = 6.6726 \times 10^{-11}$ N m^2 kg^{-2}, and R is the distance between the centers of the two masses. This is sometimes called the inverse-square law, since the force of gravity falls off as the inverse square of the distance.

The massive Sun is the gravitational center of our solar system, attracting any other body with its gravity. At a distance, R, it produces a gravitational acceleration, g, given by:

$$g = \frac{GM_\odot}{R^2} \text{ m s}^{-2},$$

where the mass of the Sun $M_\odot = 1.989 \times 10^{30}$ kg. Every planet feels this gravitational pull. At the Sun's radius, $R_\odot = 6.955 \times 10^8$ m, it has the value $g_\odot = 274$ m s^{-1}.

Why doesn't the immense solar gravity pull the entire solar system into the Sun? Motion opposes the force of gravity. So, the reason that the planets do not fall into the Sun is that each planet is also moving in a direction perpendicular to the line connecting it to the Sun. That crucial speed, called the escape velocity, is given by the expression (derived in Section 1.6, Focus 1.6):

$$V_{escape} = \left[\frac{2GM_\odot}{R}\right]^{1/2},$$

which depends only on the Sun's mass and the planet's distance, but is independent of the planet's mass. If a planet moved with a faster speed it would move away from the Sun's gravitational embrace, and if the planet traveled at a slower speed it would be pulled toward the Sun. The escape velocity at the solar radius is $V_{escape, \odot} = 6.178 \times 10^5$ m s^{-1}.

If the planets were moving in exactly circular orbits around the Sun, then the velocity would equal the circumference, $2\pi R$, divided by the orbital period, P, and by substituting $V_{escape} = 2\pi R/P$ into the previous equation and collecting terms we obtain

$$P^2 = \frac{2\pi^2}{G}\frac{R^3}{M_\odot}$$

which shows that the square of the orbital periods scales as the cube of the distances. Of course the planets move in elliptical orbits rather than circular ones, so the constant is slightly different.

the central role in governing the orbits of planets and in defining the overall structure of the Universe.

The Sun's gravitational pull holds the solar system together. That is why we call it a solar system, governed by the central Sun. The Earth and other planets are strongly attracted to the Sun's huge mass. This gravitational attraction keeps these bodies in orbit around the Sun. That is, the orbits are determined by the Sun's gravity, and therefore the mass of the Sun.

The Sun doesn't just lie at the heart of our solar system, it dominates it. Some 99.8 percent of all the matter between the Sun and halfway to the nearest star is contained in the Sun. All the objects that orbit the Sun (the planets and their satellites, the comets and the asteroids) add up to just 0.2 percent of the mass in our solar system. As far as the Sun is concerned, the planets are insignificant specks, left over from its formation and held captive by its massive gravity.

Newton used his concept of universal gravitation to derive Kepler's third law in the form:

$$P_p^2 = \frac{4\pi^2}{G}\frac{a_p^3}{(m_p + M_\odot)} = 5.9165 \times 10^{11}\frac{a_p^3}{M_\odot}$$

where a_p is the semi-major axis of the planet's orbital ellipse in meters, P_p is the orbital period in seconds, and m_p and $M_\odot$, respectively, denote the mass of the planet and the mass of the Sun in kilograms. In this expression, G is the universal gravitational constant, and $4\pi^2/G = 5.9165 \times 10^{11}$. Since the masses of the planets are so much less than the mass of the Sun, the sum $(m_p + M_\odot)$ is effectively a constant equal to the Sun's mass, $M_\odot$, regardless of the planet under consideration, at least for the accuracy available in Kepler's day.

Once we know the Earth's orbital period and mean distance from the Sun, we can weigh the Sun from a distance, determining its mass from Newton's formulation of Kepler's third law:

$$M_\odot = 5.9165 \times 10^{11}\frac{(AU)^3}{P_E^2} = 1.989 \times 10^{30} \text{ kg}$$

where the semi-major axis of the Earth's orbit is $a_E = 1$ AU $= 1.495\,978\,7 \times 10^{11}$ meters and the orbital period of the Earth is $P_E = 1$ year $= 3.1557 \times 10^7$ seconds.

For nearly two-and-a-half centuries, the solar system appeared to behave according to the laws of Newton. The paths of planets seemed to be predictable with great precision. In fact the planet Neptune was discovered by comparing the predicted positions of Uranus, using Newton's law of gravitation, with its observed place. The two sets of positions disagreed, and could only be reconciled if a large unknown planet, located far beyond Uranus, was producing a gravitational tug on Uranus. In 1845 two astronomers, John Couch Adams and Urbain Jean Joseph Leverrier, specified the location of the unknown world from a mathematical analysis of the wanderings of Uranus; it was found by Johann Galle and his student Heinrich D'Arrest in 1846, using Leverrier's prediction, and subsequently named Neptune.

But there were problems with Mercury's motion. Newton's laws failed to provide the expected connection between old and new measures of the planet's position, and a detailed analysis showed that the orientation of Mercury's elliptical orbit was rotating in space at a rate of 43 seconds of arc per century. Leverrier attributed it to the gravitational pull of an unknown planet orbiting the Sun inside Mercury's orbit and moving ahead of it. The hypothetical planet was named Vulcan, and extensive searches were conducted for it. No such planet was ever reliably detected, and Mercury's

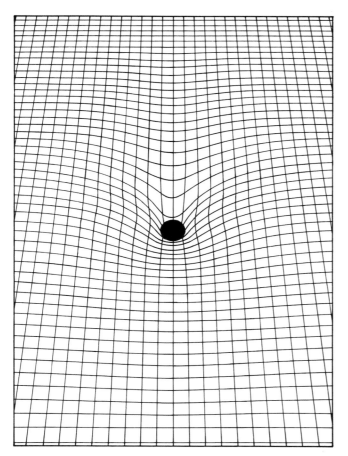

anomalous motion was not explained until 1915 when Albert Einstein extended Newton's theory of gravity to include the Sun's curvature of nearby space (Fig. 1.13).

According to Einstein's General Theory of Relativity, space is distorted in the neighborhood of matter, and this distortion is the cause of gravity. The result is a gravitational field that departs slightly from the exact inverse-square law. This departure produces planetary orbits that are not exactly elliptical. Instead of returning to its starting point to form a closed ellipse after one orbital period, the planet moves slightly ahead in a winding path that can be described as a rotating ellipse. This produces an advance of the perihelion, and the amount predicted by Einstein for Mercury was 43 seconds of arc per century – exactly the observed amount. Because the amount of space curvature produced by the Sun falls off with increasing distance, the perihelion advances for the other planets are much smaller than that of Mercury.

Fig. 1.13 **Space curvature** A massive object creates a curved indentation upon the flat Euclidean space that describes a world which is without matter. Notice that the amount of space curvature is greatest in regions near the object, while further away the effect is lessened.

1.5 Spectroscopy and the ingredients of the Sun

Nowadays scientists use special instruments called spectrographs to spread out the visible portion of the Sun's radiation into its spectral components and separate colors. Such a display of the intensity of radiation as a function of wavelength is called a spectrum. Spectroscopy is the study and interpretation of spectra, especially with a view to determine the chemical composition of, and physical conditions in, the source of radiation.

When the spectrum of sunlight is examined carefully, with very fine wavelength resolution, numerous fine dark gaps are seen crossing the rainbow-like display (Fig. 1.14). When coarser resolution is used, the separate colors of sunlight are somewhat blurred together and the dark places are no longer found superimposed on its spectrum.

These dark gaps were first noticed by William Wollaston in 1802, and investigated in far greater detail by Joseph von Fraunhofer. By 1815 Fraunhofer had catalogued the wavelengths of more than 300 of them, assigning Roman letters to the most prominent (Table 1.4).

The dark gaps of missing colors are now called absorption lines. When a cool, tenuous gas is placed in front of a hot, dense one, atoms or ions in the cool gas absorb radiation at specific wavelengths, thereby producing the dark absorption lines. They are called lines because they look like a line in the spectrum. The term Fraunhofer absorption line is also used, recognizing his investigations of them.

The detailed explanation for the Sun's absorption lines

Wavelength (nanometers)	Fraunhofer letter (Color)	Element name, symbol
393.368	K	Ionized calcium, Ca II
396.849	H (extreme violet)	Ionized calcium, Ca II
410.175	h	Hydrogen, Hδ, delta transition
422.674	g	Neutral calcium, Ca I
431.0 ± 1.0	G (violet)	CH molecule
434.048		Hydrogen, Hγ, gamma transition
438.356	d	Neutral iron, Fe I
486.134	F (blue)	Hydrogen, Hβ, beta transition
516.733	b_4	Neutral magnesium, Mg I
517.270	b_2	Neutral magnesium, Mg I
518.362	b_1	Neutral magnesium, Mg I
526.955	E (green)	Neutral iron, Fe I
588.997	D2 (yellow)[b]	Neutral sodium, Na I
589.594	D1 (yellow)[b]	Neutral sodium, Na I
656.281	C (red)	Hydrogen, Hα, alpha transition
686.719	B (red)	Molecular oxygen, O_2, in our air
759.370	A (extreme red)	Molecular oxygen, O_2, in our air

[a] The wavelengths are in nanometer units, where 1 nanometer = 10^{-9} meters. Astronomers have often used the ångstrom unit of wavelength, where 1 ångstrom = 1 Å = 0.1 nanometers. The letters were used by Joseph von Fraunhofer around 1815 to designate the spectral lines before they were chemically identified. Fraunhofer did not resolve the numbered components of the D and b lines. A Roman numeral I after an element symbol denotes a neutral, or unionized atom, with no electrons missing, whereas the Roman numeral II denotes a singly-ionized atom with one electron missing. The lines A and B are produced by molecular oxygen in the terrestrial atmosphere.

[b] Un-ionized helium, He I, was discovered in emission during a solar eclipse in 1868 at a wavelength of 587.56 nanometers, near the two sodium lines. The helium line is sometimes designated D3.

Table 1.4 **Prominent absorption lines in sunlight**[a]

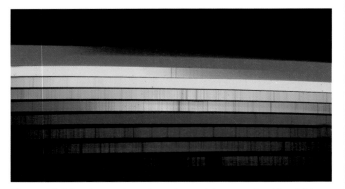

Fig. 1.14 **Visible solar spectrum** A spectrograph has spread out the visible portion of the Sun's radiation into its spectral components, displaying radiation intensity as a function of wavelength. When we pass from long wavelengths to shorter ones (*left to right* and *top to bottom*), the spectrum ranges from violet through blue, green, yellow, orange and red. Dark gaps in the spectrum, called Fraunhofer absorption lines, represent absorption by atoms or ions in the Sun. The wavelengths of these absorption lines can be used to identify the elements in the Sun, and the relative darkness of the lines establishes the relative abundance of these elements. (Courtesy of the National Solar Observatory/Sacramento Peak, NOAO.)

was provided in the mid-19th century in a chemistry laboratory in Heidelberg, Germany. There the chemist Robert Bunsen, inventor of the Bunsen burner, and his physicist colleague Gustav Kirchhoff unlocked the chemical secrets of the Universe. When they vaporized an individual element in a flame, and heated it to incandescence, the hot vapor produced a distinctive pattern of bright lines whose unique wavelengths coincided with some of the dark absorption lines in the Sun's spectrum. Moreover, when the light produced by a hot incandescent solid, such as a tungsten lamp, was passed through the cooler vaporized gas, dark lines were produced at exactly the same locations.

Kirchhoff generalized this into a law stating that the powers of emission and absorption of a body at any particular wavelength were the same at a given temperature. He also concluded that the visible solar disk was hot and incandescent, producing a continuous spectrum (the sort without lines), that became crossed by the dark Fraunhofer lines when passing through cooler overlying gas.

As Bunsen wrote in 1859:

> At present Kirchhoff and I are engaged in an investigation that doesn't let us sleep. Kirchhoff has made a wonderful, entirely unexpected discovery in finding the cause of the dark lines in the solar spectrum. He can produce these lines artificially intensified both in the solar spectrum and in the continuous spectrum of a flame, in exactly the same position as the corresponding Fraunhofer lines. Thus a means has been found to determine the composition of the Sun and fixed stars with the same accuracy as we can determine strontium chloride with our chemical reagents.

Each chemical element, and only that element, produces a unique set, or pattern, of wavelengths at which the dark lines fall. It is as if these were a characteristic barcode that can be used to identify each element, as a fingerprint might identify a criminal. The close pair of dark lines in the yellow, specified by the letter D, was attributed to sodium; they produce the distinctive yellow color of sodium-vapor street lights. The lines designated by Fraunhofer with the letters C and F were associated with hydrogen, and other conspicuous dark solar lines were attributed to magnesium, letter b, calcium, letters H and K, and iron, letter E.

The Fraunhofer lines designated A and B actually have nothing to do with the composition of the Sun. They only appear on spectra gathered beneath the Earth's atmosphere. Molecular oxygen in the terrestrial atmosphere absorbs sunlight at the wavelengths of the A and B Fraunhofer lines, creating dark lines on the Sun's spectrum. A spectrum gathered above the Earth's atmosphere would not have these lines.

The most conspicuous dark solar lines come from hydrogen, sodium, magnesium, calcium and iron (Table 1.4), but iron accounts for more lines than any other element. Since abundant heavy iron explains the Earth's high average density, and also since most of the other solar lines correspond to elements known on Earth, it was initially supposed that the Sun is made out of the same material as the Earth, but this is only partly true.

Since a greater number of atoms will absorb more light, the relative darkness of the absorption lines establishes the relative abundance of the elements in the Sun. That is, darker absorption lines generally indicate greater absorption and therefore larger amounts of the element. Studies of the absorption lines in the Sun's spectrum showed, in the 1920s, that hydrogen is the most abundant element in the visible solar gases. Since the Sun was most likely chemically homogenous, a high hydrogen abundance was implied for the entire star, and this was confirmed by subsequent calculations of its luminosity. Hydrogen accounts for 92.1 percent of the number of atoms in the Sun, and it amounts to 70.68 percent by mass.

We also now know that hydrogen is the most abundant element in most stars, in interstellar space, and in the entire Universe. The Earth is a dirty speck in the cosmos, an

Element	Symbol	Atomic number, Z	Abundance^a	Year of discovery on Earth
Hydrogen	H	1	2.79×10^{10}	1766
Helium	He	2	2.72×10^9	1895^b
Carbon	C	6	1.01×10^7	(ancient)
Nitrogen	N	7	3.13×10^6	1772
Oxygen	O	8	2.38×10^7	1774
Neon	Ne	10	3.44×10^6	1898
Sodium	Na	11	5.74×10^4	1807
Magnesium	Mg	12	1.07×10^6	1755
Aluminum	Al	13	8.49×10^4	1827
Silicon	Si	14	1.00×10^6	1823
Phosphorus	P	15	1.04×10^4	1669
Sulfur	S	16	5.15×10^5	(ancient)
Chlorine	Cl	17	5.24×10^3	1774
Argon	Ar	18	1.01×10^5	1894
Potassium	K	19	3.77×10^3	1807
Calcium	Ca	20	6.11×10^4	1808
Chromium	Cr	24	1.35×10^4	1797
Manganese	Mn	25	9.55×10^3	1774
Iron	Fe	26	9.00×10^5	(ancient)
Nickel	Ni	28	4.93×10^4	1751

^{a} Normalized to an abundance of Silicon = 1.00×10^6.

^{b} Helium was discovered on the Sun in 1868, but it was not found on Earth until 1895.

Table 1.5 **The twenty most abundant elements in the Sun**

anomaly, for it is primarily made out of heavy elements that are relatively uncommon in the Sun and the rest of the Universe. Hydrogen is, for example, about one million times more abundant than iron in the Sun, but iron is one of the main constituents of the Earth which does not have sufficient gravity to retain hydrogen in its atmosphere (Section 1.6, Focus 1.6).

Helium, the second most abundant element in the Sun, is so rare on Earth that it was first discovered on the Sun. The precise wavelength, or position, of a line in the yellow region of the spectrum, detected in the Sun's rarefied atmosphere during the solar eclipse on 18 August 1868, had no known Earthly counterpart. This meant that a new chemical element had been discovered, which Norman Lockyer named helium after the Greek Sun god *Helios* (Section 5.2). Helium was not found on Earth until 1895, when William Ramsay discovered it as a gaseous emission from a mineral called clevite.

Today, helium is used on Earth in a variety of ways, including inflating party balloons with helium gas and in its liquid state keeping sensitive electronic equipment cold. Though plentiful in the Sun, helium is almost non-existent on the Earth. It is so terrestrially rare that we are in danger of running out of helium during this century.

Altogether, 92.1 percent of the number of atoms in the

Sun are hydrogen atoms, 7.8 percent by number are helium atoms, and all other heavier elements make up only 0.1 percent. The abundances of the 20 most-abundant elements in the Sun are given in Table 1.5, normalized to an abundance for silicon of 1.00×10^6. They include elements with an abundance as low as one-10-millionth, or 10^{-7}, times that of hydrogen.

Atoms are mostly empty space, just as the room you may be sitting in is mainly empty. A tiny, heavy, positively-charged nucleus lies at the heart of an atom, surrounded by a cloud of relatively minute, negatively-charged electrons that occupy most of an atom's space and govern its chemical behavior.

The nucleus is itself composed of positively-charged protons and neutral particles, called neutrons; the proton is 1836.153 times heavier than the electron, and the neutron is 1838.684 times heavier than the electron. Ernest Rutherford established the existence of a small, positive nucleus at the center of the atom in 1911; the neutron was discovered by James Chadwick in 1932. The charge on a proton is exactly equal to that on an electron, so the complete atom, in which the number of electrons equals the number of protons, is electrically neutral.

The electrons orbit the nucleus at relatively large distances. A nucleus is therefore much smaller than the atom, about 40 000 times smaller. To put the components into perspective, if the nucleus of an atom were the size of an orange, an atom would be as large as some mountains.

Hydrogen is the simplest atom, consisting of a single electron circling around a single proton. The nucleus of helium contains two neutrons and two protons, and so has two electrons in orbit.

The electron in a hydrogen atom revolves about the nucleus according to very specific rules, and is restricted to orbits with definite, quantized values of energy. If an electron wants to stay inside an atom, it has to obey the rules and can only occupy well-defined orbits with fixed energies. The allowed quantized orbits for the single electron in a hydrogen atom are illustrated in Figure 1.15. Observations of the hydrogen lines in the solar spectrum helped describe the internal behavior of the atom (Focus 1.5).

As it turns out, almost all of the electrons in the Sun do not follow any orbits at all. This is because atoms do not exist in most of the Sun, except in the cool visible layer in which dark absorption lines are formed. It is too hot everywhere else for whole atoms to survive. Innumerable collisions fragment the abundant hydrogen atoms into their constituent pieces. Their protons and electrons have been set free from their atomic bonds, wandering throughout the solar material unattached to each other. The fragmentation, or separation, process is termed "ionization", and the resulting ionized gas is called plasma.

An atom that is missing electrons is called an ion, and the proton is the simplest, lightest ion. Most of the ions in the solar plasma are protons, with lesser amounts of other positively-charged ions of heavier atoms. However, a

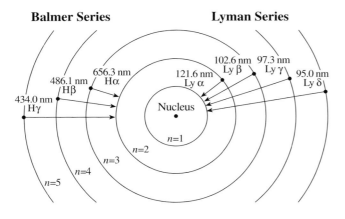

Balmer Series   **Lyman Series**

486.1 nm
Hβ
434.0 nm
Hγ
656.3 nm
Hα
121.6 nm
Ly α
102.6 nm 97.3 nm
Ly β   Ly γ
95.0 nm
Ly δ
Nucleus
n=1
n=2
n=3
n=4
n=5

Fig. 1.15 **Electron orbit transitions** A hydrogen atom's lone electron revolves around the hydrogen nucleus, a single proton, in well-defined orbits described by the integer n = 1, 2, 3, 4, 5, ... An electron absorbs or emits radiation when it makes a transition between these allowed orbits. The electron can jump upward, to orbits with larger n, by absorption of a photon of exactly the right energy, equal to the energy difference between the orbits; the electron can jump down to lower orbits, of smaller n, with the emission of radiation of that same energy and wavelength. Upward jumps produce absorption spectra; downward jumps, emission spectra. Transitions that begin or end on the n = 2 orbit define the Balmer series that is observed at visible wavelengths. They are designated by Hα, Hβ, Hγ, ... The Lyman series, with transitions from the first orbit at n = 1, is detected at ultraviolet wavelengths. The orbits are not drawn to scale for the size of their radius increases with the square of the integer n. The n = 1 orbit has a radius of 0.052 9177 nm; the n = 2 orbit has four times this radius; the n = 3 orbit is nine times larger than the n = 1 orbit, and so on.

Focus 1.5 Messages from inside the atom

The Sun's dark absorption lines help us determine how electrons orbit the nucleus of an atom. Radiation is emitted or absorbed by atoms when an electron jumps from one orbit to another, each jump being associated with a specific energy and a single wavelength, like one pure note. If an electron jumps from a low-energy orbit to a high-energy one, it absorbs radiation at this wavelength; radiation is emitted at exactly the same wavelength when the electron jumps the opposite way. This unique wavelength is related to the difference between the two orbital energies.

Adjacent lines of the same atom exhibit a strange regularity – they systematically crowd together and become stronger at shorter wavelengths. These regular spacings can only be explained if electrons occupy orbits with very specific, quantized energies (Fig. 1.15). They are described by the quantum number n which takes on integer values of $n = 1, 2, 3, 4, ..., $ (infinity) ∞.

The radius, R_n, of an electron orbiting a nucleus of charge Ze, where the atomic number, Z, is the number of protons in the nucleus and e is the charge on a proton, can be derived by assuming that the angular momentum of the electron of mass m_e and velocity V is quantized with

$$m_e V R_n = \frac{hn}{2\pi},$$

and equating the Coulomb force of attraction of the nucleus on the electron to the force from centripetal acceleration, or:

$$\frac{Ze^2}{R_n^2} = \frac{m_e V^2}{R_n},$$

to obtain:

$$R_n = \frac{h^2}{4\pi^2 m_e e^2} \frac{n^2}{Z} = a_0 \frac{n^2}{Z} = 0.529\,177 \times 10^{-10} \frac{n^2}{Z} \text{ m}.$$

The constant $a_0 = 0.529\,177 \times 10^{-10}$ m $= 0.052\,917\,7$ nm is known as the first (n = 1) Bohr orbit of hydrogen (Z = 1). The mass of the electron is $m_e = 9.1094 \times 10^{-31}$ kg, the charge of the electron is $e = 1.6022 \times 10^{-19}$ C, and Planck's constant $h = 6.6261 \times 10^{-34}$ J s.

The energy, E_n, of the nth orbit is given by:

$$E_n = -\frac{Ze^2}{R_n} + \frac{1}{2} m_e V^2 = \frac{-2\pi^2 m_e e^4 Z^2}{h^2 n^2} = \frac{-e^2 Z^2}{2a_0 n^2}.$$

The frequency, ν_{mn}, of the transition between the mth and nth orbits is given by equating the photon energy $h\nu_{mn}$ to the difference in the energies of the two orbits, or:

$$\nu_{mn} = \frac{1}{h} |E_m - E_n| = cR_\infty \left| \frac{1}{n^2} - \frac{1}{m^2} \right|,$$

where the Rydberg constant for infinite mass is:

$$R_\infty \frac{2\pi^2 m_e e^4}{ch^3} - 1.097\,37 \times 10^7 \text{ m}^{-1}$$

The wavelength λ_{mn} of the m to n transition can be inferred from the frequency from $\lambda_{mn} \times \nu_{mn} = c = 2.9979 \times 10^8$ m s^{-1}, the velocity of light.

The important hydrogen transitions between low–n orbits have been given the last names of the persons who first observed them. They are called the Lyman transitions for n = 1, the Balmer transitions for n = 2, the Paschen transitions for n = 3, the Brackett transitions for n = 4, and the Pfund transitions for n = 5, first observed in 1906, 1885, 1908, 1922, and 1924, respectively. The wavelengths of these m to n transitions are given in Table 1.6 for m between 1 and 10.

Series m	Lyman (n = 1)	Balmer (n = 2)	Paschen (n = 3)	Brackett (n = 4)	Pfund (n = 5)
2	121.567				
3	102.572	656.280			
4	97.2537	486.132	1875.10		
5	94.9743	434.046	1281.81	4051.20	
6	93.7803	410.173	1093.81	2625.20	7457.8
7	93.0748	397.007	1004.94	2165.50	4652.5
8	92.6226	388.905	954.598	1944.56	3739.5
9	92.3150	383.538	922.902	1817.41	3296.1
10	92.0963	379.790	901.491	1736.21	3038.4

[a] The wavelengths are given in nanometers where 1 nanometer $= 10^{-9}$ meters.

Table 1.6 **Wavelengths of the m to n transitions of hydrogen for n = 1 to 5 and m = 2 to 10**[a]

hydrogen atom can occasionally latch on to an electron and create a negative hydrogen ion. Overall, a plasma is electrically neutral with the negative charge of the electrons and any negative ions canceling the positive charge of the protons and heavier ions.

After solid, liquid and gas comes the fourth state of mater, plasma. The Sun is a huge incandescent ball of plasma. The stars are plasma, and indeed most of the Universe is plasma.

1.6 The planets are inside the expanding Sun

Just half a century ago, most people visualized our planet as a solitary sphere traveling in a cold, dark vacuum around the Sun. But we now know that the wide open spaces in our solar system are not empty. They are filled with pieces of the Sun! Our star constantly blows itself away, sending a billion (10^9) kilograms, or a million tons, of electrons and protons into space every second.

The tenuous solar atmosphere is expanding out in all directions, filling interplanetary space with a ceaseless wind that is forever blowing from the Sun. This solar wind is mainly composed of electrons and protons, set free from the Sun's abundant hydrogen atoms, but it also contains heavier ions and magnetic fields. The Earth and other planets are immersed within the seemingly eternal, stormy wind, so we are actually living in the outer part of the Sun.

The solar gale blows past the planets and engulfs them, carrying the Sun's rarefied atmosphere out to the space between the stars. The radial, supersonic outflow thereby creates a huge bubble of rarefied plasma, with the Sun at the center and the planets inside, called the heliosphere (Fig. 1.16), from *Helios* the Greek word for the Sun.

Within the heliosphere, conditions are regulated by the Sun. Its domain extends out to about 150 times the distance between the Earth and Sun, marking the outer boundary or edge of the solar system. Out there, the solar wind has

Fig. 1.17 **Comet tails** This photograph shows the curved dust tail and the straight ion tail of Comet Mrkos. Both comet tails point away from the Sun. The electrified solar wind deflects the charged ions and accelerates them to high velocities, creating the relatively straight ion tails. The radiation pressure of sunlight suffices to blow away the un-ionized comet dust particles, forming a broad arc that resembles a scimitar. (Courtesy of Lick Observatory.)

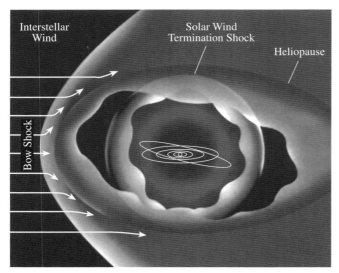

Fig. 1.16 **The heliosphere** With its solar wind going out in all directions, the Sun blows a huge bubble in space called the heliosphere. The heliopause is the name for the boundary between the heliosphere and the interstellar gas outside the solar system. Interstellar winds mold the heliosphere into a non-spherical shape, creating a bow shock where they first encounter it. The orbits of the planets are shown near the center of the drawing.

become so weakened by expansion that it can no longer repel interstellar forces (Section 7.4).

Our first clue to the existence of the solar wind came decades ago from comets. Touched by the Sun's warmth, a comet's ice and dust boil off, forming two tails that always point away from the Sun (Fig. 1.17). One is a yellow tail of dust and dirt, that litters the comet's curved path. The dust is pushed away from the Sun by the pressure of sunlight. The other tail is electric blue, shining in the light of ionized particles. The straight ion tail acts like a wind sock that is blown in the direction of the solar wind, and demonstrates the existence of this continuous electrified flow, streaming radially out in all directions from the Sun. Both comet tails always point away from the Sun, so they travel head-first when approaching the Sun and tail-first when departing from it.

The German astronomer, Ludwig Biermann, noticed that the ions in a comet's tail move with velocities many times higher than could be caused by the weak pressure of sunlight. So he proposed, in the 1950s, that a perpetual space-filling flow of electrically-charged particles pours out of the Sun at all times and in all directions, colliding with the

Focus 1.6 Escape from the Sun, Earth and Moon

When the kinetic energy of motion of an object or a particle of mass, m, moving at velocity, V, is just equal to the gravitational potential energy exerted on it by a larger mass, M, we have the relation:

$$\text{kinetic energy} = \frac{mV^2}{2} = \frac{GmM}{D} = \text{gravitational potential energy},$$

where the Newtonian gravitational potential is $G = 6.6726 \times 10^{-11}$ N m^2 kg^{-2}, and D is the distance in meters between the centers of the two masses. When we solve for the velocity, we obtain:

$$V_{escape} = \left[\frac{2GM}{D}\right]^{\frac{1}{2}}$$

where the subscript escape has been added to show that the small mass must be moving faster than V_{escape} to leave a larger mass, M. This expression is independent of the value of the smaller mass, m.

At the visible solar disk, where D becomes the solar radius of 6.955×10^8 m, the equation gives

$$V_{escape}\,(\text{Sun}) = 6.178 \times 10^5 \text{ m s}^{-1},$$

where the Sun's mass is 1.989×10^{30} kg.

By way of comparison, the escape velocity at the Earth's surface is:

$$V_{escape}\,(\text{Earth's surface}) = 1.12 \times 10^4 \text{ m s}^{-1}$$

where the mean radius of the Earth is 6.371×10^6 m and the mass of the Earth is 5.9742×10^{24} kg. The escape velocity from the surface of the Moon is:

$$V_{escape}\,(\text{Moon's surface}) = 2.37 \times 10^3 \text{ m s}^{-1}$$

The Moon's mass is 7.348×10^{22} kg and its mean radius is 1.738×10^6 m.

A rocket must move faster than 11.2 thousand meters per second if is to move from the Earth into interplanetary space, and if it travels at a slower speed the rocket will crash back down into the Earth. A lunar craft only needs to be propelled at about one-fifth of this speed to leave the Moon. There is no atmosphere on the Moon because it has a very low escape velocity, and molecules can therefore easily leave it.

An atom, ion or molecule moves about because it is hot. Its kinetic temperature, T, is defined in terms of the thermal velocity, $V_{thermal}$, given by the expression equating the thermal energy to the kinetic energy of motion,

$$\text{thermal energy} = \frac{3}{2}kT = \frac{1}{2}mV^2_{thermal} = \text{kinetic energy},$$

or, solving for the thermal velocity:

$$V_{thermal} = \left[\frac{3kT}{m}\right]^{\frac{1}{2}},$$

where Boltzmann's constant $k = 1.38066 \times 10^{-23}$ J K^{-1}, the temperature of the particle is denoted by T, and its mass by m. We see right away that at a given temperature, lighter particles move at faster speeds. Colder particles of a given mass travel at slower speed. Anything will cease to move when it reaches absolute zero on the kelvin scale of temperature.

Everything in the Universe moves, and there is nothing completely at rest. You might say that motion seems to define existence. When you stop moving it is all over.

For the lightest-known element, hydrogen, our expression gives

$$V_{thermal}\,(\text{hydrogen atoms}) = 157 \, T^{1/2} \text{ m s}^{-1},$$

where the mass of a hydrogen atom is $m = 1.6735 \times 10^{-27}$ kilograms. So, hydrogen atoms in the visible solar disk, with an effective temperature of 5780 K, move at about 12 thousand meters per second. Since this velocity is way below the Sun's escape velocity, hydrogen and any other heavier element in the bright solar disk are not hot enough or moving fast enough to leave the Sun by themselves. Even in the higher, rarefied solar atmosphere, where the temperature is about 2 million degrees, the thermal velocity of a hydrogen atom is 222 thousand meters per second. To leave the Sun, this hydrogen has to be given an extra push out to a distance of a few solar radii, where the solar gravity and escape velocity have become smaller. The same conclusion applies to protons that have essentially the same mass as a hydrogen atom. Since a free electron is 1836 times lighter than a proton, it has a thermal velocity that is 42.8 times faster at a given temperature.

comet's ions and imparting momentum to them. This would accelerate the ions to high speeds and push them radially away from the Sun in straight tails.

The comet tails serve as probes of the solar particles streaming away from the Sun; it is something like putting a wet finger in the wind to show it is there. Biermann estimated a speed of 500 thousand to one million meters per second from moving irregularities and the directions of comet tails.

But why is the Sun continuously blowing itself away? Its outer atmosphere, called the corona, is exceedingly hot, with a temperature of a few million degrees kelvin (Section 5.3). The million-degree corona is so hot that it cannot stand still. Indeed, the solar wind consists of an overflow corona, which is too hot to be entirely constrained by the Sun's inward

gravitational pull. The hot gas creates an outward pressure that tends to oppose the inward pull of the Sun's gravity. At great distances, where the solar gravity weakens, the hot protons and electrons overcome the Sun's gravity and accelerate away to supersonic speed (Focus 1.6).

In 1957 the English geophysicist Sydney Chapman noticed that the free electrons in the million-degree corona make it a very good thermal conductor, even better than a metal. The electrons in the corona would therefore carry its intense heat far into space, somewhat like an iron bar that is heated at one end and therefore becomes hot all over. This meant that the outer solar atmosphere must spread out to the Earth's orbit and beyond; and that the temperatures just outside the Earth's atmosphere must be about 100 000 degrees kelvin. According to Chapman, this extended, non-expanding gas would block any outward stream of charged particles from the Sun. Biermann disagreed, arguing that the solar particles would sweep any stationary gas out of the solar system.

These conflicting ideas, of a static atmosphere and a flow of charged particles, were reconciled by Eugene N. Parker of the University of Chicago. In 1958, he added dynamic terms to Chapman's equations, showing how a relentless flow might work, and dubbing it the solar wind.

Parker's theoretical conclusions were very controversial, and were initially received with a great deal of skepticism. The Sun might be sporadically ejecting material from localized regions, but it was difficult to envisage a continual ejection over the entire corona. Critics also wondered how the solar atmosphere could be hot enough to sustain such a powerful wind. Then scientists realized that there are better ways than comets to see the solar wind in space. Spacecraft can be sent out to show exactly how the solar wind is blowing, measuring its density, velocity and magnetic fields.

All reasonable doubt concerning the existence of the solar wind was removed when spacecraft were sent out to touch and feel it, first by the Soviet Luna 2 spacecraft in 1959 and then by the American Mariner 2 in 1962–63. The Luna 2 measurements indicated a solar wind flux of 2 million million (2×10^{12}) ions (presumably protons) per square meter per second. This is in rough accord with all subsequent measurements.

More than one hundred days of Mariner 2 data, obtained as the spacecraft traveled to Venus, were used to show that charged particles are continuously emanating from the Sun, for at least as long as Mariner observed them. The velocity of the solar wind was accurately determined by the Mariner instruments, with an average speed of 500 thousand meters per second, in rough accord with Biermann's and Parker's predictions.

The average wind ion density was shown by Mariner to be 5 million (5×10^6) protons per cubic meter near the distance of the Earth from the Sun. We now know that such a low density close to the Earth's orbit is a natural consequence of the wind's expansion into an ever greater volume, but that variable wind components can gust with higher densities.

The Mariner data unexpectedly indicated that the solar wind has a slow and a fast component. The slow one moves at a speed of 300 thousand to 400 thousand meters per second; the fast one travels at twice that speed. The low-velocity wind was identified with the perpetual expansion of the million-degree solar atmosphere. The high-velocity component swept past the spacecraft every 27 days, suggesting long-lived, localized sources on the rotating Sun. When viewed from the Earth, the solar equator rotates once every 27 days. Moreover, peaks in geomagnetic activity were correlated with the arrival of these high-speed streams at the Earth, indicating a direct connection between some

Parameter	Approximate mean value
Particle density, N	10 million particles per cubic meter (5 million electrons and 5 million protons)
Velocity, V slow $\qquad\qquad V$ fast	4×10^5 m s^{-1} 8×10^5 m s^{-1}
Flux, F	10^{12} to 10^{13} particles per square meter per second.
Temperature, T	1.2×10^5 K (protons) to 1.4×10^5 K (electrons)
Particle thermal energy, kT	2×10^{-18} J $\approx$ 12 eV
Proton kinetic energy, $0.5\, m_p V^2$	10^{-16} J $\approx$ 1000 eV = 1 keV
Particle thermal energy density, NkT	10^{-11} J m^{-3}
Proton kinetic energy density, $0.25\, Nm_p V^2$	10^{-9} J m^{-3}
Magnetic field strength, H	6×10^{-9} T = 6 nT = 6 x 10^{-5} G

[a]These solar-wind parameters are at the mean distance of the Earth from the Sun, or at one astronomical unit, 1 AU, where 1 AU = 1.496×10^{11} m; the Sun's radius, R$_\odot$, is R$_\odot$ = 6.955×10^8 m. Boltzmann's constant k = $1.380\ 66 \times 10^{-23}$ J K^{-1} relates temperature and thermal energy. The proton mass m$_p$ = $1.672\ 623 \times 10^{-27}$ kg.

Table 1.7 **Mean values of solar-wind parameters at the Earth's orbit**[a]

unknown region on the Sun and disturbances of the Earth's magnetic field.

Interplanetary space probes have been making in-situ (Latin for "in original place", or literally "in the same place") measurements of the solar wind for decades, both within space near the Earth and further out in the Earth's orbital plane (Table 1.7). Unlike any wind on Earth, the solar wind is a rarefied plasma or mixture of electrons, protons and heavier ions, and magnetic fields that streams radially outward in all directions from the Sun at supersonic speeds of hundreds of thousands of meters per second.

This perpetual solar gale brushes past the planets and engulfs them, carrying the Sun's atmosphere out into interstellar space at the rate of almost a million tons (10^6 tons = 10^9 kilograms) every second. As the corona disperses, it must be replaced by gases welling up from below to feed the wind. Exactly where this material comes from is an important subject of contemporary space research, subsequently discussed in greater detail in Section 7.2.

1.7 How the solar system came into being

The origin of the solar system is one of the most fundamental problems of science. It helps us understand where we came from, and whether or not life might be possible on other planets. Important constraints on any origin theory include the ages of the satellites and planets, as well as the Sun, and the extraordinary regularity of the planetary orbits. The composition of the planets and the distribution of mass and angular momentum in the solar system must also be explained by any successful theory. In this section we describe these boundary conditions, and show how the nebular hypothesis can explain them. According to the nebular hypothesis the Sun and planets formed together during the collapse of a rotating interstellar cloud, called the solar nebula. It suggests that planetary systems might be relatively common, and that they will be found around other stars.

Age of the solar system

Precisely when did the solar system originate? Its precise age is determined by examining primitive meteorites, ancient rocks returned from the Moon, and deep ocean sediments. These relics have remained unaffected by the erosion that removed the primordial record from most terrestrial rocks.

Their ages can be determined by measuring the relative amounts of radioactive materials and their non-radioactive products. When this ratio is combined with the known rates of radioactive decay, the time since the rock solidified and "locked-in" the radioactive atoms is found. The isotopes used for dating are given in Table 1.8 together with the half-life, the time for half of the radioactive substance to decay into a stable element.

The method is known as radioactive dating, and it works this way. Certain types of nuclei, known as unstable parent isotopes, decay at a constant rate into stable lighter isotopes known as daughters. By measuring the amount of daughter material and knowing the rate of decay, the age of the rock can be estimated. The detailed mathematical treatment is given in Focus 1.7. The method is something like determining how long a log has been burning by measuring the amount of ash and watching a while to determine how rapidly the ash is being produced.

The daughter isotopes must be trapped in the rock and not escape or the estimated age will be too short. In fact, the daughters can escape quite easily when the rock is molten; only when it cools and solidifies do the daughters start to accumulate. For this reason, the ages determined for the rocks are really the times since the rock became solid. And if the rock is remelted, say by the impact of a meteorite, its radioactive clock is reset, and the age will measure the time since the last solidification.

The radioactive dating method has been used to study primitive meteorites, known as carbonaceous chondrites, to obtain an age of 4.566 billion years, with an uncertainty of about 0.002 billion years (Focus 1.7). These meteorites are thought to date back to the earliest days of the solar system. Radioactive dating of the oldest rocks returned from the Moon indicate an age of about 4.5 billion years, and deep sediments in the Earth's oceans are dated at 4.55 billion years. Astronomers sometimes use the term gigayear, abbreviated Gyr, to denote cosmic ages, where $1 \text{ Gyr} = 10^9 =$ one billion years, but here we will date ages in the equivalent billions of years.

Rounding off the numbers and allowing for possible systematic errors, we can say that the Earth, Moon and meteorites solidified at the same time some 4.6 billion years ago, with an uncertainty of no more than 0.1 billion years. If the solar system originated as one entity, then this should also be the approximate age of the Sun and the rest of the solar system.

The Sun had to be shining for all or at least most of this time, to keep the ocean waters from freezing and to make the Earth a suitable place for life. Sedimentary rocks found on Earth, which must have been deposited in liquid water, date

Radioactive parent [Name (Symbol) Mass no.]	Stable daughter [Name (Symbol) Mass no.]	Half-life (millions of years)
Rubidium (Rb) 187	Strontium (Sr) 87	48 800
Rhenium (Re) 187	Osmium (Os) 187	44 000
Lutetium (Lu) 176	Hafnium (Hf) 176	35 700
Thorium (Th) 232	Lead (Pb) 208	14 050
Uranium (U) 238	Lead (Pb) 206	4 470
Potassium (K) 40	Argon (Ar) 40	1 270
Samarium (Sm) 146	Neodymium (Nd) 142	100
Uranium (U) 235	Lead (Pb) 207	704
Plutonium (Pu) 244	Thorium (Th) 232	83
Iodine (I) 129	Xenon (Xe) 129	16
Palladium (Pd) 107	Silver (Ag) 107	6.5
Manganese (Mn) 53	Chromium (Cr) 53	3.7
Aluminum (Al) 26	Magnesium (Mg) 26	0.72

Table 1.8 **Radioactive isotopes used for dating**

Focus 1.7 The age of rocks

Radioactive elements can be used to clock the age of rocks on the Earth's surface, meteorites, and lunar rock samples. The number, N, of radioactive atoms in a rock changes with the time, t, since its solidification according to the differential equation:

$$\frac{dN}{dt} = -\lambda N,$$

where λ is the decay rate. This equation integrates to give the number of radioactive atoms, N_t, at time t:

$$N_t = N_0 \exp(-\lambda t) = N_0 \exp\left(-0.693 t / \tau_{1/2}\right),$$

where N_0 is the number of atoms at time t = 0, the time of solidification. The radioactive decay constant $\lambda = 0.693/\tau_{1/2} = \ln(2)/\tau_{1/2}$ and $\tau_{1/2}$ is the half-life of the radioactive species. Half-lives for the decay of radioactive isotopes are given in Table 1.8.

The number of radioactive atoms in the rock will be halved in a time equal to the half-life. Radioactive uranium, ^{238}U, decays, for example, into lead, ^{206}Pb (which is stable), with a half-life of about 4.47 billion years; so every 4.47 billion years the amount of uranium-238 in a rock will be halved. We can apply the equations to ^{238}U, and express the abundance in terms of another kind of lead, ^{204}Pb which is not a radioactive decay product. If a terrestrial rock, lunar sample or a non-terrestrial meteorite became a closed system at time t = 0, then the present abundances of lead and uranium are related by the equation:

$$\left(\frac{^{206}Pb}{^{204}Pb}\right)_t = \left(\frac{^{238}U}{^{204}Pb}\right)_t [\exp(\lambda_{238}t) - 1] + \left(\frac{^{206}Pb}{^{204}Pb}\right)_0,$$

where the subscripts t and 0 denote the present and initial abundances, respectively.

If all of the rock samples have the same initial $^{206}Pb / ^{204}Pb$ abundance, and if all of them have the same age, t, then a plot of $(^{206}Pb/^{204}Pb)$t against $(^{238}U/^{204}Pb)$t should result in a straight line of slope $[\exp(\lambda_{238}t) - 1]$. Such a plot is called an isochron. If a system formed t years ago and initially contained no lead, then a curve of the ratios $^{207}Pb / ^{206}Pb$ and $^{238}U / ^{206}Pb$ also provides the age t.

These and similar methods have been used to determine an age of carbonaceous chondrite meteorites of t = 4.566 ± 0.002 billion years, where one billion years = 10^9 yr = 1 Gyr. (Also abbreviated as Ga, where "a" is short for "annum", latin for "year".)

Radioactive dating of samples returned from the lunar highlands indicate that the Moon formed about 4.5 billion years ago and that an intense bombardment created the large impact basins and most of the highland craters about 4 billion years ago. An era of volcanism, from 4.2 to 3.1 billion years ago, flooded the great impact basins with lava and produced the dark circular Maria that can be seen today.

The oldest known rocks on earth are 3.9 billion years old, but terrestrial crystals of zircon are as old as 4.4 billion years. Erosion by wind, water and geological processes has wiped out the oldest terrestrial rocks. The deep ocean sediments, that are least affected by continuing geological activity on Earth, have ages of about 4.55 billion years.

from 3.8 billion years ago, and there is fossil evidence in these rocks for the emergence of life at least 3.5 billion years ago.

Although a star with the Sun's mass can continue to shine for more than 10 billion years, the observed regularities in the solar system strongly suggest that the Sun formed together with the planets about 4.6 billion years ago.

Regular planetary orbits and the nebular hypothesis

Any successful theory for the origin of the solar system must account for the regular pattern of the orbits of the planets. The ancients noted that:

1 The planets move in a narrow band around the sky: the band of the "zodiac". This implies that the orbits all lie in nearly the same plane.
2 Viewed from the northern hemisphere of the Earth, the orbits of the planets move in the same direction from right to left across the sky, corresponding to prograde or counterclockwise revolution about the Sun. (Temporary interludes of apparent backward motion as viewed from the Earth, the retrograde loops of Mars, Jupiter and Saturn, are explained by Sun-centered orbits in which more-distant planets move around the Sun with slower speeds.)

Telescopes have revealed several other regularities:

3 The orbits are nearly circular.
4 The equator of the Sun's rotation nearly coincides with the plane of the planets' orbits.
5 The orbits of most of the satellites imitate the planets in being confined to the planet's equatorial plane and following prograde motion.
6 With the exception of Venus, Uranus and Pluto, the Sun and the planets all rotate about their axes in the same prograde direction.
7 The orbital radii of the planets follows a regular pattern, called the Titius–Bode law (Section 1.4) in which a planet's orbit has roughly twice the radius of its inward neighbor.

This regular orbital arrangement of the planets is not accidental. Even if a million million million (10^{18}) solar systems were made haphazardly and the planets and satellites were thrown into randomly oriented orbits, only one of these solar systems would be expected to look like our own. So, it is exceedingly unlikely that the planets became aligned by chance.

Although Newton's and Kepler's laws describe the present behavior of the solar system, they cannot explain the remarkable arrangement. Some additional constraints are required that describe the state of affairs before the planets were formed and set in motion. These initial conditions are provided by the nebular hypothesis, in which the Sun and planets formed out of a single collapsing, rotating cloud of interstellar gas and dust, called the solar nebula (Fig. 1.18).

The nebular hypothesis has been around for a long time, with two historical versions. In the version proposed by the German philosopher Immanuel Kant, in 1755, the Sun

Fig. 1.18 **Formation of the solar system** In this artist's impression, unformed planets circle a nascent star, our Sun, before its nuclear fires burst forth. According to the nebular hypothesis, the Sun and planets were formed at the same time during the collapse of an interstellar cloud of gas and dust that is called the solar nebula. (Courtesy of Helmut K. Wimmer, Hayden Planetarium, American Museum of Natural History.)

formed at the center of the spinning solar nebula, while the planets formed from swirling condensations in a flattened disc revolving around it. According to Pierre Simon Laplace's modification, suggested in 1796, the shrinking Sun shed a succession of gaseous rings, and each ring condensed into a planet. Then each planet, in turn, became a small rotating nebula in which its own family of rings and satellites was born.

Modern versions of the nebular hypothesis provide additional caveats, but the basic tenets of the original idea are still valid. Billions of years ago the spinning solar nebula, attracted by its own gravity, fell in on itself, getting denser and denser, until its middle became so packed, so tight and hot, that the Sun began to shine. The planets formed at the same time, within a flattened rotating disk centered on the contracting proto-Sun.

This is the essence of the original nebular hypothesis, which explained qualitatively the fact that the planets and their satellites all revolve in the plane that coincides with the equator of the rotating Sun. The highly regular pattern, which cannot be accidental, is a natural consequence of the rotation and collapse of a solar nebula composed of gas and dust from which the planets were produced.

Terrestrial planets	Mercury	Venus	Earth	Mars
Equatorial radius, R_p (m)	2 439 000	6 051 000	6 378 140	3 397 000
Equatorial radius, (R_E = 1.000)	0.382	0.949	1.000	0.533
Reciprocal mass (M_o/M_p)	6 023 600	408 525.1	332 946.043	3 098 710
Mass, M_p (kg)	3.3022×10^{23}	4.8690×10^{24}	5.9742×10^{24}	6.4191×10^{23}
Mass, (M_E = 1.000 00)	0.055 27	0.814 99	1.000 00	0.107 45
Mean mass density (kg m^{-3})	5430	5250	5520	3930
Giant planets	Jupiter	Saturn	Uranus	Neptune
Equatorial radius, R_p (m)	71 492 000	60 268 000	25 559 000	24 764 000
Equatorial radius, (R_E = 1.0)	11.19	9.46	3.98	3.81
Reciprocal mass (M_o/M_p)	1047.3492	3497.91	22902.94	19434
Mass, M_p (kg)	1.8992×10^{27}	5.6865×10^{26}	8.6849×10^{25}	1.0235×10^{26}
Mass, (M_E = 1.0000)	317.894	95.1843	14.5373	17.1321
Mean mass density (kg m^{-3})	1330	710	1240	1670

[a] Pluto has a radius of 1.123×10^6 m and a mass of 1.36×10^{22} kg.

Table 1.9 **Physical properties of the planets**[a]

Explaining the details

If the nebular hypothesis is correct, and the whole solar system originated at the same time, then you might expect the planets to have a similar chemical composition to the Sun. The abundances of the elements in the giant planet Jupiter do indeed mimic those of the Sun, with a predominance of the lightest element, hydrogen. Unlike the Sun, the Earth is mainly composed of heavier elements, but this can be explained by the loss of volatile gases in the regions of the solar nebula near the central, hot, forming Sun.

The planets fall into two major compositional groups, the compact "terrestrial", or Earth-like, planets of low mass and high mass density; and the extended giant planets with a larger mass and size but lower mass density (Table 1.9). Dense rocky substances dominate the four terrestrial planets that are nearest the Sun (Mercury, Venus, Earth and Mars), while the lighter icy and gaseous substances dominate the outer giant planets (Jupiter, Saturn, Uranus and Neptune).

These compositional differences appear to result from the fact that the terrestrial planets formed close to the bright young Sun. In the inner regions of the solar nebula, the higher temperatures would vaporize icy material that could not condense, leaving only rocky substances to coalesce and merge together to form the terrestrial planets. Also the modest masses of these planets and their proximity to the Sun did not allow them to capture and retain the abundant

---

Focus 1.8 Angular momentum, gravitational collapse and the halting spin of old age

According to the nebular hypothesis, the Sun and planets formed together as the result of the gravitational collapse of a rotating interstellar cloud, called the solar nebula. During such a contraction, the angular momentum is conserved. For a body of mass, M, radius, R, and rotation period, P, this means that:

$$\text{angular momentum} = \frac{2\pi MR^2}{P} = \text{constant},$$

where the rotation velocity $V = 2\pi R/P$.

Since the Sun contains 99.87 percent of the mass of the solar system, we can assume that the mass remains constant during the collapse of the solar nebula to form the Sun. The nearest star to the Sun lies at a distance of 4.29 light-years = 4.06×10^{16} m. We might therefore speculate that the solar nebula had an initial radius, R_{sn}, of about one-thirtieth of this distance, or

$$\text{radius of solar nebula} = R_{sn} = 1.35 \times 10^{15} \text{ m}.$$

We can also conservatively assume that the interstellar gas cloud was initially turning with the general rotation of the Milky Way at the Sun's distance (Section 2.1) with the rotation period given by:

$$\text{rotation period of solar nebula} = P_{sn} = 2.4 \times 10^8 \text{ yr} = 7.6 \times 10^{15} \text{ s},$$

where 1 yr = 3.1557×10^7 s. It is hard to imagine a slower rotation period.

By the time the Sun collapsed to its present size of $R_\odot = 6.955 \times 10^8$ m, its rotation period ought to be:

$$\text{rotation period of the young Sun} = P_{sn}\left(\frac{R_\odot}{R_{sn}}\right)^2 = 2 \times 10^3 \text{ s} = 0.55 \text{ h}.$$

By way of comparison, Jupiter has a rotation period of 9 h 55 min, and we would expect objects of similar density, such as the Sun and Jupiter, to have about the same rotation periods if they formed under similar conditions. However, the Sun now has an equatorial rotation period of 25.7 days, or 2.22×10^6 s, more than 1000 times longer than expected. Some process other than gravitational contraction must have slowed the Sun's spin after its birth, perhaps by magnetic braking or as the result of mass and momentum transfer between the proto-Sun and its surrounding disk.

The Earth was also born with a more rapid rotation than it has now. Our planet has been slowed down by the rhythmic ebb and flow of the ocean tides, driven by the Moon's gravitational pull on the Earth's ocean water. As the Earth's rotation carries the continents past the tidal ocean bulges, the water at beaches will rise and fall. These tides create eddies in the water, producing friction and dissipating energy at the expense of the Earth's rotation. The tides therefore act as brakes on the spinning Earth, slowing it by friction in much the same way that the brakes of a car slow its wheels.

Tidal friction is still slowing the rotation of the Earth, and the day is becoming longer at the rate of 2 milliseconds (0.002 seconds) per century, or one second every 50 000 yrs. That doesn't sound like much, but it adds up over the aeons. If the present rate for the lengthening of the day is extrapolated back into the distant reaches of time, the Earth's day would be only 4 hours long about four billion years ago.

Since angular momentum is conserved in the Earth–Moon system, the decrease in the Earth's rotational angular momentum is transferred to the Moon, increasing its orbital angular momentum by an equivalent amount. This causes the Moon to swing away from the Earth at the rate of about 0.04 meters per year, which is only a fraction of the 384.4 million, or 3.844×10^8, meters. Small as it is, the increasing distance of the Moon has been measured with laser beams that bounce light off reflectors planted on the Moon by the Apollo astronauts. By measuring the round-trip travel time of laser-pulses, and multiplying the time by the velocity of light, the laser-ranging experiments have determined the distance to the Moon to an astonishing accuracy of 0.0007 m, or about 2 parts in a million million. Such measurements, made over the years, indicate that the Moon is now moving away from the Earth at the rate of 0.0382 meters per year.

The planets Mercury and Venus have exceptionally slow rotation periods, of 58.646 and 243 Earth-days, respectively. The youthful energy of their fast initial rotation has probably been tempered by tidal interaction with the massive nearby Sun. These would be tides in the solid body of the planets, for there are no oceans on Mercury or Venus. Such a tidal interaction is suggested by the fact that Mercury spins on its axis exactly three times during two full revolutions about the Sun, so its rotation period is exactly two-thirds of Mercury's orbital period of 87.969 Earth-days.

lighter gases, hydrogen and helium, directly from the solar nebula.

At larger distances where the solar nebula was cooler, icy substances could condense and combine with heavier ones to form the large, massive cores of giant planets. These cores eventually became sufficiently massive to gravitationally accrete, accumulate and capture the surrounding hydrogen and helium. The low temperatures at remote distances from the Sun thus enabled the giant planets to retain the abundant light gases and grow even bigger, with large masses and low mass densities.

The terrestrial planets close to the Sun might have acquired some gas from the solar nebula, but could not hold on to it because of their low mass and high temperatures. The powerful winds of the new-born Sun would also most likely blast away any nebular gas in the primordial atmospheres of the terrestrial planets.

The nebular hypothesis also has to be adjusted to explain the present distribution of mass and angular momentum in the solar system. Most of the mass is in the central Sun; the total mass of all the planets is a fraction of one percent of the Sun's mass. Yet, the angular momentum of the solar system is concentrated in the orbital angular momentum of Jupiter, and the rotating Sun has less than one percent of the amount carried by this giant planet. In other words, a very small fraction of the mass of the solar system has significant angular momentum, while most of the mass has relatively little angular momentum.

According to the law of conservation of angular momentum, the rotation of a shrinking object will speed up as the radius decreases, just as a skater will when her arms are pulled in (Focus 1.8). The young Sun should therefore be rotating very rapidly, with a rotation period of perhaps one hour, but it is now spinning at an unexpectedly slow rate with an equatorial rotation period of 25.7 days.

The spinning Sun must have slowed down as it aged. One possibility is the action of magnetic brakes that could connect the Sun to the distant, slowly rotating material in the surrounding disk. Another possibility is that frictional forces caused mass to move inwards from the disk to the central Sun while transporting angular momentum outward. Whatever the exact explanation, it seems to apply to other new-born stars, for it is the youngest stars that rotate at the fastest speeds.

The observations of rotating flattened disks of gas and dust around many proto-stars in nearby molecular clouds provides strong supporting evidence for the nebular hypothesis. Moreover, planets have recently been discovered orbiting several nearby stars, which would be expected if the nebular hypothesis is extended to the formation of other stars and planetary systems. This therefore brings us to a consideration of the Sun's relationship to other stars.

CHAPTER TWO

The Sun as a star

A star dies while new ones are born A massive star ended its life in a tremendous supernova explosion, creating shock waves that are seen plowing through the interstellar medium. Sulfur atoms are emitting the red light, oxygen atoms radiate the blue light, and hydrogen atoms emit the green light. This image of a supernova remnant, known as the Cygnus Loop, also shows dense clumps of gas as small as our solar system, just as they are overrun by the shock waves; they may mark the location of future stars and planets. The bluish ribbon of light stretching from left to right across the image may be a knot of gas ejected by the supernova. Traveling at about 3 million meters per second, this interstellar "bullet" is just catching up to the slower-moving shock front. Supernova remnants play an important role in stellar evolution by enriching space with heavy elements and by triggering new star formation by compressing interstellar gas. (Courtesy of Jeff J. Hester, the Hubble Space Telescope and NASA.)

2.1 The Sun's place in the Milky Way

On a clear moonless night, away from city lights and pollution, we can see a broad, diffuse band of light stretching across the sky. This is the Milky Way (Fig. 2.1). The ancients imagined it to be milk spilled from the breast of a goddess, but we now know that it is a flattened disk-shaped collection of stars, seen edgewise from inside.

Our Sun is a star, and all the other stars are distant Suns, kin to our own daytime star. Some of the ancient Greeks apparently suspected this, as did Isaac Newton, but we only became certain that the Sun was a typical star when we could measure the distances of nearby stars, and compare their luminosity to that of the Sun.

Yet, the Sun remains a special star. Together with the Earth, we mortals are bound to the Sun, and since time immemorial it has supplied us with light and heat. As Francis William Bourdillon stated in 1878:

> The night has a thousand eyes,
> And the day but one;
> Yet the light of the bright world dies,
> With the dying Sun.

Altogether there are about 5000 stars visible to the unaided eye in the entire celestial sphere. With a good pair of binoculars, the hazy band of the Milky Way can be resolved into a host of tiny stars, perhaps 200 000 of them. Countless multitudes of stars can be detected with powerful telescopes.

There are hundreds of billions of stars in the Milky Way, with a roughly equal mass of interstellar gas and dust. This immense collection of stars and interstellar matter is called our Galaxy, with a capital G. The luminous band of stars, known as the Milky Way, describes the disk and plane of our Galaxy.

How far away are the stars in the Milky Way? Their distances are often specified in light years. The light year is the distance that a ray travels in one year at the speed of light, 299 792 458 m s^{-1}. Astronomers often prefer to specify distances between stars in parsecs, or pc for short, while distances comparable to the extent of our Galaxy are measured in units of kiloparsecs, abbreviated kpc and equal to one thousand parsecs.

To set the units straight in your mind:

one parsec = 1 pc = 3.0857×10^{16} meters.

Fig. 2.1 **The Milky Way at visible wavelengths** A panoramic view of the Milky Way as seen by optical telescopes. Luminous concentrations of bright stars and dark intervening dust clouds extend in a band across the celestial sphere. This Milky Way defines the plane and main disk of our Galaxy. We live in this disk, and look out through it. Our view is eventually blocked by the build-up of interstellar dust, and the light from more distant regions of the disk cannot get through. The center of the Galaxy is located at the center of the picture, in the direction of the constellation Sagittarius. Although the disk appears visibly wider in that direction, the galactic center is not visible through the dust. The Large and Small Magellanic Clouds, small companion galaxies to the Milky Way, can be seen as bright swirls of light below the plane to the right of center. (This map of the Milky Way was hand-drawn from many photographs by Martin and Tatjana Keskula under the direction of Knut Lundmark; courtesy of the Lund Observatory, Sweden.)

We can compare this to the average distance between the Sun and the Earth, the astronomical unit, or AU, with:

one parsec = 1 pc = 206 265 AU,

and to the light-year:

one parsec = 1 pc = 3.2616 light-years.

Proxima Centauri, is the nearest star other than the Sun, at a distance of 1.314 parsec or 4.29 light-years. The light you see from this star left it 4.29 years ago. It is 271 000 times farther away from us than the Sun.

Stellar distances can be gauged by comparing their intrinsic brightness to the amount of their radiation reaching the Earth. This method depends on the inverse square law, in which the absolute luminosity, L, is reduced by the square of the distance, D, as light travels outward and is dispersed into space. It states that the

$$\text{apparent luminosity} = l = \frac{L}{D^2} = \frac{\text{absolute luminosity}}{\text{square of distance}}$$

or that:

$$\text{distance} = D = \left(\frac{L}{l}\right)^{1/2}.$$

This explains why the lights of an oncoming car brighten when it approaches you and its distance decreases.

If you have a separate method of establishing the distance of a star, you can use the inverse-square law to determine its intrinsic brightness, or absolute luminosity. For example, the distances to nearby stars have been measured by triangulation (Focus 2.1) and combined with their observed luminosity to show that their intrinsic brightness is often comparable to that of the Sun, and sometimes vastly exceeds it. This means that the other stars are truly distant Suns.

The unit of absolute luminosity, L, is a watt, or equivalently a joule per second. The unit of apparent luminosity is watts per square meter. For the Sun we have $L_\odot = 3.85 \times 10^{26}$ watts; this number was derived in Section 1.3, Focus 1.2.

Astronomers often describe the apparent and intrinsic brightness of celestial objects in terms of a magnitude scale invented by the ancient Greeks. Magnitude allows astronomers to rank how bright, comparatively, different stars appear to the unaided eye. Because of the way our eyes detect light, a star ten times more luminous than a second star will appear less than ten times brighter to human eyes. In the magnitude system the measure of brightness is logarithmic and the scale factor is such that a difference of five magnitudes corresponds to an intensity ratio of 100.

The fourteen brightest stars visible from Greece were called stars of "first magnitude", while the faintest stars visible to the unaided eye were said to be of "sixth magnitude". Even today, we retain this peculiar convention in which fainter celestial objects have larger magnitudes. The lower a star's magnitude, the brighter it is. Stars with negative magnitudes are the brightest of all. The apparent magnitude, m, is that measured from the Earth on the basis of direct observation by eye or telescope. The absolute

Focus 2.1 Distance to the stars

How far away are the stars? To triangulate the distances of nearby stars we need a wide baseline, the Earth's annual orbit around the Sun. The trigonometric, or annual, parallax, π_t, is defined as half the apparent angular displacement of a star when measured at intervals of six months from opposite sides of the Earth's orbit. Thus:

$$\pi_t = \frac{AU}{D} \text{ radians},$$

where one AU = $1.495\ 979 \times 10^{11}$ m is the astronomical unit and D is the distance of the star. When the trigonometric parallax is specified in seconds of arc, then

$$D = \frac{1}{\pi_t} \text{ parsecs}.$$

To convert between the two methods of measuring angle, one radian is equivalent to $2.062\ 648 \times 10^5$ seconds of arc.

If the direction of a star lies in the plane of the Earth's orbit, or in the ecliptic, the star will move backwards and forwards in this plane during the year by the amount of π_t. When the star's direction is perpendicular to this plane, it moves in a circle of angular radius π_t, and at other ecliptic latitudes the star appears to move in an ellipse with semi-major axis π_t.

Parallaxes are very difficult to measure, and can take years of observations to obtain. The stars are so far away, and the parallaxes so tiny, that the first reliable measurement of a star's trigonometric parallax did not occur until 1838, when Friedrich Wilhelm Bessel succeeded with the star 61 Cygni. Over the course of a year, the apparent position of 61 Cygni shifted back and forth over a total angle of almost two-thirds of a second of arc, to give a parallax of half that amount, or $\pi_t = 0.29$ seconds of arc, and a distance of 3.4 parsecs.

Trigonometric parallaxes have now been obtained using Earth-based telescopes for about 2000 stars that are closer than 22 pc away. More distant stars have parallaxes smaller than 0.05 seconds of arc, and they cannot be measured from the ground because of atmospheric distortions that limit the angular resolution of optical telescopes (Section 9.1). However, instruments aboard the *Hipparcos* satellite, which orbits the Earth above its confusing atmosphere, have obtained parallaxes for more than a hundred thousand stars with an unprecedented accuracy of 0.001 seconds of arc, leading to dramatic improvements in our estimates of the distances of nearby stars.

The bright variable stars, known as Cepheid variables, provide another tool for measuring remote stellar distances. The intrinsic luminosity of these stars, or their absolute magnitude, M, increases with their period, P, of brightness fluctuation; stars with longer variation periods are brighter. For classical Cepheids in our Milky Way, the period–luminosity relation is:

$$M = -1.371 - 2.986 \log P,$$

where M is in magnitudes and P is in days.

magnitude, M, is defined as the apparent magnitude, a star or other celestial object would have if it was at an arbitrary

distance of 10 pc. Any apparent magnitude can be converted to absolute through the simple formula:

$$\text{absolute magnitude} = M = m + 5 - 5 \log D,$$

where D is the distance in parsecs. The Sun has an apparent magnitude of $m_{\odot} = -26.74$ and an absolute magnitude of $M_{\odot} = +4.83$.

Astronomers sometimes make a distinction between the apparent magnitude observed in blue light, centered at 440 nm and designated m_B, and that detected in a so-called visual passband, m_V centered at 550 nm. They together define a color index $B - V = m_B - m_V$. The total radiant flux integrated over all wavelengths, visible and invisible defines the apparent bolometric magnitude, m_{bol}, and the absolute bolometric magnitude, M_{bol}.

To find the absolute luminosity of a star, in units of the Sun's luminosity, $L_{\odot}$, just use:

$$\log \left(\frac{L}{L_{\odot}} \right) = 0.4 \, (M_{\odot} - M),$$

or

$$L = 10^{0.4 \, (M_{\odot} - M)} L_{\odot}$$

where the absolute magnitude of the Sun is $M_{\odot} = 4.83$, the absolute magnitude of the other star is M, and the absolute luminosity of the Sun is $L_{\odot} = 3.85 \times 10^{26}\,\text{W} = 3.85 \times 10^{26}\,\text{J s}^{-1}$.

So, all you need to know is the true brightness of a star, either its absolute luminosity or its absolute magnitude, and you can determine its distance from the star's observed, or apparent, brightness. You might, for example, suspect that all stars are intrinsically as bright as the Sun, but this is a very bad assumption (Section 2.3). The great star Deneb, also known as α Cyg, shines brightly in the night sky with an apparent magnitude of m = 1.25. It is at a distance of 552 pc,

Fig. 2.2 **Our place in space** In this painting, the Milky Way and its massive core are depicted from slightly above the plane of our Galaxy. Bright stars are concentrated within spiral arms that wind out from a central galactic bulge. Since the Sun lies within one of the spiral arms, we look into them edgewise and see the nearby ones as the luminous band of the Milky Way. (Courtesy of Jon Lomberg.)

Fig. 2.3 **Globular star cluster** A million stars are crowded together in this globular star cluster, known as M13 and NGC 6205, which is only 17 pc across. It has an angular extent to 16.6 minutes of arc and a distance of 6900 pc or 6.9 kpc. (Courtesy of the United States Naval Observatory.)

and has an absolute magnitude of M = − 7.5. Deneb is therefore about 85 000 times more powerful and intrinsically more luminous than the Sun.

Cepheid variable stars, named after their prototype δ Cephei, provide us with an extra measurement that can be used to determine their intrinsic brightness and internal power. These stars shine with a brightness that varies periodically with time, at a period of 2 to 150 days depending

Fig. 2.4 **Edge-on view of our Galaxy** This infrared image of our Galaxy was obtained from the *InfraRed Astronomical Satellite* (*IRAS*) Point Source Catalogue. The infrared observations, obtained at 0.12 and 0.25 $\times$ 10^{-4} m, or 12 and 25 μm, can penetrate the obscuring veil of interstellar dust that hides the Milky Way from full view at visible wavelengths. The central nuclear bulge of our Galaxy shines brightly at the center of this picture. The narrow bands of the galactic disk extend to each side of the center, marking regions of star formation and newborn stars. (Courtesy of the Leiden Observatory.)

on the star. Their absolute luminosity increases with the period according to a well-defined period-luminosity relation (Focus 2.1). Cepheid variables are also very bright stars with absolute magnitudes between − 2 and − 7. Since they are so luminous, these variable stars can be seen to exceptional distances.

Once we know the all-important stellar distance, we can determine the shape and form of our Galaxy. The bright stars and interstellar gas are concentrated within spiral arms that wind out from a massive galactic center. The Sun is an undistinguished star located within one of these arms at a distance of 8500 pc, or about 27 700 light-years, from the center of our Galaxy (Fig. 2.2). That is:

Sun's distance from galactic center = $D_{\odot}$ = 8500 pc = 8.5 kpc.

This distance is 1.75 billion (1.75×10^9) times the distance between the Earth and Sun.

The first clue to the immensity of our Galaxy came from observations of globular star clusters that lie outside the Milky Way plane of our Galaxy, rather than from stars within the Milky Way. The globular star clusters are round, densely packed concentrations of up to millions of stars that are gravitationally bound together (Fig. 2.3). In 1918, Harlow Shapley used variable stars in globular star clusters to show that they are distributed within a large, roughly-spherical halo around a remote galactic center.

Since interstellar dust absorbs the light of distant stars within the Milky Way, the galactic center is hidden from view at visible wavelengths. The longer infrared and radio waves can more easily slip past the dusty obscuring regions of the Milky Way, penetrating the obscuring veil of interstellar dust to detect the center (Fig. 2.4). It is located in the constellation Sagittarius, best seen from the southern hemisphere in June.

The highly flattened disk of our Galaxy is about 15 000 pc in radius and about 1000 pc thick, depending on the type of stars used to map its extent. The bright, massive young stars are found in a disk that is only 120 pc or so thick, despite being over 30 000 pc across. Other types of stars define thicker disks; some of them form one that is 2000 pc thick. Assuming a thickness of about 1000 pc, the disk of our Galaxy has a volume of about 700 billion (7×10^{11}) cubic parsecs, pc^3.

How many stars are there in the Milky Way? The distribution of stars mapped by *Hipparcos* indicates that the total mass density in the galactic plane near the Sun is:

nearby galactic mass density =
$\rho(0) = 0.515 \times 10^{-20} \text{ kg m}^{-3} = 0.076 \text{ M}_{\odot} \text{ pc}^{-3}$,

which is comparable to the mass density of visible stars in our vicinity. Here, $M_{\odot}$ is the mass of the Sun. Assuming a uniform distribution of stars in the disk and multiplying this density by the disk volume, we obtain a lower limit of at least 50 billion stars with a mass equal to that of the Sun. Since the stars are more concentrated toward the central regions of our Galaxy, there is at least twice this amount, so the galactic disk contains roughly 100 billion (10^{11}) stars like the Sun.

Observations of the motions of nearby stars in the Milky Way indicate that the entire stellar system is rotating like a gigantic pinwheel about a remote, massive galactic center.

Focus 2.2 Cosmic motions and the Doppler effect

Just as a source of sound can vary in pitch or wavelength, depending on its motion, the wavelength of electromagnetic radiation shifts when the emitting source moves with respect to the observer. This Doppler shift is named after Christiaan Doppler who discovered it in 1842. If the motion is <u>toward</u> the observer, the shift is to shorter wavelengths, and when the motion is away the wavelength becomes longer. You notice the effect when listening to the changing pitch of a passing ambulance siren. The tone of the siren is higher while the ambulance approaches you and lower when it moves <u>away</u> from you.

If the radiation is emitted at a specific wavelength, $\lambda_{emitted}$, by a source at rest, the wavelength, $\lambda_{observed}$, observed from a moving source is given by the relation:

$$\text{redshift} = z = \frac{\lambda_{observed} - \lambda_{emitted}}{\lambda_{emitted}} = \frac{V_r}{c},$$

where V_r is the radial velocity of the source along the line of sight <u>away</u> from the observer and c is the velocity of light. The parameter z is called the redshift since the Doppler shift is to the longer, redder wavelengths in the visible part of the electromagnetic spectrum. When the motion is <u>toward</u> the observer, V_r is negative and there is a blueshift to shorter, bluer wavelengths.

The Doppler effect applies to all kinds of electromagnetic radiation, including X-rays, visible light and radio waves. The Doppler effect is a very important tool for astronomers, determining the radial velocity of all kinds of cosmic objects, including the oscillating solar disk, the stars, interstellar gas and galaxies.

Our equation is strictly valid for radial velocities that are much smaller than the velocity of light, or for V_r less than c, where $c = 2.9979 \times 10^8$ m s^{-1}. The largest redshifts measured for the most distant galaxies (Section 2.2) reach $z = 5$ or more, but this does not mean that the radial velocity exceeds c. Nothing can move faster than the velocity of light. For objects that move at speeds comparable to that of light, somewhat more complicated expressions describe the Doppler effect:

$$1 + z = \left[\frac{c + V_r}{c - V_r}\right]^{1/2}$$

and equivalently $\dfrac{V_r}{c} = \dfrac{(z+1)^2 - 1}{(z+1)^2 + 1}$.

Everything in the Universe moves, and the stars are no exception. We can measure the radial, or line-of-sight, component of their motion by observing the Doppler shift in the wavelength of a star's spectral lines (Focus 2.2). Moreover, radio astronomers can use the same Doppler effect with the spectral line of interstellar hydrogen atoms, emitted at a wavelength of 0.21 m, to trace out the motions of the gas located between the stars. Both techniques indicate that the Sun and nearby gas and stars are revolving about the galactic center at a speed of:

Sun's velocity around galactic center = $V_\odot$ = 220 thousand meters per second.

By way of comparison, stars randomly dart here and there, like bees in a swarm, each with a velocity of about ten

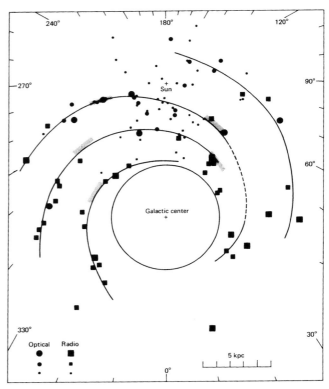

Fig. 2.5 **Spiral arms of our Galaxy** Luminous emission nebulae, known as ionized hydrogen, or H II, regions, act like beacons that mark out the spiral structure of our Milky Way Galaxy. The galactic longitude is indicated along the edge of the figure, the Sun and solar system are located at the upper center, and the figure is centered at the center of our Galaxy. The scale is shown at the lower right, for a linear extent of 5 kpc.

thousand meters per second, superimposed on the much larger circular motion.

The flattened, rotating disk contains relatively young stars and interstellar clouds arranged in a multi-arm spiral structure. Sections of these arms, that wind out from the galactic center, are delineated by the luminous emission nebulae (H II regions) that envelop very bright, massive stars (Fig. 2.5). Such stars have not yet had time to move away from the interstellar material in which they were born.

Radio observations of the 0.21–m transition of neutral, unionized (H I) hydrogen have provided a more extensive delineation of our Galaxy's spiral arms. The distances of the interstellar hydrogen are inferred from the dependence of the circular rotational velocity on the distance from the galactic center, and the density of the hydrogen is inferred from the strength of the transition. The hydrogen atoms are found concentrated in several elongated, circular features that trace out the spiral arms of our Galaxy; the Sun is located within one of these arms, in the remote, outer parts of the Galaxy (Fig. 2.6).

The Sun, which is moving around the center at a velocity of 220 thousand meters per second, takes 240 million years to complete one trip around the center of the Galaxy. That is, the period, $P_\odot$, for one rotation of the Sun around the galactic center is:

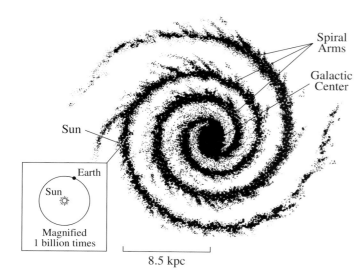

Fig. 2.6 **Structure of our Galaxy** This drawing depicts our Galaxy as viewed from above the Milky Way or galactic plane. The stars and interstellar material in our Galaxy are concentrated within spiral arms. The Sun lies within one of these spiral arms at a distance of 8,500 parsec, 8.5 kpc, or 27 700 light-years, from the center of our Galaxy. This distance is 1.75 billion times the distance between the Earth and the Sun.

Sun's rotation period about galactic center =

$$P_\odot = \frac{2\pi D_\odot}{V_\odot} = 2.4 \times 10^8 \text{ years},$$

where 1 year $= 3.1557 \times 10^7$ s. The Sun has circled the Galaxy more than 19 times during its 4.6-billion-year lifetime. So, the spiral arms should have been wrapped around the galactic center many times during the lifetime of the Sun, and a persistent dilemma has been understanding why they have not wound up to form a featureless ball of gas, dust and stars. The explanation seems to be density waves that control the concentrations of stellar and interstellar material, somewhat like highways that direct the motions of cars and trucks.

The stars in our Milky Way Galaxy, and the gas and dust between them, are rotating differentially about the massive galactic center, following Kepler's third law (Section 1.4). Stars nearer the center move around it at faster speeds, and those further away from the center revolve at a slower pace. We can use the measurements of the Sun's distance and velocity with Kepler's third law to infer the central mass that gravitationally controls the Sun's orbit:

Central galactic mass =

$$\frac{4\pi D_\odot^3}{G P_\odot^2} = \frac{D_\odot V_\odot^2}{G} = 1.9 \times 10^{41} \text{ kg} = 0.95 \times 10^{11} \text{ M}_\odot.$$

Here, the gravitational constant $G = 6.6726 \times 10^{-11}$ N m^2 kg^{-2} and the mass of the Sun is $M_\odot = 1.989 \times 10^{30}$ kg. So there is a mass equivalent to about 100 billion stars like the Sun within the Sun's galactic orbit, and most of it is presumably concentrated toward the central bulge of our Galaxy.

A Galaxy mass equivalent to 100 billion Suns is consistent with that obtained from the local mass density and volume of the Galaxy.

The absolute luminosity of our Galaxy, L_{BG}, in blue light is about 25 billion times that of the Sun, or $L_{BG} = 2.5 \times 10^{10}$ $L_{B\odot}$, where the absolute blue luminosity of the Sun is $L_{B\odot} = 3.0 \times 10^{26}$ W. So, the stars in our Galaxy shine with the light of 25 billion Suns. Their combined absolute magnitude is $M_{BG} = -20.5$.

The very center of our Galaxy is thought to harbor a black hole, probably associated with a compact, non-thermal radio source called Sagittarius A*, or Sgr A* for short. Observations of the motions of nearby ionized gas and stars indicate that a mass equivalent to about 2 million Suns is concentrated within the innermost parsec of Sgr A*. So, a small but massive black hole is most likely present in the core of our Galaxy. However, there is much more mass distributed in stars and material throughout the Milky Way, and a good deal of invisible matter surrounds it.

The flattened, rotating disk of our Galaxy contains about 100 billion visible stars, and a comparable mass is found in interstellar molecular clouds and other interstellar material. A spherical halo surrounds the disk, extending out in all directions from the galactic center to distances of at least 20 000 pc. The halo of our Galaxy contains globular star clusters whose stars are older than those found in the disk, but most of the halo mass is not visible to the human eye. The halo contains large amounts of hot gas with temperatures approaching half-a-million degrees kelvin. It is thought to be heated and replenished by stars that exploded long ago.

Distant stellar systems, located above and below the disk of the Milky Way, are circling our Galaxy so rapidly that they have to be immersed in an extensive dark halo of material to stay in orbit around the Milky Way. If the orbits of small, distant satellite galaxies are dynamically constrained by the mass of our Galaxy, then it contains a total mass of about 500 billion Suns within 50 000 pc of its center. Under the assumption that the dwarf spheroidal companions of our Galaxy are gravitationally bound to it, an even larger mass of about one thousand billion solar masses is inferred to exist within about 200 000 pc from the galactic center. The total mass is obtained by noting that the observed velocities of the remote stellar systems must be less than the escape velocity of the Galaxy at that distance. There may be about 10 times more dark material than luminous stars hidden within the vast, unseen places outside the visible disk of the Milky Way.

Thus, although the Sun is extremely important to Earth and to our solar system, it is just an average star, one of about one hundred billion stars in our Galaxy. Moreover, the observable Universe is not limited to our stellar system, but instead populated by more than one hundred billion galaxies, each composed of about one hundred billion stars. These galaxies stretch as far as the largest telescope can see – and perhaps beyond.

2.2 The Sun's place in the Universe

The nearest large galaxy is Andromeda, designated as M31 or NGC 224, is located at a distance of about 800 000 parsec, or 0.80 Megaparsec, written 0.80 Mpc. The designation M31 means that this object is the 31st one in Charles Messier's list of nebulous objects, first published in 1771, while NGC is an acronym for the New General Catalogue published by J. L. E. Dryer between 1888 and 1905. M31 is the most distant object seen by the unaided eye, visible as a faint, fuzzy glow in the constellation Andromeda.

Elegant Andromeda is a flattened spiral galaxy that looks oval because we view it from an angle (Fig. 2.7). It is the near twin of our Galaxy. The combined mass and the combined luminosity of all the visible stars in both galaxies are within a factor of two of each other. Both Andromeda and our Galaxy have dark invisible matter of up to 10 times that contained in

their stars, and massive black holes are found at the centers of both of them.

Andromeda is escorted through space by two small companions, M32 and NGC 205, that are elliptical galaxies without disks or spiral arms. They are devoid of significant interstellar material and young luminous stars. The elliptical and spiral galaxies are the two main types of optically visible galaxies. Examples of spiral galaxies, observed both edgewise and face on, are shown in Figures 2.8, 2.9, and 2.11.

All the spirals and ellipticals are remote galaxies, or island universes of stars, which are scattered throughout space as far as we can see. They are typically separated by a few million parsecs, or a few Mpc, and about 100 galaxy diameters. Thus the Universe appears to be largely empty space as far as galaxies are concerned. Yet, the Universe isn't

Fig. 2.7 **The Andromeda Nebula**
The nearest spiral galaxy, the Andromeda Nebula, also known as M31 and NGC 224, is located at a distance of about 800 kpc, so its light takes about 2.6 million years to reach us. Both the Andromeda Nebula and our Galaxy are spiral galaxies with total masses of about a million million, or 10^{12}, solar masses, and roughly a hundred billion, or 10^{11}, visible stars. Two smaller galaxies are also shown in this picture. They are M32, or NGC 221, at the edge of the Andromeda Nebula, and NGC 205, that is located somewhat further away. These are elliptical systems at about the same distance as M31, but with only about one-hundredth of its mass. (Courtesy of Dr Siegfreid Mark, Central Institute of Astrophysics of the Academy of Science, GDR, and the Karl-Schwarzschild-Observatorium, Tautenburg.)

Fig. 2.8 **Spiral galaxy** A spiral galaxy NGC 4603 in the Centaurus cluster of galaxies. The bright, blue spiral arms outline the sites where star formation has been triggered recently in the disk of this galaxy. The bright nuclear bulge is an aggregate of stars believed to have formed more than 10 billion years in the past during the earliest evolutionary stages in the development of NGC 4603. (© AURA, The Cerro Tololo Interamerican Observatory.)

Fig. 2.9 **Galaxy cluster and gravitational lens** A *Hubble Space Telescope* image of a rich cluster of galaxies named Abell 2218, number 2218 in George Abell's catalogue. A typical rich cluster will span 1 to 10 Mpc, and may contain hundreds and even thousands of galaxies, each composed of hundreds of billions of stars. The powerful gravitation of the compact, massive cluster Abell 2218 acts as a gravitational lens, spreading the light of very distant galaxies, that lie behind it, into a spider-like array of faint arcs. (Courtesy of NASA.)

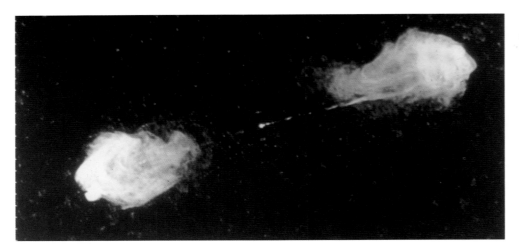

Fig. 2.10 **Radio galaxy Cygnus A** A radio map of the source Cygnus A, or 3C 405. Its features include the compact core in the center of the galaxy; the jets emanating from the core and carrying energy and particles to the lobes; and the radio lobes themselves with the back-flowing cocoons of wispy regions and enhanced emission. Barely visible in the overexposed lobes are the "hot spots" where the jets are terminated. (Courtesy of Richard A. Perley, John W. Dreher, and the NRAO.)

quite as empty as the solar system; the mean distance from the Earth to the Sun is about ten thousand times the Earth's diameter. When compared to the separations of planets, in units of their size, the galaxies are comparatively close to each other, even though their real distances are enormous.

The galaxies are gathered into clusters that extend across hundreds of millions of parsecs (Fig. 2.9). The nearest one spreads across the constellation of Virgo. It lies at a distance of about 18 million parsecs, or 18 Mpc, gauged from Cepheid variable stars using the *Hubble Space Telescope*. The clusters of galaxies are themselves gathered into superclusters with a typical extent of about 50 million parsecs and a combined mass of about 100 thousand galaxies, or 10 million billion (10^{16}) solar masses.

Powerful radio sources can be associated with distant elliptical galaxies. They are characterized by non-thermal radiation and by the expulsion of energy in two oppositely directed beams. These jets remain extraordinarily straight and surprisingly stable across distances as large as a Megaparsec. Such a radio galaxy often consists of two extended lobes containing magnetic fields and high-speed electrons moving at nearly the speed of light (Fig. 2.10). They are anchored in a gigantic spinning black hole that acts as a gyroscope and feeds energy along the jets to the lobes for hundreds of millions of years.

The quasars lie further out, beyond most radio galaxies. Quasars are extremely distant and very bright objects giving off hundreds, even thousands, of times as much energy in visible light as ordinary nearby galaxies. Yet, their brightness in visible light can vary enormously over the course of a single day. Since nothing can move faster than the velocity of light, a variation over this time indicates that some quasars are no larger than the solar system.

Both radio galaxies and quasars are thought to be powered by supermassive black holes at their center. The enormous gravity of the black hole, with a mass of up to a billion Suns, pulls in surrounding stars and gas. The infalling material forms a flat, orbiting accretion disk that spirals into the black hole, energizing twin radio jets that stream out in opposite directions along the spin axis of the supermassive black hole.

Compelling evidence for a dark, supermassive object has been found at the core of M87, one of the biggest and brightest elliptical galaxies in the nearby Virgo cluster of galaxies. An inner disk has been detected orbiting a supermassive black hole, whose estimated mass is three billion solar masses, apparently feeding a one-sided optical and radio jet that emerges nearly perpendicular to the disk.

The incredible aspect of the Universe of normal galaxies, radio galaxies and quasars is its motion. As long ago as 1917, before anyone knew what galaxies were, Vesto M. Slipher used the Doppler effect (see Focus 2.2) to show that many spiral-shaped nebulae are receding from our Galaxy at remarkable speeds, some greater than a million meters per second. As often happens in astronomy, this was an entirely serendipitous discovery, for Slipher was looking for their expected rotations under the belief that spiral nebulae, as they were then called, are primitive solar systems in the making.

It was hard to believe that objects with such enormous speeds could long remain a part of the Milky Way system, since their velocities exceed the escape velocity of our Galaxy. The escape velocity is defined in Focus 1.6, and amounts to about a million meters per second at the edge of our Galaxy. According to Slipher, the observations indicated "a general fleeing from us or the Milky Way". By 1925 Edwin Hubble had used Cepheid variable stars in Andromeda to determine its distance, showing that it lies well outside the confines of our Galaxy and is itself a spiral galaxy.

Perhaps the most profound discovery of optical astronomy has been the discovery of the expanding Universe. After measuring the distances of nearby galaxies, showing that they are all outside the Milky Way, Edwin Hubble found that the galaxies are flying apart at speeds that increase with distance. The further away the galaxy, the faster it is fleeing from us. According to the now-famous Hubble's law, the radial velocity, V_r, of a galaxy, as measured by the Doppler effect, is given by the linear relation:

$$V_r = cz = H_0 \times D,$$

where the velocity of light $c = 2.9979 \times 10^8$ m s^{-1}, z is the redshift (Focus 2.2), and the Hubble constant, H_0, is the ratio of the speed with which distant galaxies are receding from us to their distance, D. The relation is linear. If the distance doubles, so does the speed. It is as though the expanding Universe started with a gigantic explosion, with the fastest-moving parts having traversed the greatest distances.

The current rate of expansion of the Universe is quantified by the Hubble constant, H_0, often expressed in the form:

$$H_0 = 100h \text{ km s}^{-1} \text{ Mpc}^{-1},$$

where 1 km s^{-1} = 1000 m s^{-1} and 1 Mpc = 10^6 pc. The Hubble constant sets the physical scale of the Universe, with the distance to any galaxy given by:

$$D = \frac{cz}{H_0} = 2997.9 \left(\frac{z}{h}\right) \text{Mpc}.$$

There is an ongoing controversy over the exact value of this important constant, with estimates during the past decades ranging between 50 and 100 km s^{-1} Mpc^{-1}, and current observational constraints of h lying between 0.50 and 0.85.

Since distances can only be measured for relatively nearby extragalactic objects, Hubble's law for remote objects is sometimes expressed by the redshift-magnitude relation in which the apparent magnitude, m, is given by:

$$m = 5 \log\left(\frac{cz}{H_0}\right) + M + 25,$$

where M is the absolute magnitude and the Hubble constant H_0 is given in units of km s^{-1} Mpc^{-1}.

Astronomers are now detecting galaxies out to redshifts as great as $z = 5$, and distances on the order of 10 000 Megaparsec. Altogether there are roughly a billion galaxies in the volume of space that modern telescopes can detect (Focus 2.3), and there is no end in sight. So far, there has been no observable edge to the Universe.

Telescopes are time machines. When we look farther out into space, we travel back in time and see objects as they

Fig. 2.11 **Hubble Deep Field** Numerous spiral and elliptical galaxies are found in this *Hubble Space Telescope* image that looks deep into space and far back in time. This patch of sky is just 1.0 minutes of arc on a side, or about 1/30 of the diameter of the full Moon. The narrow field of view enables the telescope to look between the foreground stars of the Milky Way, and detect remote galaxies as faint as magnitude 30, or nearly four billion times fainter than human vision. The brightest few-dozen galaxies in the image are at a redshift, *z*, of about 0.5; the light now reaching the Earth from these galaxies was emitted about six billion years ago. For most of the galaxies in the image we are looking back more than half the age of the Universe. The light detected from some of them may have been emitted less than one billion years after the big-bang explosion that created the expanding Universe. (Courtesy of Robert Williams, the Hubble Deep Field Team and NASA.)

were. The radiation that we now detect from an astronomical object was generated one light-travel-time ago. This time is therefore sometimes called the look-back time, for we are looking back to a time when the object was younger than it is now. Moving at the speed of light, it takes 2.6 million years for light to travel from the closest spiral galaxy, Andromeda, to the Earth. That is, the light we see today left Andromeda 2.6 million years ago. We can only see the galaxy as it was then and not as it might be now. The radiation we now detect from the most distant objects in the Universe was emitted soon after the Universe was created!

Some galaxies are so remote that a telescope captures the light that was emitted by them 10 billion years ago. We see these objects as they were before the Sun came into existence and life began on Earth. Some of the most distant galaxies that we see now may no longer exist, but they were embryonic galaxies when the light now reaching the Earth began its journey.

This brings us to an important property of light – its permanence. As long as a ray of starlight passes through empty space and encounters no atoms or electrons it will survive unchanged. A ray of light has no way of marking

---

Focus 2.3 Density and total number of galaxies

The average volume density of galaxies, n_g, is:

volume density of galaxies = n_g = 0.005 52h^3 Mpc^{-3},

where Hubble's constant H_0 = 100h km s^{-1} Mpc^{-1}.

From Hubble's law, the distance, D, of a galaxy at redshift, z, is D = cz/H_0, where the velocity of light c = 2.9979 × 10^5 km s^{-1}. So the total number of galaxies, N_G, out to redshift z is:

total number of galaxies to redshift $z = \frac{4\pi}{3}\left(\frac{cz}{H_0}\right)^3 n_g = 0.6 \times 10^9 z^3$

which is independent of h and the exact value of the Hubble constant.

Since galaxies have been observed out to redshifts greater than unity, there are roughly a billion, or 10^9, galaxies in the observable Universe. The exact shape and form of the Universe complicates the precise calculations, since the curvature of space changes the distances at large redshifts. Still, our estimate should be correct to within an order of magnitude, or a factor of ten.

time, and it will persist forever, bringing its message forward to the end of the Universe.

The *Hubble Space Telescope*, named after the discoverer of the expanding Universe, has been used to study galaxies formed early in the Universe. Deep imaging with this telescope has opened a gateway to at least halfway back to when the expansion began (Fig. 2.11).

For an object at redshift z, we can specify the look-back time, t_L, at which the radiation was emitted. Looking at objects at larger and larger redshift is looking further and further into the past. The look-back time can be expressed in terms of the Hubble time, t_H, which is the reciprocal of the Hubble constant, H_0. For small redshifts, very much less than unity, we have

$$\text{look-back time} = t_L = zt_H = \left(9.778\,\frac{z}{h}\right) 6.6 \text{ billion years,}$$

where the Hubble constant $H_0 = 100h$ km s^{-1} Mpc^{-1}.

We cannot tell exactly how far away the distant galaxies are, or precisely how long ago their radiation began its journey, until we understand the curvature of space. If space is curved, the path through space might be noticeably longer than that expected in flat, uncurved space, with larger distances and greater look-back time. As Einstein showed in dealing with the Sun's gravitational interaction with Mercury (Section 1.4), space is curved by mass, and the curvature of the Universe is determined by its total mass. The detailed expressions for the distance and look-back time therefore depend upon the amount of space curvature, or the mass

Focus 2.4 Age, shape and fate of the Universe

A hot explosive origin of the Universe is inferred from a backward extrapolation of the expansion of the galaxies, which must have been propelled from a common origin in the big bang explosion. The big bang has been confirmed by the discovery of the cosmic microwave background, a relic of the initial fireball, and by the observed abundance of the light elements that must have been synthesized during the big bang.

Precise estimates for the time of the big bang, called the expansion age of the Universe, depend on assumptions about the mean mass density of the Universe, the possibility of a cosmological constant that opposes gravity, and the exact value of the Hubble constant. In a matter-dominated Universe, the age, fate and shape of the Universe are all interrelated. They depend upon the value of the expansion age, t_0, in comparison with the reciprocal of the Hubble constant, or the:

$$\text{Hubble time} = \frac{1}{H_0} = \frac{9.778}{h} \times 10^9 \text{ yr,}$$

where $H_0 = 100h$ km s^{-1} Mpc^{-1}. Measurements of the Hubble constant limit the Hubble time to somewhere between 10 and 20 billion years.

The Hubble constant defines a critical mass density, ρ_c, that must be exceeded if gravity overcomes expansion. Imagine the most distant galaxy with mass, m_g, distance D_g, and velocity $V_g = H_0 D_g$. Gravity will just balance the expansion of this galaxy if its kinetic energy of expansion is equal to the gravitational potential energy of all of the rest of the Universe, or if:

$$\text{kinetic energy} = \frac{m_g V_g^2}{2} = \frac{m_g H_0^2 D_g^2}{2} = \frac{G m_g M_U}{D_g}$$

$$= \text{gravitational potential energy,}$$

where M_U is the total mass of all the rest of the Universe inside distance D_g. In other words, the velocity, V_g, of the most distant galaxy is just equal to the escape velocity, V_{escape}, of the entire Universe, $V_{escape} = (2GM_U/D_g)^{1/2}$ (Focus 1.6).

Collecting terms we obtain:

$$\text{Critical mass density} = \rho_c = \frac{M_U}{\frac{4\pi}{3}D_g^3} = \frac{3H_0^2}{8\pi G}$$

or

$$\rho_c = 1.879 \times 10^{-26} h^2 \text{ kg m}^{-3},$$

where the Hubble constant $H_0 = 100h$ km s^{-1} Mpc^{-1}.

If the observed mass density of the Universe exceeds the critical value, then the total mass of the Universe will be enough to halt its expansion in the future. The expansion age, t_0, is then given by:

$$0 < t_0 < 2t_H/3 \text{ for a closed Universe in spherical space.}$$

The expansion is decelerating, or slowing down, as time goes on, and space is curved inward with a positive, spherical curvature.

If there is not enough matter around to stop the expansion in the future, then the expansion age is given by :

$$2t_H/3 < t_0 < t_H \text{ for an unbound Universe in open hyperbolic space.}$$

This open, unbound Universe continues to expand, accelerating and thinning out forever as space curves outward.

For the special case of a flat, Euclidean Universe without space curvature, the matter precisely balances the expansion indefinitely, and the expansion age is given by:

$$t_0 = 2t_H/3 \text{ for an unbound Universe in flat Euclidean space.}$$

The oldest globular star clusters in our Galaxy have a median age of 14.56 billion years, with a lower limit of 12.07 billion years. That does not include a formation time of between 0.1 and 2.0 billion years. The Hubble constant has to be on the low side of its possible values to be consistent with the lower limits to the ages of the globular clusters. Any larger value of the Hubble constant, and/or greater star-cluster ages would be inconsistent with any scenario in which the Universe stops its expansion in the future.

As a matter of fact the observational evidence has come down in favor of an open, hyperbolic, ever-expanding Universe for more than half-a-century. The mass density in visible galaxies is less than one percent of the critical mass density required to eventually stop the expansion. The observed light-element abundance, formed in the immediate aftermath of the big-bang explosion, requires that the density of ordinary matter, in both invisible and visible form, be less than 10 percent of the critical mass density. There simply is not enough ordinary seen or unseen matter in the Universe to stop the expansion of the Universe and eventually close it.

density of the Universe, as well as the redshift.

The exact mathematical equations can be complicated, and the effects increase at larger redshifts. Nevertheless, the uncertainties in the distances and look-back times that are introduced by the unknown mass density are no larger than those caused by our imprecise knowledge of the Hubble constant. Even at a redshift $z = 5.0$, corresponding to the largest ones observed, the look-back time ranges from 10 to 15 billion years depending on the choice of the Hubble constant and the mass density of the Universe. So we can ignore the effects of space curvature in most practical computations.

At first we seem to be at the center of the expansion, since all the galaxies appear to be fleeing from us; but if we traveled to another galaxy, all the galaxies would apparently be fleeing from it. The galaxies therefore act like dots painted on the surface of an expanding balloon. So the Universe has no preferred, observable center.

A backward extrapolation of the present expansion indicates that all the matter in the Universe was once highly compressed and extremely hot. This was the time of the big bang explosion that gave rise to the expanding Universe. The big bang marks the beginning of time.

A wide range of observational evidence, including the cosmic abundance of the light elements and the precise black-body character of the microwave background radiation, provide strong evidence for the big-bang model. The light elements had to be synthesized in the dense, hot early stages of the big bang and the background radiation is the cooled, left-over relic of the primeval fireball.

We cannot specify the location from which the expanding Universe began, but we can identify the time at which it started. A rough estimate for the age of the expanding Universe is the time it took any one galaxy to get from us to its present position at the measured velocity. That age must be the same for all galaxies, and it is the time of the big-bang explosion. It's approximate value, called the Hubble time, is the reciprocal of the Hubble constant (see Focus 2.4).

Everyone is agreed that the Hubble time lies somewhere between 10 and 20 billion years, but the experts have been quibbling about the exact value for decades. The controversy involves the precise estimate for the Hubble constant. A faster rate of expansion, or a larger value of H_0, means that less time has elapsed since the expansion began with the big bang, and a slower expansion with a small value of the Hubble constant suggests a greater age.

The Hubble time is an upper limit to age because gravity can slow the expansion, and the exact value for the time of the big bang depends on the mass density of the Universe as well as on the uncertainties in the Hubble constant (Focus 2.4). The mass of the Universe curves its shape, establishes its geometry and determines its fate.

What is the ultimate destiny of the Universe? If there is enough mass in the Universe, its gravitational pull will eventually halt the expansion and the Universe will fall back on itself. On the other hand, there might not be enough mass to stop the expansion, and the Universe will then continue to expand forever. The amount of mass in optically visible matter, like stars and galaxies, is much too small to stop the presently-observed expansion. In fact, the density of ordinary matter in the Universe, in both visible and invisible forms, is ten times too small to ever stop the expansion (Focus 2.4). Thus all the available observational evidence indicates that the Universe will continue to expand forever, eventually becoming dark and cold.

To sum up, the stars are not distributed uniformly, or even randomly, through the Universe. They are grouped together in huge assemblages called galaxies. The stars and galaxies outline the structure and motion of our Universe, revealing its enormous extent and showing that it has been expanding for 10 to 20 billion years. They also define our place in space and time. The Sun and planets are located in the nondescript, outer parts of our Galaxy, and they are less than half as old as the expanding Universe. Like humans, the stars are born, live and die, describing an evolution that we consider next.

2.3 How stars evolve

The visible Universe is composed of condensations of matter, called stars, that emit large amounts of light. We can examine regions of future star formation, witness stars that are now being born, and observe stars that originated long ago in the distant reaches of time. Close scrutiny of the light from existing stars has provided us with an understanding of how stars evolve and has also answered one of the most profound questions in nature. Where did the chemical elements come from? Most of them were manufactured inside former stars that threw them into space during their explosive deaths. The Sun and planets contain these elements, so the Sun is a relative newcomer in the recurring life cycle of stars. It is a second-generation star formed after other stars have lived and died.

Comparisons of the Sun with other stars in our Galaxy

To understand stellar evolution we must first investigate stars of different physical properties and ages. It is somewhat like looking at all the trees in a forest to see how they grow. In comparing the light production, mass and size of the Sun to other stars, astronomers find that the Sun is a perfectly ordinary star typical of its size, doing all the things a star like it should.

This does not mean that all stars are alike. They differ enormously in intrinsic brightness, or in the amount of energy that they radiate each second. There are stars that are ten thousand times fainter than the Sun, and there are a few very rare stars that are ten million times brighter than the Sun. The solar luminosity, $L_\odot = 3.85 \times 10^{26}$ W serves as the unit of comparison, with the absolute luminosity, L_s, of other stars ranging from 0.0001 $L_\odot$ to 10 000 000 $L_\odot$.

In comparison with the large range in stellar brightness, there is relatively little variety in the masses of stars. These masses may be determined from observations of binary-star systems. When the sizes and periods of their orbital motion are determined, the mass can be inferred from Kepler's third law (Section 1.4). The Sun is in the middle of the stellar mass range, with a mass of $M_\odot = 1.989 \times 10^{30}$ kg, and the majority of observed stars have masses between 0.3 and 3.0 times the mass of the Sun, or between 0.3 and 3.0 $M_\odot$. Stars with a larger mass have been observed, up to about 27 times the mass of the Sun, but the most massive stars are relatively rare.

Theoretical calculations indicate that stars can exist with masses between 0.08 and 120 times that of the Sun. At 0.08 $M_\odot$ we have reached the lower limit for a gaseous body to become a star. Its central regions are too rarefied and too cool to sustain the nuclear fusion reactions that energize a star and

make it shine. As the mass increases, so does the central temperature; this is because the mass must be supported by the internal pressure that increases with temperature (Focus 2.5). Nevertheless, the temperatures inside stars cannot become too high, and therefore the masses cannot exceed an upper bound of 120 $M_\odot$. Since the outward pressure increases with internal temperature, and therefore with mass, a very massive star can be blown apart.

Although there is not much variation between the masses of stars, a little difference can become very important. It turns out that the mass of a star determines its brightness, lifetime and ultimate fate.

A more massive star is hotter inside, and it ought to be more luminous. Observations indicate that the absolute luminosity, L_S, of a star increases with the fourth power of

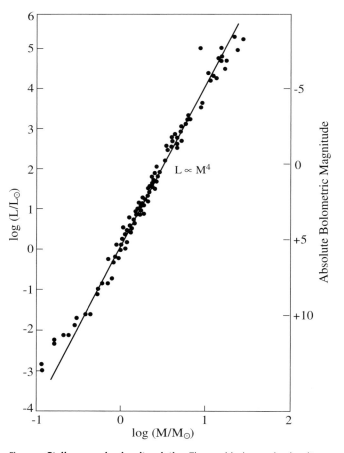

Fig. 2.12 **Stellar mass–luminosity relation** The empirical mass–luminosity relation for main-sequence stars of absolute luminosity, L, in units of the solar luminosity, $L_\odot = 3.85 \times 10^{26}$ W, and mass, M, in units of the solar mass, $M_\odot = 1.989 \times 10^{30}$ kg. The straight line corresponds to a luminosity proportional to the fourth power of the mass.

Focus 2.5 Temperature at the center of a star

Since the most abundant element in the Universe is hydrogen, and because the inside of a star is too hot for atoms to exist, we can assume that the interiors of most stars are composed of protons and electrons, the subatomic constituents of hydrogen atoms. The protons are 1836 times more massive than the electrons, so they dominate the gravitational effects inside the star. The temperature, T_C, at the center of a star like the Sun can be estimated by assuming that each proton down there is hot enough and fast enough to counteract the gravitational compression it experiences from the rest of the star. That is:

$$\text{thermal energy} = \frac{3}{2}kT_C = \frac{Gm_pM_s}{R_s} = \text{gravitational energy}$$

where Boltzmann's constant $k = 1.380\ 66 \times 10^{-23}$ J K^{-1}, the gravitational constant $G = 6.6726 \times 10^{-11}$ N m^2 kg^{-2}, the mass of the proton is $m_p = 1.6726 \times 10^{-27}$ kg, and M_s and R_s respectively denote the mass and radius of the star.

Solving for the central temperature we obtain:

$$T_C = \frac{2Gm_pM_s}{3kR_s} = 1.5 \times 10^7 \left(\frac{M_s}{M_\odot}\right)\left(\frac{R_\odot}{R_s}\right) \text{K},$$

where the mass and radius of the star are given in solar units in the numerical approximation, and $M_\odot = 1.989 \times 10^{30}$ kg and $R_\odot = 6.955 \times 10^8$ m. Thus, the temperature at the center of the Sun is 15 million degrees kelvin. If stars have about the same mass density inside, then the mass will increase with the cube of the radius, or the radius will scale as the cube-root of the mass, and the central temperature will be given by:

$$T_C = 1.56 \times 10^7 \ (M_s)^{2/3} \text{K}.M_\odot$$

So, the central temperature increases with mass, and the more-massive stars are hotter inside.

We have assumed that the outward gas pressure of the moving protons supports the inward pull of gravity, a condition known as hydrostatic equilibrium. The gas pressure, P_g, is given by the ideal gas law:

$$\text{gas pressure} = P_g = NkT,$$

where $N = M_s/m_p$ is the total number of particles in the star.

Gas pressure does indeed balance gravity in the Sun and most other stars, called dwarfs, but it does not apply for much larger, and relatively rare, stars, termed giants. The giants are enormously distended stars with low mass density and high luminosity. If we assumed that the inner temperature of giant stars remains high enough to generate a gas pressure sufficient to balance gravitation, then their luminosity greatly exceeds that actually observed.

Giant stars are supported by a different kind of pressure, known as radiation pressure, P_r, which, at temperature, T, is given by:

$$\text{radiation pressure} = P_r = aT^4/3,$$

where the radiation constant $a = 7.5659 \times 10^{-16}$ J m^{-3} K^{-4}. The radiation pressure is much less than the gas pressure at the center of the Sun.

There can be too much of a good thing. Since the radiation pressure increases with the fourth power of the temperature, and the temperature has to increase with the mass, the radiation pressure can overcome the gravity of an exceptionally massive star. That is, a giant star cannot remain in equilibrium if the central temperature and mass become too high, and this occurs for masses greater than about 120 solar masses.

---

the star's mass, M_S, or that $L_S = constant \times M_S^4$ (Fig. 2.12). So, a small increase in a star's mass implies a big increase in its brightness. Less-massive stars have less weight pressing down on their cores, so the core is cooler, and the stars are dimmer.

The life span of stars also depends on their mass. The more massive a star is, the shorter it lives. A massive star uses up its available energy at a greater rate, and it therefore cannot shine as long. It is as if bigger fires burn brighter and last a shorter time.

It turns out that the energy supply is proportional to the mass, from Einstein's famous expression $E = Mc^2$ (Section 3.3 − E is the energy, M the mass, and c is the velocity of light). The rate at which energy is being radiated away, the absolute luminosity L_S, also increases with the mass, but as the fourth power of the mass. So, the length of time, t, that a star shines, is another dramatic function of the mass. It decreases with increasing mass, and scales roughly as the inverse third power of mass.

More than a century ago, astronomers noticed that stars of different colors exhibit different spectral lines. The spectra of white stars like Vega and Sirius are dominated by strong absorption lines of hydrogen, while some blue ones have noticeable helium absorption lines. Yellow stars like the Sun

Class	Dominant Lines	Color	Color Index	Effective Temperature (K)	Examples
O	He II	Blue	−0.3	28 000 – 50 000	χ Per, ε Ori
B	He I	Blue–White	−0.2	9900 – 28 000	Rigel, Spica
A	H	White	0.0	7400 – 9900	Vega, Sirius
F	Metals; H	Yellow–White	0.3	6000 – 7400	Procyon
G	Ca II; Metals	Yellow	0.7	4900 – 6000	Sun, α Cen A
K	Ca II; Ca I	Orange	1.2	3500 – 4900	Arcturus
M	TiO; Ca I	Orange–Red	1.4	2000 – 3500	Betelgeuse

[a] An H denotes hydrogen, He is helium, Ca is calcium, and TiO is a titanium oxide molecule. A neutral, unionized atom is denoted by the Roman numeral I, the number II describes an ionized atom missing one electron.

Table 2.1 **The spectral classification of stars**[a]

have strong absorption lines of calcium and metals in their spectra.

A system of stellar classification based on their spectra was developed in the early 20th century, and is still in use today. Working under the direction of Edward C. Pickering, astronomers at the Harvard College Observatory examined the spectra of hundreds of thousands of stars. Pickering's co-worker, Annie Jump Cannon, distinguished the stars on the basis of the absorption lines in their spectra, and arranged most of them in a smooth and continuous spectral sequence, designated O B A F G K M (Table 2.1).

A star has a certain color because most of its radiation is emitted at the wavelengths corresponding to that color. The wavelength of maximum starlight intensity varies inversely with temperature, according to the Wien displacement law (Section 1.2). Thus, bluer stars are hotter and redder stars are cooler. This means that the spectral sequence is also one of decreasing effective temperature for the visible stellar disk (Table 2.1). Cannon further refined each spectral class by adding numbers from 0 to 9, running from hot to cold. As an example, the hottest F star is designated F0, and the coolest as F9, followed by G0. In this system our Sun is classified as G2.

Once the absolute luminosity of individual stars was obtained by parallax measurements and apparent luminosity, astronomers were able to show that the intrinsic brightness of most stars exhibits a systematic decrease with the progress of spectral class from O to M. The brightness drop is illustrated in the famous Hertzsprung–Russell, or H–R, diagram of absolute luminosity or absolute magnitude plotted against the spectral class or effective temperature. It is named after Ejnar Hertzsprung and Henry Norris Russell, who drew the first ones in the early-20th century (Fig. 2.13), and thousands of nearby stars have subsequently been examined in this way. Most stars lie along the main-sequence part of the H–R diagram, which goes from the upper left to the lower right (Fig. 2.14).

The general characteristics of the H–R diagram are described by the Stefan–Boltzmann law:

$$L_s = 4\pi\sigma R_s^2 T_{es}^4,$$

where L_s is the absolute luminosity of the star, R_s is its radius, T_{es} is the effective temperature of the visible stellar disk, $\pi = 3.141\ 59$ and the Stefan–Boltzmann constant $\sigma = 5.670\ 51 \times 10^{-8}\,J\,m^{-2}\,K^{-4}\,s^{-1}$. This expression tells us that for

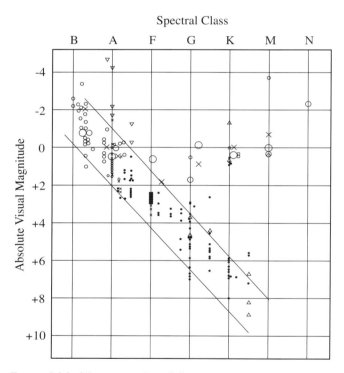

Fig. 2.13 **Original Hertzsprung–Russell diagram** When data on stellar parallax enabled the distances and absolute luminosities of stars to be determined, Henry Norris Russell published, in 1914, this plot of the absolute visual magnitudes of stars in moving, or open, star clusters – the Hyades (*filled circles*), the Ursa Major group (*small crosses*), the large group in Scorpius (*small open circles*) and the 61 Cygni group (*triangles*) – against their spectral class. The two diagonal lines mark the boundaries of Enjar Hertzsprung's sequence, inferred in 1911 for the Pleiades and Hyades open star clusters; this is now known as the main sequence along which most stars, including the Sun, are located. The giant stars are located at the upper right. The large circles and crosses represent points calculated from the mean parallaxes and magnitudes of other groups of stars. In the same 1914 paper, Russell published a very similar diagram for individual bright stars whose distances had been measured. It contained just one anomalous point in the lower left part of the diagram, at spectral class Ao and apparent visual magnitude 10.3. It belonged to the first white-dwarf star, o Eridani B, discovered by Walter Adams, also in 1914.

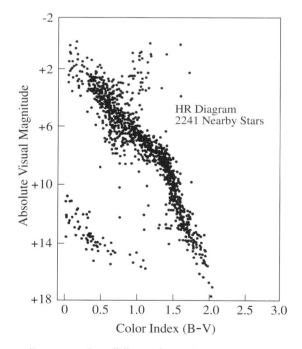

Fig. 2.14 **Hertzsprung–Russell diagram for nearby stars** The absolute visual magnitudes, M_V, are plotted against the color, $B–V$, for 2241 nearby stars. The vast majority of the stars occupy the main sequence, running diagonally from the hot luminous upper left to the cooler, less-luminous lower right. The Sun is a main-sequence star with $M_V = 4.8$ and $B–V = 0.68$. The white-dwarf stars, in the lower left, are low-luminosity stars about the size of the Earth.

a fixed radius the brightness of a star increases with the fourth power of the effective temperature, so cooler stars are less luminous. That is just what happens along the main sequence.

Although the radius of main-sequence stars does not vary much, certain giant stars, that are not on the main sequence, are very big. If a supergiant is 25 000 times brighter than the Sun, with the same temperature, it follows from the Stephan–Boltzmann law that it will be 158 times bigger than the Sun. A supergiant can be big enough to swallow up not just the Sun and Earth, but the entire Earth's orbit around the Sun.

Most of the stars in the Milky Way are main-sequence stars, and most of them are real dwarfs - smaller, cooler and dimmer than the Sun. The effective temperature, mass, luminosity, radius and lifetime of main-sequence stars are given in Table 2.2. The lifetimes are based upon the amount of time to consume all of the hydrogen fuel in the stellar cores, and are described in more detail later, in Section 3.5 and Focus 3.8. Very massive and bright stars will survive only a few million years. That is why they are so rare and hard to find. Some of these stars have come and gone since humans began to walk the Earth. The very luminous, massive stars that we can now see were not shining when the dinosaurs ruled the Earth; had they been shining then, more than 65 million years ago, they would have burned out long ago.

Stars of intermediate mass, such as the Sun, will shine for 10 to 20 billion years. Some of them were born when our Galaxy originated; they have been around ever since. The Sun formed about 4.6 billion years ago, roughly 10 billion years after the Galaxy began. In another 7 billion years, the Sun is expected to end its life on the main sequence (Section 3.5). Stars with very low mass, below 0.8 solar masses, have lifetimes that exceed the age of our Galaxy. Every one that ever was is still there, and none of these stars has yet had time to perish. Thus, to understand stellar evolution we have to examine the upper part of the H–R diagram, where the more-massive stars are located.

Spectral type	Effective temper-ature (K)	Mass (M/M$_\odot$)	Luminosity (L/L$_\odot$)	Radius (R/R$_\odot$)	Lifetime (years)
O5	44 500	60	7.9×10^5	12	3.7×10^6
B0	30 000	17.5	5.2×10^4	7.4	1.1×10^7
B5	15 400	5.9	8.3×10^2	3.9	6.5×10^7
A0	9 520	2.9	5.4×10	2.4	2.9×10^8
F0	7 200	1.6	6.5	1.5	1.5×10^9
G0	6 030	1.05	1.5	1.1	5.1×10^9
K0	5 250	0.79	0.42	0.85	1.4×10^{10}
M0	3 850	0.51	0.077	0.60	4.8×10^{10}
M5	3 240	0.21	0.011	0.27	1.4×10^{11}

[a]The mass, M, is in units of the Sun's mass $M_\odot = 1.989 \times 10^{30}$ kg; the absolute luminosity, L, is in units of the Sun's absolute luminosity, $L_\odot = 3.85 \times 10^{26}$ W; and the radius, R, in units of the Sun's radius, $R_\odot = 6.955 \times 10^8$ m. The lifetime is the amount of time required to exhaust the nuclear hydrogen fuel that supplies the energy of star on the main sequence (Section 3.5, Focus 3.8).

Table 2.2. **The Main-Sequence Stars**[a]

Fig. 2.15 **Dark interstellar cloud** This dark nebula, known as Barnard 86, is located in the constellation Sagittarius. When Edward Barnard discovered these dark places, he thought they might be black holes in the sky, but they are actually interstellar clouds of gas and dust. The dust absorbs the light of distant stars, making this region appear dark in contrast to the numerous background stars in the Milky Way. Interstellar molecules and newborn stars are found in these dark, interstellar clouds. Barnard 86 might be the leftover residue of the formation of nearby, bright blue stars in the young star cluster named NGC 6520, found just to the left of the dark cloud. (Courtesy of David F. Malin, © 1980 Anglo Australian Telescope Board.)

Stars are born, live and die

The space between the stars in our Galaxy may look dark and empty to the eye, but it contains vast clouds of gas and dust (Fig. 2.15). There is now about as much material in interstellar clouds as there is in visible stars. These clouds provide the raw stuff from which stars and planets can be made. They are the future nurseries of new-born stars.

The dust is composed of heavy elements concentrated into tiny solid particles resembling smoke. It absorbs the light of all the distant stars lying behind it, and hides the interior of the clouds from view. Interstellar dust also shields nearby gas from energetic starlight, keeping the clouds cold and permitting molecules to form.

The present-day birth-place of stars is within the dark, giant clouds of molecular hydrogen whose total mass is between one hundred thousand and one million times the mass of the Sun, or $10^5 - 10^6$ M$_\odot$. These clouds are extremely cold, about 10 degrees kelvin; water freezes at 273 degrees on the kelvin scale. The molecular clouds can also be very compact, about 10 pc across, and very dense when compared to other interstellar material. As many as 200 million, or 2×10^8, hydrogen molecules can be packed into every cubic meter. In contrast, there are only about a million hydrogen atoms in a cubic meter of the interstellar material outside the molecular clouds.

These clouds are now in the process of forming future stars. If a cloud becomes sufficiently massive and dense, the mutual gravitation of its parts will overcome the internal pressure, and the cloud will start falling in on itself. Clouds with a mass smaller than giant molecular clouds have a more difficult time. They require external compression to get the collapse underway. The shocks and expanding remnants of a nearby exploding star can trigger the collapse, but such events do not occur very often. For the most part, the interstellar gas and dust is too tenuous, too hot, and too agitated to spontaneously condense into stars. This is why there is now still plenty of material around to form stars after billions of years of continual star formation.

Any interstellar cloud which becomes spontaneously unstable to gravitational collapse has a mass of roughly a million Suns. Once collapse is underway, the giant cloud will fragment, and the pieces can become stars like the Sun. Each material fragment will fall toward its center under its own self-gravity, and a massive object will grow there. Once gravitational collapse begins, it takes almost a million years to build up a star with the Sun's mass at the center from a cloud fragment.

Fig. 2.16 **The Orion Nebula** In the gas and dust of the Orion Nebula, designated M42 or NGC 1976, gravity has already pulled in the debris of long-dead stars, igniting the celestial fires of new-born stars. The hot, massive stars ionize the nearby interstellar gas, causing it to fluoresce with red light. The Orion Nebula is about 1 degree in angular extent; it is located at a distance of 500 pc and is about 13 pc across. This bright crucible of star creation is immersed in a giant molecular cloud that is 4 by 10 degrees in angular dimensions. This photograph was taken in red light during a 90-minute exposure on the 1.2-m United Kingdom Schmidt Telescope. The original photographic plate has been copied using the unsharp masking technique for controlling contrast, thereby permitting fine detail to be resolved in both the bright central region and the fainter outlying filaments. (Royal Observatory Edinburgh © 1978, prepared by David F. Malin.)

When the local concentrations of hydrogen gas are pulled together by the force of gravity, the core becomes more compressed and the temperature rises. The infalling gas accrues energy, called gravitational potential energy, much the way a hammer or a waterfall gains energy when it moves toward the ground. Heated by this energy, the collapsing core eventually becomes hot enough to ignite nuclear fusion, burning hydrogen to make the star shine (Section 3.3). For the Sun, this occurred after almost a million years of gravitational collapse, when the temperature at the center reached 15.6 million degrees kelvin. The outward pressure of the hot gas prevents the star from collapsing further. It has settled down for a long, rather uneventful life as a main-sequence star, the longest stop in its life history.

The most massive stars that are formed in this way light up the surrounding material, forming colorful regions of ionized hydrogen in the same general region that giant molecular clouds are found (Fig. 2.16).

Rotation gives spin to the infalling material, so the collapse is not spherical. A spinning "accretion" disk is created around the central proto-star, the future home of planetary systems. A stellar wind also breaks out along the rotation axis of the system, creating a bipolar outflow of cold molecular gas. Circumstellar disks and jet-like outflows are often found in regions of star formation. Large planets have even been detected in orbit about some Sun-like stars.

Both the total number of stars and their lifetime rapidly increase down the main sequence from the high-mass, luminous upper end to the low-mass dimmer lower end. Ninety percent of all main-sequence stars have a mass below $0.8\,M_\odot$, and have been on the main sequence ever since they were born. They thus provide us with no information about stellar evolution.

Some of the massive stars, that were born long ago, have had enough time to burn up their available hydrogen fuel and advance to the next stage of the stellar life cycle. What happens next was hinted at in the original H–R diagram, where a few stars were found up and to the right of the main sequence (see Fig. 2.13). If these stars collectively describe the course of stellar evolution, then they have become brighter and cooler than their main-sequence counterparts. This could only happen if the stars became larger as their outer layers cooled, and if the swelling was enough to offset any dimming caused by a drop in temperature. Such stars have become the giants.

The path of stellar evolution is most clearly portrayed in the H–R diagram of a globular star cluster that can contain millions of stars. We can assume that all of the stars in the cluster were born about the same time from the same giant cloud. Thus they have the same ages and initial chemical composition.

The stars in a globular star cluster are born with a full range of stellar masses occupying the entire main sequence. As the cluster ages, the massive stars perish first, and the main sequence starts to disappear from the top down. Since the globular star clusters contain the oldest known stars in our Galaxy, the massive ones have had time to move off the upper end of the main sequence, delineating the evolution to the giant stage of stellar life (Fig. 2.17).

The age of a globular cluster can be determined from (i) the luminosity of the main-sequence turn-off point, or from (ii) the cluster's detailed evolutionary path in the H–R diagram. Method (i) involves theoretical calculations of just how long a main-sequence star's central fuel supply can last; Method (ii) incorporates models of what happens when the fuel is used up. These methods indicate that the oldest globular star clusters in our Galaxy are between 10 and 20 billion years old, or between two and five times the age of the Sun, a mere 4.6 billion years old.

The evolutionary information contained in the various kinks and bends of the H–R diagrams has been deciphered using complex theoretical models and high-speed computers. After spending most of its life on the main sequence, at essentially constant effective temperature and luminosity, an aging star consumes the available hydrogen near its center. All the hydrogen in its core has been

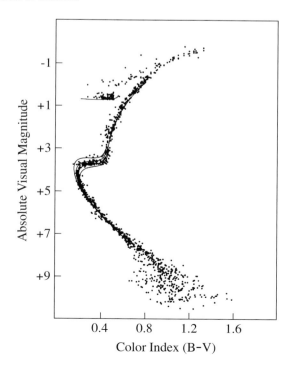

Fig. 2.17 **Hertzsprung–Russell diagram of a globular star cluster** A plot of the absolute visual magnitudes, M_V, against the intrinsic color, $B - V$, for the stars in the southern globular star cluster 47 Tucanae, or NGC 104. Although low-mass, relatively faint stars are still on the main sequence (*diagonal line from middle left to bottom right*), the massive, bright stars in the cluster have left the main sequence and are evolving into giant stars (*top right*). The turn-off point from the main sequence is at the left side of the horizontal arm at about $M_V = +4$. Theoretical tracks, called isochrones, show the evolutionary distributions at different ages of 10, 12, 14 and 16 billion years. The best fit corresponds to an age of between 12 and 14 billion years for this star cluster, which allows us to date the formation of the cluster and the halo of our Galaxy. (Courtesy of James E. Hesser.)

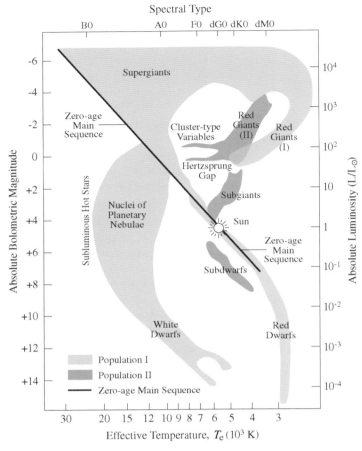

converted into helium. The core then contracts, liberating gravitational energy, and the envelope expands into a giant star. The star's outer layers cool and become red or yellow, but its central temperature increases until it is hot enough to burn helium. The bright red star Betelgeuse in the constellation of Orion, the Hunter, is one of these red giant stars, five hundred times larger than the Sun and over

Fig. 2.18 **Composite Hertzsprung–Russell diagram** Stars of different absolute luminosity, L (*right axis, in units of the Sun's absolute luminosity* $L_\odot = 3.85 \times 10^{26}$ W) and absolute bolometric magnitude, M_{bol}, (*left axis*) are plotted as a function of the effective temperature, T_e, of their visible disk (*bottom axis*). The vast majority, roughly 90 percent, of stars occupy the main sequence of the Hertzsprung–Russell diagram, running diagonally from the hot, luminous upper left to the cooler, less-luminous lower right. The hot, massive supergiants (*top left*) are the most luminous stars. The white-dwarf stars (*lower middle*) are low-luminosity stars about the size of the Earth.

Fig. 2.19 **The Helix planetary nebula** The planetary nebula NGC 7293 in Aquarius is almost half a degree across, or about the angular size of the Moon and the Sun, and is located at a distance of 500 pc. It is called the Helix nebula after its helical form that resembles a coiled spring. The Helix planetary nebula is expanding into space from a central white-dwarf star, at a velocity of about 14×10^3 m s⁻¹. The light of the central star excites atoms and ions of oxygen and nitrogen that emit the green light within the circular envelope and the red light in the radial structures. This color photograph is a combination of photographs made in red, green and blue light using the 3.9-m Anglo Australian Telescope at Siding Spring, New South Wales, Australia. (Anglo Australian Telescope Board © 1979, Photo by David J. Malin.)

Fig. 2.20 **The Crab Nebula supernova remnant** The Crab Nebula (M1 or NGC 1952) is an expanding cloud of interstellar gas that was ejected by the supernova explosion of a massive star, about nine solar masses, in 1054 A.D. The filamentary gases are expanding at a velocity of 1.5 million meters per second. The red-yellow wisps of gas shine primarily in the light of hydrogen, while the blue-white light is the non-thermal radiation of high-speed electrons spiraling in magnetic fields. The south-westernmost (*bottom right*) of the two central stars is the remnant neutron star of the supernova explosion, and a radio pulsar with a period of 0.33 s. This supernova remnant has angular dimensions of 4.5×7 minutes of arc; it is located at a distance of 1930 ± 110 pc and has a diameter of 2.9 pc. This famous supernova remnant is also a source of intense emission at radio and X-ray wavelengths. (Courtesy of Rudolph Schild, Harvard-Smithsonian Center for Astrophysics.)

ten thousand times more luminous. If our Sun were the size of Betelgeuse, it would extend all the way out to Mars. There will come a time, in roughly seven billion years, when the aging Sun will swell up to become a giant star; though not quite as big as Betelgeuse (Section 3.5).

Most of the life history of stars can be traced out in the composite H–R diagram that includes intrinsically bright, distant and relatively rare stars. It shows two well-defined classes of stars, called dwarfs and giants on the basis of their size. The main-sequence stars are the dwarfs. The giant stars are located in a broad horizontal band near the top of the H–R diagram; the giants are much bigger stars with higher luminosity than most main-sequence stars (see Fig. 2.18). This follows from the Stephan–Boltzmann law that indicates that a star of a certain color and effective temperature will have a much larger luminosity if it has a bigger radius.

Very-massive stars near the top of the main sequence become supergiants; those with masses comparable to the Sun become red giants. Aging then accelerates and progresses rapidly, as it can for humans. The supergiants quickly consume their successive internal fuel sources at ever-increasing central temperatures and rates; the helium is converted into carbon, and then the carbon serves as fuel, and so on up the chain of the nuclear species. The red giants are not sufficiently massive to burn anything but helium. In the meantime, intense winds in the outer atmosphere of these stars blows some of the stuff out of the giants back into interstellar space.

The ultimate fate of a star also depends upon its mass. When giant stars with a moderate, Sun-like mass use up their helium fuel, they collapse until they are about the size of the Earth, or only about one-hundredth, or 0.01, of the Sun's radius. The collapsed star is called a white-dwarf star, for it has a small size and it is initially white in color. Such stars are

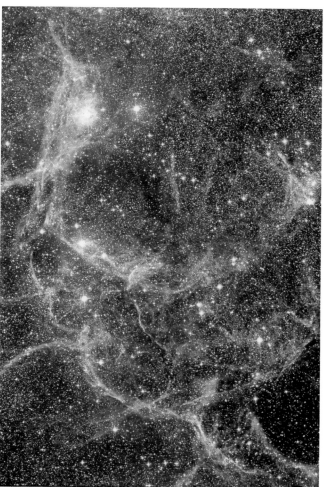

Fig. 2.21 **Supernova remnant in Vela** The most-massive stars end their lives in violent explosions called supernovae. These tenuous red wisps mark the expanding debris of the supernova that exploded in the constellation Vela, perhaps 12 thousand years ago. Its gases are still expanding outward as a result of the explosion. The Vela supernova remnant now subtends an angle of about 255 minutes of arc; it is located at a distance of about 500 pc and is now about 30 pc across. (Royal Observatory Edinburgh © 1979. Photo prepared by David F. Malin.)

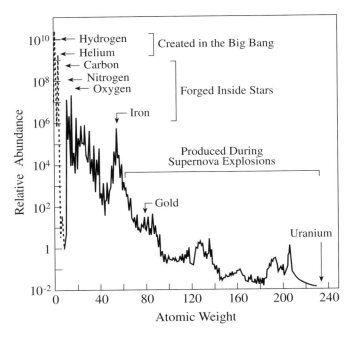

Fig. 2.22 **Abundance and origin of the elements in the Sun** The relative abundance of the elements in the Sun plotted, as a function of their atomic weight, on a logarithmic scale that spans twelve orders of magnitude. The abundance is normalized to a value of 1.0×10^6 for silicon, Si, and the atomic weight or mass number, is the number of protons and neutrons in the nucleus of the element. Hydrogen, the lightest and most-abundant element in the Sun, was formed 10 to 20 billion years ago in the immediate aftermath of the big-bang explosion that led to the expanding Universe. Most of the helium now in the Sun was also created then. All the elements heavier than helium were synthesized in the interiors of stars that no longer shine, and then wafted or blasted into interstellar space where the Sun subsequently originated. Carbon, nitrogen, oxygen and iron were created over long time intervals during successive nuclear-burning stages in former stars. Because any nuclear reaction involving the iron group must absorb energy rather than release it, these elements cannot serve as fuel in further chains of nuclear reactions inside stars. Elements heavier than iron, such as gold and uranium, were produced by neutron-capture reactions during the supernova explosions of massive stars that lived and died before the Sun was born. The exponential decline of abundance with increasing mass can be explained by the rarity of stars that have evolved to later stages of life.

to be found near the bottom of the composite H–R diagram (Fig. 2.18), below the main sequence.

The visible disk of a white dwarf can be as hot as one hundred thousand degrees kelvin when it is created, but its light comes from the leftover heat of a former life. There is no fuel left in its core to power a nuclear furnace, so a white dwarf slowly cools and fades into darkness as it ages. This, then, will be the final resting place of our Sun. It will ultimately fade from sight, like a dying ember in a fireplace once the fire has gone out.

Dying stars with moderate mass can blow off a spherical puff of gas before settling down to their final state. These shells are called planetary nebulae, so-called because their round shapes, when viewed through early telescopes, resembled planets (Fig. 2.19).

More-massive stars have an explosive death. After the core has become hot enough to produce iron, the star has reached the end of the line. It has become bankrupt, completely spending all its nuclear resources, and there is nothing left to do but collapse under the sheer weight of it all. In a matter of milliseconds the central core is crushed into a ball of neutrons about 10 000 m across, no bigger than a city. The outer layers also plunge in toward the center, but they rebound like a tightly coiled spring. The pent-up energy generated in the collapsing center blasts the outer layers out in a violent explosion called a supernova, littering space with its cinders. These ashes will join the debris from countless other explosions, providing the raw material for the next generation of stars.

In a galaxy the size of the Milky Way, a supernova explodes on average once every hundred years or so. The energy released in a supernova is immense. For a few weeks it can be brighter than the combined brightness of all the other stars in a galaxy. Then, as the debris expands outwards, it cools and

becomes fainter. Astronomers use the name supernova remnant for this expanding shell of gas (Figs. 2.20, 2.21). This material moves out into interstellar space, seeding it with heavy elements that were forged inside the former star.

Where did the chemical elements come from?

The majority of atoms that we see today were formed at the dawn of time before the stars even existed, in the immediate aftermath of the big-bang explosion that produced the expanding Universe. All of the most-abundant element, hydrogen, that is now in the Universe was created back then, about 10 or 20 billion years ago, and so was most of the helium, the second most-abundant element. The hydrogen and helium were synthesized in the earliest stages of the infant Universe, within just a wink of the cosmic eye. As the Universe expanded, it cooled and thinned out, prohibiting primordial nucleosynthesis after the first few hundred seconds of the big bang.

The first stars could not have had rocky planets like the Earth, because there was initially nothing but hydrogen and helium. The only possible planets would have been balls of gas. Without carbon, life as we know it could not evolve in these planets or on their, possible frozen, surfaces.

Observations have shown that very old stars, born long ago when the Universe was young, have practically no elements except hydrogen and helium, produced in the early Universe. We see these survivors of the earliest times in globular star clusters; they have probably existed ever since our Galaxy formed 10 to 20 billion years ago.

Stars that contained only hydrogen and helium are called first-generation stars. Middle-aged stars like the Sun are second-generation stars, meaning that some of their material came from previous stars. Sun-like stars contain heavy elements that were formed inside massive first-generation stars or at the time of their explosions (Fig. 2.22).

During the billions of years before the Sun was born, massive stars reworked the chemical elements, fusing lighter elements into heavier ones within their nuclear furnaces. Carbon, oxygen, nitrogen, silicon, iron, and most of the other heavy elements were made this way. The enriched stellar material was then cast out into interstellar space by the short-lived massive stars, gently blowing in their stellar winds or explosively ejected within supernova remnants.

The Sun and its retinue of planets condensed from this material about 4.6 billion years ago. They are partly composed of heavy elements that were synthesized long, long ago and far, far away, in the nuclear crucibles of stars that lived and died before the Sun was born. The Earth and everything on it have been spawned from this recycled material, the cosmic leftovers and waste products of stars that disappeared long ago.

Perhaps the most fascinating aspect of stellar alchemy is its implications for life on Earth. Most of the chemical elements in our bodies, from the calcium in our teeth to the iron that makes our blood red, were created billions of years ago in the hot interiors of long-vanished stars. We are true children of the stars, for we are all made of star stuff. If the Universe were not very, very old, there would not have been time enough to forge the necessary elements of life in the ancient stellar cauldrons. The same nuclear powerhouse that synthesized heavy elements from light ones, and made these stars shine, is now at work inside the Sun.

CHAPTER THREE

What makes the Sun shine?

Wild geese in sunlight Geese flying south in the northern winter, following the Sun's warmth. In this V-shaped pattern of flight, the lead bird deflects currents of air and makes flying easier for those that follow in its wake. The Earth's magnetic field similarly deflects the Sun's wind. (Courtesy of James Tallon.)

3.1 Awesome power, enormous times

The Earth intercepts only a modest fraction of the energy being pumped out in all directions from the Sun. When we measure the total amount of sunlight that illuminates and warms our globe, and extrapolate back to the Sun, we find that it is emitting a power of 385 million million million million, or 3.85×10^{26}, watts. An enormous amount of energy is being expended. In just one second the energy output of the Sun equals the entire energy consumption of the United States for a million years.

The astonishing thing is the Sun's durability; the Sun has managed to last billions of years despite radiating such awesome amounts of energy. In looking back at Earth's history, we find that the Sun has been shining steadily and relentlessly for aeons, with a brilliance that could not be substantially less than it is now. The radioactive clocks in rock fossils indicate, for example, that the Sun was hot enough to sustain primitive creatures on Earth 3.5 billion years ago (Section 1.7).

What has kept the Sun hot for so long? The seemingly constant and unchanging Sun must eventually run out of fuel. After all, nothing can stay hot forever, and all things wear out in time.

There is no ordinary fire within the Sun. If the Sun was made entirely of burning coal, and had an adequate supply of oxygen to sustain combustion, it could only generate as much heat and light as the Sun now does for a few thousand years. So, chemical burning cannot explain the fires that keep the Sun hot.

In 1892, the physicist William Thompson, later Lord Kelvin, showed that gravity might supply the Sun's energy for a much longer time than an ordinary fire. If the Sun was gradually shrinking, the matter falling inward would collide and heat the solar gas. Thompson showed that this slow contraction could sustain the Sun's heat for about 100 million years. Giant stars can only shine for about 30 000 years if gravitational contraction supplies their energy (Focus 3.1). Moreover, radioactivity was discovered in 1895, and within a decade it was used to show that Earth's rocks are older than Thompson's time limit for sustaining the Sun's fires.

By the early-20th century, when your parents and/or grandparents might have lived, no one had any clue as to why the Sun, or any other star, could shine so brightly for billions of years. That understanding had to await the discovery of the ingredients of the atom, observations that the Sun is mainly composed of hydrogen (Section 1.5), and the realization that sub-atomic energy can be released within the extraordinary conditions inside stars.

Focus 3.1 How long can gravity fuel a star?

If a star is slowly contracting, the shrinkage produces gravitational potential energy that can be converted into heat. For a star of mass, M_s, and radius, R_s, the total amount of that energy is:

$$\text{gravitational potential energy} = \Omega = \frac{GM_s^{\,2}}{R_s}$$

where the gravitational constant $G = 6.6726 \times 10^{-11}$ N m^2 kg^{-2}. The change, $\Delta\Omega$, in gravitational potential energy created by a decrease in radius, ΔR_s, is:

$$\Delta\Omega = \frac{GM_s^2}{R_s^2}\Delta R_s = 0.55 \times 10^3 \left(\frac{M_s}{M_\odot}\right)^2 \left(\frac{R_\odot}{R_s}\right)^2 \Delta R_s \ \text{J},$$

where the radius decrease is in meters, and we have normalized the mass and radius in terms of the Sun's mass $M_\odot = 1.989 \times 10^{30}$ kg and the Sun's radius $R_\odot = 6.955 \times 10^8$ m.

If the energy change provides an absolute luminosity, L_s, in a time interval, Δt, then $L_s = \Delta\Omega/\Delta t$, and the rate of change in radius is:

$$\frac{\Delta R_s}{\Delta t} = 7.05 \times 10^{-7} \left(\frac{L_s}{L_\odot}\right)\left(\frac{M_\odot}{M_s}\right)^2 \left(\frac{R_s}{R_\odot}\right)^2 \ \text{m s}^{-1},$$

where the Sun's absolute luminosity is $L_\odot = 3.85 \times 10^{26}$ W. Since one year is 3.1557×10^7 seconds, this shows that a contraction of only 22.3 m yr^{-1} will power the Sun at the present rate. That is a very small change considering the very large radius of the Sun.

The problem with this mechanism is the long duration of the Sun and other stars. If the source of a star's present luminosity is gravitational potential energy, then the current radius would shrink to zero in a total gravitational lifetime, τ_S , of:

$$\text{gravitational lifetime} =$$

$$\tau_s = \frac{R_s}{(\Delta R_s/\Delta t)} = \frac{\Omega}{L_s} = 3.1 \times 10^7 \left(\frac{M_s}{M_\odot}\right)^2 \left(\frac{R_\odot}{R_s}\right)\left(\frac{L_\odot}{L_s}\right) \text{yr}.$$

Thus, if the Sun shines by converting gravitational potential energy into heat, it will disappear in only 31 million years.

The problem is much worse for a giant star that has both a larger radius and a greater luminosity than the Sun. Of course, a brilliant giant star might be more massive than the Sun, since the stellar luminosity increases with stellar mass, but the lifetime will still be inversely proportional to the radius and a low power of the luminosity. If we assume that the luminosity increases with the mass cubed, a giant star that is 100 times bigger than the Sun and has an absolute luminosity 1000 times the solar value, could only be powered by its gravitational potential energy for about 30 000 years.

3.2 A hot, dense core

Whole atoms are only found in the Sun's relatively cool, visible layers and they do not exist in most of the Sun. The solar atoms are stripped bare and lose their identity inside the Sun. The temperature is so high, and the particles are moving so fast, that innumerable collisions tear the atoms apart into their sub-atomic constituents.

The most-abundant atom in the Sun is hydrogen (Section 1.5). Each hydrogen atom is made up of a nucleus that contains a positively-charged proton, and a remote, negatively-charged electron that orbits the nucleus. In the Sun, collisions rip the hydrogen atoms apart, and separate the electron from the nucleus. Both the electron and the proton are liberated from their atomic bonds, and are set free to move and wander throughout the solar interior unattached to each other.

What is left is a plasma, a seething mass of electrically-charged particles – electrons, protons and heavier ions. Helium and heavier elements are ionized inside the Sun, with some of their atomic electrons set free, but these ions are much less abundant than the protons. The electrical charge of the protons and other positively-charged ions balances and cancels that of the electrons removed from atoms to make them, so the plasma has no net electric charge.

The Sun is just one huge mass of incandescent plasma, compressed on the inside and more tenuous further out. With their electrons gone, hydrogen nuclei (protons) can be packed much more tightly than complete atoms. This is because all atoms are largely empty space, with their

---

Focus 3.2 The equations of solar structure

The four differential equations that determine a stars initial position on the main sequence of the Hertzsprung–Russell diagram and its subsequent evolutionary history are:

The equation of hydrostatic equilibrium. This equation states that the inward force of gravity caused by the mass, $M(r)$, within a distance, r, from the stellar center is just balanced by the outward gas pressure, $P_G(r)$, at radius, r, so that:

$$\frac{dP_G(r)}{dr} = \frac{\rho G M(r)}{r^2} \text{ or equivalently } = -\frac{dP_G(r)}{dM(r)} = \frac{GM(r)}{4\pi r^4},$$

where the gravitational constant $G = 6.6726 \times 10^{-11}$ N m^2 kg^{-2}, the mass-density is denoted by ρ, and the gas pressure is given by the ideal gas law (Section 2.3, Focus 2.5).

The mass equation. This equation specifies the mass, $M(r)$, contained within radius, r, in terms of the mass density, ρ, by:

$$\frac{dM(r)}{dr} = 4\pi r^2 \rho \text{ or equivalently } \frac{dr}{dM(r)} = \frac{1}{4\pi r^2 \rho}.$$

This equation is subject to the boundary conditions of zero mass at zero radius, or $M(r) = 0$ at $r = 0$, and a mass that is now equal to the total mass of the star, M_s, at the visible stellar radius, R_s, or $M(R_s) = M_s$. For the Sun, $M_s = M_\odot = 1.989 \times 10^{30}$ kg and the radius $R_s = R_\odot = 6.955 \times 10^8$ m.

The equation for energy balance. This equation states that the energy generated per unit mass per unit time in the star's core, denoted by ε, supplies the energy flux, $L(r)$ carried across radius, r, or that:

$$\frac{dL(r)}{dr} = 4\pi r^2 \rho \varepsilon \text{ and equivalently } \frac{dL(r)}{dM(r)} = \varepsilon.$$

This equation has the boundary condition provided by the current luminosity, L_s, for a star of total mass, M_s, and radius, R_s. For the Sun, we have $L_s = L_\odot = 3.85 \times 10^{26}$ W. Hydrogen-

burning is responsible for the energy generation of stars on the main sequence (Section 3.3). The energy generated, ε, is a function of the initial composition, mass density and temperature.

The equation for radiation energy transfer. This equation relates the temperature, $T(r)$, at radius, r, to the amount of energy being transferred by radiation to that distance. It is related to the opacity to radiation, κ, which measures the resistance of the material to energy transport by radiation. Opacity is like the dirt on a window that keeps some of the sunlight from getting through. The equation is:

$$\frac{dT(r)}{dr} = \frac{-3\kappa\rho L(r)}{4ac\,[T(r)]^3 4\pi r^2} \text{ or equivalently } \frac{dT(r)}{dM(r)} = \frac{-3\kappa L(r)}{64\pi^2\, ac[T(r)]^3 r^4},$$

where the radiation density constant $a = 7.5659 \times 10^{-16}$ J m^{-3} K^{-4}, and the velocity of light $c = 2.9979 \times 10^8$ m s^{-1}. The total luminosity of a star, L_s, of radius, R_s, is given by the Stefan–Boltzmann law $L_s = 4\pi\sigma R_s^2 T_e^4$ where the Stefan–Boltzmann constant $\sigma = ac/4 = 5.6705 \times 10^{-8}$ W m^{-2} K^{-4} and T_e is the effective temperature.

The chemical composition of a zero-age main sequence star is assumed to be homogenous, and is often specified by X = fraction by mass of material in the form of hydrogen, Y = fraction by mass of material in the form of helium, and $Z = 1 - X - Y$ = fraction by mass of material heavier than helium. Observations of sunlight and meteorites indicate that $X = 0.706\,83 \pm 0.025$, $Y = 0.274\,31 \pm 0.06$ and $Z = 0.018\,86 \pm 0.000\,85$ for the Sun. The mass density, ρ, for a fully ionized plasma of total number density, N, is given by:

$$\rho = \left(2X + \tfrac{3}{4}Y + \tfrac{1}{2}Z\right)^{-1} N m_H,$$

where the mass of the hydrogen atom is $m_H = 1.673\,534 \times 10^{-27}$ kg.

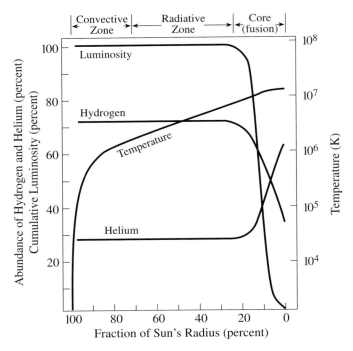

Fig. 3.1 **Internal compression** The Sun's luminosity, temperature, and composition all vary with depth in its interior, from the Sun's visible disk (*left*) to the center of the Sun (*right*). The nuclei are squeezed into a smaller volume by the pressure of the material above, becoming hotter and more densely concentrated at greater depths. At the Sun's center, the temperature has reached 15.6 million degrees kelvin, and the pressure is 233 billion times that of the Earth's atmosphere at sea level. Nuclear fusion in the Sun's energy-generating core synthesizes helium from hydrogen, so this region contains more helium and less hydrogen than the primordial amounts detected in the light of the visible Sun.

orbiting electrons located at remote distances from their nucleus.

At great depths inside the Sun, the pressure of overlying material is enormous, the protons are squeezed tightly together, and the material is very hot and extremely dense. To understand how this works, imagine a hundred mattresses stacked into a pile. The mattresses at the bottom must support those above, so they will be squeezed thin. Those at the top have little weight to carry, and they retain their original thickness. The gas at the center of the Sun is similarly squeezed into a smaller volume by the overlying material, so it becomes hotter and more densely concentrated.

Since we cannot see inside the Sun, or any other star, astronomers use mathematical models to determine the internal structure of stars; these models have recently been confirmed by measurements of sound waves in the Sun (Section 4.2). The models use the laws of physics to describe a self-gravitating sphere of highly ionized hydrogen, helium and small quantities of heavier elements (Focus 3.2). At every point inside the Sun, the force of gravity must be balanced precisely by the gas pressure, which itself increases

with the temperature. The energy released by nuclear fusion in the Sun's core heats the gas, keeps it ionized, and creates the pressure. This energy must also make its way out of the Sun to provide its luminosity and keep it shining.

After arrival on the main sequence (Section 2.3), which is called the time of zero age, the internal structure of a star and its subsequent history can be determined by just four equations (Focus 3.2), provided that the age, mass and chemical composition of the star are specified. Stars with the same composition will have radii, luminosities, effective temperatures, and mean densities that are determined solely by the stars' ages and masses. The crucial equations can be solved without any knowledge of properties of the star before arrival on the main sequence. They describe: nuclear energy generation by burning hydrogen in the central core of the star; hydrostatic equilibrium that balances the outward force of gas pressure and the inward force of gravity; energy transport by radiative diffusion; and an opacity determined from atomic physics calculations.

These equations are integrated over 4.6 billion years, the age of the Sun, to obtain the present luminosity of the Sun, for a star of its mass and radius. This results in a Standard Solar Model that specifies the current mass-density, temperature and pressure as a function of depth within the Sun (Fig. 3.1).

The model indicates just how hot and dense it is down at the core of the Sun (Fig. 3.1). At the Sun's center the temperature is 15.6 million degrees (Section 2.3, Focus 2.5). The central mass density is 1.5×10^5 kg m^{-3}, or more than 13 times that of solid lead. Yet, the protons are so small that they can move about freely as a gas, even at this high density.

The visible solar disk is relatively cool, at 5780 K, and extremely rarefied, about ten thousand times less dense than the air we breathe. The pressure of this tenuous gas is less than that beneath the foot of a spider. Just as the pressure on your body increases as you dive deeper and deeper in the sea, so does the pressure increase with greater depth into the Sun. The pressure at the center of the Sun is 233 billion times Earth's air pressure at sea level (Fig. 3.1); all that outward gas pressure is needed to support the mass of the Sun.

When the Sun arrived on the main sequence, at zero age, it was assumed to have a homogeneous chemical composition, with a hydrogen abundance of about 71 percent by mass and helium at about 27 percent by mass. Nuclear fusion reactions convert hydrogen into helium within the Sun's hot, dense core, providing the Sun's energy and explaining its luminosity. The amount of helium in the core of the Sun has therefore increased over the past 4.6 billion years, due to the ongoing synthesis of helium from hydrogen, and the amount of hydrogen in the core has decreased (Fig 3.1).

3.3 Nuclear fusion reactions in the Sun

The ultimate source of the Sun's energy is nuclear fusion in its core. The intense pressures and searing temperatures at the Sun's core are fusing together the nuclei of the most-abundant element in the Universe, hydrogen, to form nuclei of the second most-abundant element, helium. The mass lost in the nuclear fusion reactions supplies the energy that makes the Sun shine. The nuclear fusion begins when two of the fastest-moving protons collide head on, and very occasionally tunnel through the electrical barrier that almost always keeps them apart. When two protons merge and come into each other, they initiate a chain of reactions that ends when four protons have joined together to make one helium nucleus consisting of two protons and two neutrons. This sequence of nuclear fusion reactions is called the proton–proton chain. In main-sequence stars more massive than the Sun, carbon acts as a catalyst in hydrogen–burning by the CNO cycle; less massive main-sequence stars burn hydrogen by the proton-proton chain.

Mass lost is energy gained

The only known method for keeping the Sun shining with its present intensity for billions of years involves nuclear fusion reactions at great depths within the Sun. They are termed "nuclear" because it is the interaction of atomic nuclei that powers the Sun. In nuclear fusion reactions, two or more atomic nuclei fuse together to produce a heavier nucleus, releasing energy and tiny elementary particles.

The Sun uses as fuel the simplest element, hydrogen, the lighter-than-air gas that once lifted airships to the skies. The power comes from the Sun's ability to squeeze hydrogen nuclei, the protons, together so hard that they make helium. Still, protons can only be fused together near the center of the Sun, where it is hot and dense enough for them to move into each other. Outside the solar core, where the overlying weight and compression are less, the gas is cooler and thinner, and nuclear fusion cannot occur.

Nuclear fusion, or the coming together of nuclei, has been achieved on Earth in the hydrogen bomb, showing how dangerous it can be. Scientists have been unsuccessfully trying to fuse nuclei together in a more controlled way for more than fifty years. You need temperatures about as hot as the center of the Sun, with fantastic pressures to force the particles together. The Sun squeezes its protons together with gravity; on Earth, strong magnetic fields are used to push them close enough to fuse. At present, we can only keep the fusion going for a few seconds. If and when we succeed in making it last, controlled nuclear fusion could be a source of power for future generations.

Still, replicating what the Sun does every second of every day is not an easy thing to do. For the time being, nuclear energy can only be generated relatively safely in terrestrial nuclear reactors that work not by fusing atoms, but by splitting and breaking them apart.

Energy can only be derived from energy, and the source of energy in nuclear fusion is mass loss. The foundation was provided by Albert Einstein's special theory of relativity, which included the famous formula $E = mc^2$ for the equivalence of mass, m, and energy, E. Because the velocity of light, $c = 2.997\ 9 \times 10^8$ m s^{-1}, is very large, only a tiny amount of mass is needed to produce a huge amount of energy. Even the smallest grain of sand holds an enormous quantity of energy locked up inside its atoms.

Such large quantities of energy are often expressed in the macabre unit of kilotons or megatons of TNT, or trinitrotoluene. There are 4.128×10^{12} J of energy per kiloton of TNT, and a 100 megaton hydrogen bomb contains 4.128×10^{17} J. The consumption of just one gram of hydrogen in the Sun's proton-proton chain produces an energy equivalent to 20 kilotons of TNT, about the same as the nuclear bomb that destroyed Hiroshima on 6 August 1945.

The next clue to explaining the Sun's awesome energy came from measurements in the laboratory on Earth. They showed that the helium nucleus is slightly less massive (by a mere 0.7 percent) than the sum of the mass of the four hydrogen nuclei, or four protons, that combine to make it. So, what you get out in making helium is less than what you put into it (like the usual outcome of a slot machine or other type of gambling). The part that disappears goes into energizing the Sun and other stars. The mass difference, Δm, is converted into energy, ΔE, to power the Sun, all in accordance with Einstein's equation $\Delta E = (\Delta m)\ c^2$.

Shortly after these mass measurements, the English astrophysicist Sir Arthur Stanley Eddington argued in 1920 that hydrogen might be transformed into helium inside stars, with the resultant mass difference released as energy to power the Sun. A similar idea was advanced at about the same time by the French physicist Jean Perrin, but with less impact on the astronomical community. Eddington demonstrated that the sub-atomic power would provide an almost limitless reservoir of energy, sufficient to keep the Sun shining at its present rate for 15 billion years, thereby solving the riddle of why the Sun has stayed hot for so long.

The basic idea was there, but it took decades to unravel the details. Astronomers had not yet discovered that hydrogen is the most-abundant element in the Sun (Section 1.5), and most physicists were convinced that protons could not fuse together inside the Sun.

Break on through to the other side

Two protons resist coming together because they have the same electrical charge. This repulsive electrical barrier cannot be easily overcome, even in a head-on collision of protons in the extraordinary conditions inside the Sun. Two protons would need to have a temperature of 11 billion degrees kelvin to get close enough to fuse together (Focus 3.3), but the temperature at the Sun's center is only 15.6 million degrees kelvin. The temperature at the heart of the Sun was therefore thought to be too low to permit protons to fuse together and for nuclear reactions to take place.

The paradox was resolved when the quantum theory provided physicists with a new set of tools to describe the way particles interact. In the quantum world, you can never be quite sure exactly where a particle is, for it does not have a precisely defined position. This is because the particle is also a wave, and that wave is a spread-out thing. This wave-like nature results in an uncertainty in the exact positions of the two protons, and helps them tunnel through the electrical obstruction between them.

In 1928 George Gamow showed how this uncertainty could explain the escape of particles from inside radioactive atoms. The wave-like position of a particle can extend beyond the range of the strong nuclear force that holds an atomic nucleus together, permitting the particle to tunnel through the nucleus to the outside world. The strong nuclear force is, as its name implies, very powerful, but only over very short distances.

This tunnel effect works the other way around, and explains why protons can fuse together inside the Sun. Because of their wave-like character, two very fast protons only have to move close enough for their waves to overlap; then the strong nuclear force takes over and pulls them together. Thus, protons sometimes get close enough to move into each other and fuse together, even though their average energy is well below that required to overcome their electrical repulsion.

The bizarre tunneling reaction does not happen very often. The center of the Sun is so dense that there are about 10^{32} protons packed into every cubic meter; to obtain this number density just divide the mass density at the center by the mass of the proton. This means that a proton, with a radius of 10^{-15} meters, can only move about 0.03 meters before encountering another proton; this mean free path is obtained by dividing the proton's area by the number density of protons.

Only a tiny fraction of the protons inside the Sun are ever moving fast enough to overpower the repulsive electrical force. At a central temperature of 15.6 million degrees kelvin, the protons are darting about so fast, with a mean speed of 6.2×10^5 m s^{-1}, that each one of them collides with other protons about 20 million times every second. Yet, the protons nearly always bounce off each other without triggering a nuclear reaction during a collision. Even with

Focus 3.3 Electrical repulsion

Like charges repel each other, and this sets up an electrified barrier that keeps them from coming too close together. If a proton approaches another proton, it will be repelled by a force that is proportional to the square of the electrical charge, and is inversely proportional to the distance. Yet, if a proton is moving fast enough, with enough kinetic energy, it might get close enough to another proton to overcome the electrical repulsion.

To determine the velocity that is just fast enough for one proton to move into another, we equate:

$$\text{kinetic energy} = \tfrac{1}{2} m_p V^2 = \frac{e^2}{4\pi\varepsilon_0 R} = \text{electrical potential energy},$$

and solve for the velocity:

$$\text{velocity } V = \left[\frac{2e^2}{4\pi\varepsilon_0 m_p R}\right]^{1/2} = 1.66 \times 10^7 \text{ m s}^{-1},$$

where the charge of a proton is $e = 1.6022 \times 10^{-19}$ coulomb, the permittivity of free space is $\varepsilon_0 = 8.854 \times 10^{-12}$ farad per meter, the mass of the proton is $m_p = 1.6726 \times 10^{-27}$ kg, the constant $\pi = 3.141\ 592\ 65$ and we have assumed that the separation of two protons is comparable to the size of an atomic nucleus, or $R = 10^{-15}$ m, when they touch and move into each other.

The mean speed of a proton inside the Sun is determined by equating the kinetic energy of the moving proton to the thermal energy that keeps it hot, or

$$\text{kinetic energy of motion} = \tfrac{1}{2} m_p V^2 = \tfrac{3}{2} kT = \text{thermal energy},$$

where T is the temperature and Boltzmann's constant $k = 1.380\ 66 \times 10^{-23}$ J K^{-1}, and then solve for the velocity:

$$\text{thermal velocity} = V = \left[\frac{3kT}{m_p}\right]^{1/2} = 157 T^{1/2} \text{ m s}^{-1}.$$

For the center of the Sun, where the temperature $T = 15.6$ million degrees kelvin, the speed is only $V = 6.2 \times 10^5$ m s^{-1}, lower than that required to overcome the electrical repulsion between two protons. A proton would have to be at a temperature of $T = 1.1 \times 10^{10}$, or 11 billion, degrees kelvin for the nuclear fusion to occur.

So, the classical physicists were convinced that proton fusion cannot occur in the Sun, but the astronomer Arthur Eddington was certain that sub-atomic energy fueled the stars, stating in 1926 that:

> We do not argue with the critic who urges that the stars are not hot enough for this process; we tell him to go find a hotter place.

As it turned out Eddington was right and the physicists were wrong, but the physicists corrected the mistake by inventing a new quantum theory that permitted protons to fuse together in the center of the stars. In this theory, a proton acts like a spread-out wave, with no precisely defined position; and the proton's energy fluctuates about its average, thermal value. This means that a proton near the center of the Sun has a very small but finite chance of occasionally moving close enough to another proton to overcome the barrier of repulsion and tunnel through it.

the help of tunneling, a typical proton would have to wait over 10 billion years before penetrating the electrical barrier that separates it from another proton; so the average proton has to make about 10^{25} collisions before nuclear fusion can happen. For the fusion to occur, the collision must still be almost exactly head-on, and between exceptionally fast protons.

Moreover, the discovery of tunneling still did not reveal how the reactions that produce helium from hydrogen proceed inside the Sun, or why they produce enough radiation to make the Sun shine. This required the additional discovery of several sub-atomic particles, including the neutron, the positron and the neutrino.

Hydrogen burning

The Sun and most other stars are mainly composed of hydrogen (Section 1.5), and that is the material that must energize them. If these stars were made only of heavy elements, instead of hydrogen, they could not shine. To understand how hydrogen is burned within the central furnace of stars, we need to know about the sub-atomic constituents of matter.

The most familiar sub-atomic particles, the particles that make up atoms, are protons, neutrons and electrons. The nucleus of a hydrogen atom consists of a single proton, and the nucleus of any other atom contains protons and neutrons, collectively called nucleons. The nucleus of the helium atom, for example, has four nucleons – two neutrons and two protons.

Yet, it wasn't until 1919 that Ernest Rutherford showed that a proton lies at the center of a hydrogen atom, and James Chadwick discovered the neutron in 1932. The proton and neutron have about the same size and mass, but protons have a positive electric charge, while neutrons are electrically neutral. Rutherford was awarded the 1908 Nobel Prize in Chemistry for his investigations of radioactive elements, that led to our understanding of the internal structure of the atom, and Chadwick received the 1935 Nobel Prize in Physics for the discovery of the neutron.

Our understanding of nuclear reactions additionally requires knowledge of the positron and the neutrino. The positron is the positive electron, or the anti-matter version of the electron, with the same mass and a reversed charge. The insubstantial neutrino has no charge and very little mass; it moves very fast, at nearly the velocity of light (Section 3.4). There are three types of neutrinos, and the kind of neutrino that is made inside the Sun is called an electron neutrino.

Three other Nobel Prizes in Physics resulted from these discoveries – to Paul A. M. Dirac in 1933 for his prediction of anti-matter and the positron in 1930; to Carl D. Anderson in 1936 for his discovery of the positron in cosmic rays in 1932; and to Frederick Reines in 1995 for his detection of neutrinos emitted from a terrestrial nuclear reactor in 1956.

It wasn't until the late-1930s that Hans A. Bethe and Charles L. Critchfield demonstrated how a sequence of nuclear reactions makes the Sun shine. Bethe was awarded the Nobel Prize in Physics in 1967 for this and other discoveries concerning energy production in stars. The hydrogen fusion reactions that fuel the Sun are collectively called the proton–proton chain (Fig. 3.2). They are also known as hydrogen-burning reactions, for it is hydrogen nuclei, the protons, that are being consumed to make helium; but it is a chain of nuclear reactions and not combustion in the ordinary chemical sense.

Nuclear reactions are often written in a short-hand notation using letters to denote the nuclei and other sub-atomic particles. An arrow → specifies the reaction. Nuclei on the left side of the arrow react to form products given on the right side of the arrow.

A nucleus is designated by a letter preceded by a superscript that gives the number of particles, called nucleons, in the nucleus; it is the sum of the neutrons and protons. Different isotopes of an element have the same number of protons, and the same letter symbol, but a different number of neutrons. For instance, a rare isotope of helium, designated ^{3}He, has two protons and one neutron in its nucleus, while the common form of helium, ^{4}He, has two

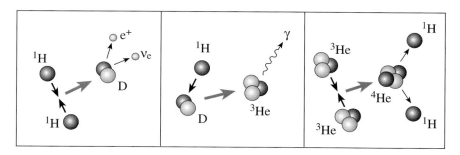

Fig. 3.2 **The proton–proton chain** The Sun gets its energy when hydrogen nuclei are fused together to form helium nuclei within the solar core. This hydrogen burning is described by a sequence of nuclear-fusion reactions called the proton-proton chain. It begins when two protons, here designated by the symbol ^{1}H, combine to form the nucleus of a deuterium atom, the deuteron that is denoted by D, together with the emission of a positron, e^+, and an electron neutrino, v_e. Another proton collides with the deuteron to make a nuclear isotope of helium, ^{3}He, and then a nucleus of helium, ^{4}He, is formed by the fusion of two ^{3}He nuclei, returning two protons to the gas. Overall, this chain successively fuses four protons together to make one helium nucleus. Even in the hot, dense core of the Sun, only rare, fast-moving particles are able to take advantage of the tunnel effect and fuse in this way.

protons and two neutrons. For historical reasons, the nuclei of the hydrogen isotopes ^{1}H, ^{2}H, and ^{3}H, are named protons, deuterons and tritons, and the nucleus of ^{4}He is called an alpha particle.

A Greek letter γ denotes energetic gamma-ray radiation, the positron is denoted by e$^+$, an electron by e$^-$, and an electron neutrino by ν_e.

The proton–proton chain describes the fusion of hydrogen into helium, which makes the Sun shine and our own existence possible. The four protons do not combine directly to form a helium nucleus, since they are moving rapidly about and are almost never in the same place at the same time. Moreover, the electrical repulsion between four protons is too great to overcome, even if they happen to come together at about the same time. Instead, the protons merge together in a series of steps to form a helium nucleus, and these steps are called the proton–proton chain.

In the first step, two exceptionally fast protons, designated by the symbol ^{1}H, meet head on and merge into each other, tunneling through the electrical barrier between them. The two protons combine to make a deuteron, D, an isotope of hydrogen that contains one proton and one neutron which is also denoted ^{2}H. A deuteron is the nucleus of a heavy form of the hydrogen atom, called deuterium, and an example of this on Earth is found in so-called heavy water.

Since a deuteron consists of one proton and one neutron, one of the protons entering into the reaction must be neutralized, being instantly turned into a neutron, with the ejection of a positron, e$^+$, to carry away the proton's charge, and a low-energy electron neutrino, ν_e, to balance the energy in the reaction. The first step of the proton-proton chain is therefore represented by

$$^1\text{H} + {}^1\text{H} \to \text{D} + \text{e}^+ + \nu_e \,.$$

In the second step of the chain, the deuteron collides with another proton to form light helium, ^{3}He, an isotope of helium which has two protons and one neutron. Less energy is needed to maintain this helium nucleus than is needed to maintain a deuteron and a proton separately. The extra energy is released as a photon, or a packet of radiation. It is energetic gamma-ray radiation, denoted by the symbol γ. The notation for this, the second step, is:

$$\text{D} + {}^1\text{H} \to {}^3\text{He} + \gamma \,.$$

In the final step of the proton–proton chain, two light helium nuclei meet and fuse together to form a nucleus of helium, ^{4}He, which has two protons and two neutrons. This reaction also returns two unattached protons to the solar gas. The notation for this, the third and last step is:

$$^3\text{He} + {}^3\text{He} \to {}^4\text{He} + {}^1\text{H} + {}^1\text{H} \,.$$

Eighty-six percent of the light helium generated takes part in this last step in the proton–proton chain. The remaining fourteen percent is involved in less-frequent terminations, reacting with normal helium to form light beryllium (Focus 3.4).

Anti-matter and matter cannot coexist. The reason why we live in a material world is simply because there is more matter than anti-matter. As soon as any anti-matter is

Focus 3.4 Secondary nuclear fusion reactions in the Sun

Fourteen percent of the light helium generated in the second step of the proton-proton chain will form beryllium, ^{8}Be. A nucleus of light helium, ^{3}He, will fuse with a nucleus of heavier helium, ^{4}He, to form a nucleus of light beryllium, ^{7}Be, according to the nuclear fusion reaction:

$$^3\text{He} + {}^4\text{He} \to {}^7\text{Be} + \gamma,$$

where γ denotes energetic gamma-ray radiation.

Most of the light beryllium will combine with an electron, e$^-$, to make a nucleus of lithium, ^{7}Li, which then joins a proton, ^{1}H, to make two nuclei of helium, ^{4}He. The reactions are:

$$^7\text{Be} + \text{e}^- \to {}^7\text{Li} + \nu_e,$$

and

$$^7\text{Li} + {}^1\text{H} \to {}^4\text{He} + {}^4\text{He},$$

where ν_e denotes an electron neutrino.

About 0.02 percent of the light beryllium combines with a proton to make boron, ^{8}B. The boron is a radioactive nucleus that decays in just one second into beryllium, ^{8}Be, together with the emission of a positron, e$^+$, and an electron neutrino. The beryllium then decays to make two nuclei of helium, completing the conversion of protons into helium. These secondary nuclear fusion reactions are:

$$^7\text{Be} + {}^1\text{H} \to {}^8\text{B} + \gamma$$

and

$$^8\text{B} \to {}^8\text{Be} + \text{e}^+ + \nu_e,$$

with

$$^8\text{Be} \to {}^4\text{He} + {}^4\text{He}.$$

They are relatively unimportant when considering the total energy produced by nuclear fusion in the core of the Sun. However, since the electron neutrino that is produced during the boron decay is about 36 times more energetic than that produced by the other nuclear reactions in the Sun, it has played a role in detecting solar neutrinos and the related solar neutrino problem (Section 3.4). The nuclear astrophysicist, William Fowler, was one of the first to realize the ramifications of this rare side reaction; he received the Nobel Prize in Physics in 1983 for his theoretical and experimental studies of nuclear reactions in the stars and the early Universe.

produced, it is immediately wiped out of existence. The positron created in the first step of the proton–proton chain almost instantly encounters a free electron, and both become pure energy. The two sub-atomic particles collide and annihilate each other in a flash of radiation at gamma-ray wavelengths. This reaction is written symbolically as:

$$\text{e}^+ + \text{e}^- \to \gamma + \gamma \,.$$

A positron, e$^+$, and an electron, e$^-$, self destruct when they meet, and their mass is transformed into two gamma-ray photons, each designated by a γ.

The net result of the proton–proton chain is that four protons have merged and fused together to make one helium nucleus, gamma rays and electron neutrinos. The overall reaction is obtained by summing the left and right sides of all the participating reactions, to obtain:

$$6\,^1H \rightarrow\, ^4He + 2\,^1H + 6\gamma + 2\nu_e\,.$$

Six protons are put into the proton–proton chain, and two of them are returned to the stellar interior at the end of the chain to be re-used later.

The helium is slightly less massive, by a mere 0.7 percent, than the four protons that combine to make it, so there is an energy, ΔE, released given by:

$$\Delta E = \Delta m c^2 = (4m_p - m_{He})\,c^2 = 0.007\,(4m_p)\,c^2$$
$$= 4.2 \times 10^{-12}\,J,$$

where the mass of the proton is $m_p = 1.6726 \times 10^{-27}$ kg, the mass of the helium nucleus is $m_{He} = 6.644 \times 10^{-27}$ kg, and $c = 2.9979 \times 10^8$ m s^{-1} is the velocity of light. This energy leaves the Sun as radiation, and the part of this radiation that constitutes visible light is what makes the Sun shine.

The Sun is consuming itself at a prodigious rate. Every second, roughly 100 trillion trillion trillion, or 10^{38}, helium nuclei are created from about 700 million tons of hydrogen, where one ton is equivalent to 1000 kilograms. In doing so, 5 million tons (0.7 percent) of this matter disappears as pure energy, which is enough to keep the Sun shining with its present, brilliant luminous output of 385 million billion billion, or 3.85×10^{26} joules per second.

Every second the Sun becomes that much less massive. However, the Sun's mass loss is trivial compared to its total mass of 1.989 billion billion billion tons, or 1.989×10^{30} kilograms, so the Sun has scarcely changed its mass even over the 4.6 billion years it has been shining. It has lost only 1 percent of its original mass in all that time.

Nuclear fusion reactions proceed at a slow and stately pace in the Sun, staying at just the right temperature to light and heat the Earth for billions of years. If the central temperature of the solar crucible was a lot colder, the reactions would occur less often and make the Sun shine feebly, like a flashlight with a worn-out battery. If the temperature at the center was a lot higher, the frequent and rapid nuclear fusion would cause the entire Sun to blow up almost instantaneously, like a giant hydrogen bomb.

Focus 3.5 Hydrogen burning in massive main-sequence stars

In 1938, the same year that hydrogen burning by the proton–proton chain was delineated, Carl Von Weizsäcker discovered the cyclic CNO chain of reactions, in which carbon acts as a catalyst in the synthesis of helium from hydrogen. The participating nuclei are heavier and have greater electrical charge than protons, so the nuclear reactions only work at the higher temperatures needed to overcome the larger electrical repulsion. The CNO chain is the dominant type of hydrogen burning for stars that have central temperatures of more than 20 million degrees kelvin. Such temperatures occur for stars that are more than 1.5 times as massive as the Sun. This method of burning hydrogen proceeds faster than the proton–proton chain that burns hydrogen in all main-sequence stars with masses less than 1.5 times the mass of the Sun.

An important distinction is that carbon must already be present in the star before the CNO cycle can begin. This could not have been the case in the early stages of the expanding Universe when nuclear fusion reactions had to be limited to those that begin from hydrogen and helium. These first-generation stars must have synthesized the carbon that is required to make massive second-generation stars burn hydrogen by the CNO cycle.

This chain of reactions starts when a carbon nucleus, ^{12}C, fuses with a proton, ^1H, to produce a nucleus of light nitrogen, ^{13}N, and gamma-ray radiation, γ. The light nitrogen decays to form a nucleus of heavy carbon, ^{13}C, a positron, e^+, and an electron neutrino, ν_e. These beginning nuclear fusion reactions are:

$$^{12}C + \,^1H \rightarrow\, ^{13}N + \gamma\,,$$

and

$$^{13}N \rightarrow\, ^{13}C + e^+ + \nu_e\,.$$

The cycle then continues when the heavy carbon combines with a proton to form nitrogen, ^{14}N, that then fuses with another proton to form light oxygen, ^{15}O. The light oxygen decays to make ^{15}N which then combines with a proton to make carbon, ^{12}C, together with a helium nucleus 4He. The nuclear fusion reactions are:

$$^{13}C + \,^1H \rightarrow\, ^{14}N + \gamma$$
$$^{14}N + \,^1H \rightarrow\, ^{15}O + \gamma$$
$$^{15}O \rightarrow\, ^{15}N + e^+ + \nu_e$$

and

$$^{15}N + \,^1H \rightarrow\, ^{12}C + \,^4He\,.$$

The positrons, e^+, annihilate with the electrons, e^-, to produce energetic gamma-ray radiation by the reactions:

$$e^+ + e^- \rightarrow 2\gamma\,.$$

Like the proton–proton chain, the net result of the CNO cycle is that four protons have been fused together to form one helium nucleus, gamma rays and electron neutrinos. By summing together the left and right sides of all the participating reactions, we obtain:

$$^{12}C + 4\,^1H \rightarrow\, ^{12}C + \,^4He + 7\gamma + 2\nu_e\,.$$

Also like the proton–proton chain, about 4×10^{-12} J are released as energy to make the star shine. The difference is that carbon acts as a catalyst in the CNO cycle. The carbon is destroyed and then recreated, and so is available for the sequence of reactions to occur over and over again. This cycle could also begin at the intermediate stages with the nitrogen or oxygen, so the entire cycle has been called the carbon–nitrogen–oxygen, or CNO, cycle.

The Sun has tamed the hydrogen bomb. Inside our star an energy equivalent to a million million hydrogen bombs is being released every second, and each of these bombs has the explosive equivalent of 100 megatons of TNT. It is enough to devastate the entire Earth; that is why the Sun is best viewed from a safe distance, though still not with the naked eye.

If there is an enormous hydrogen bomb at the center of the Sun, why isn't the star blasted apart by a tremendous explosion? We have already made mention of the force that stops the Sun from blowing itself to pieces – gravity. The sheer weight of the gas surrounding the core keeps the lid on the nuclear cauldron and prevents it from blowing up. The Sun's massive gravity pulls and holds it all together, reining in all that nuclear energy. There is an equilibrium between the inward pull of gravity and the outward pressure of the particles moving around in the Sun.

There is a wonderful balancing act in the Sun, controlled by its temperature-sensitive nuclear reactions that act like a thermostat. If the Sun's gravity pulls our star in and makes it smaller, it gets hotter inside and its nuclear fusion proceeds at a more rapid rate. The particles move faster, create more pressure and push the Sun back out. As it expands outward, the Sun cools a little and is restored to its original temperature. And if the Sun expands a little, it cools a little

inside, the nuclear reactions slow down, and the gravity pulls the star in again, restoring the equilibrium. So the pendulum keeps swinging between gravity and fusion, with no winner. That is how the Sun harnesses its nuclear energy, and it has been doing that for 4.6 billion years.

The same equilibrium applies to all stars that have settled down to long lives on the main sequence, and they all shine by hydrogen burning. The exact details differ according to the overall mass of the star, but the bottom line is always the same. Four protons combine to make one helium nucleus, releasing radiative energy and neutrinos in the process. Main-sequence stars less than 1.5 solar masses burn hydrogen by the proton–proton chain; those with masses greater than 1.5 times that of the Sun burn hydrogen by the CNO cycle (Focus 3.5).

As confident as we are about understanding the energy production in the Sun and other stars, there is at least one nagging problem marring our model. The nuclear reactions in the Sun's central furnace create prodigious quantities of electron neutrinos, but experiments detect only one-third to one-half of the particles that theory predicts, a discrepancy known as the solar neutrino problem. As we shall next see, it now seems that there is nothing wrong with our models for generating energy inside stars; we instead have an incomplete knowledge of neutrinos.

3.4 The mystery of solar neutrinos

More than half-a-century ago, scientists found that the amount of energy coming out of radioactive rocks varied by an unpredictable amount, and they resolved this paradox by inventing invisible particles called the neutrinos that carried away just the right amount of missing energy. The tiny elusive particles were discovered a few decades later, coming out of a nuclear reactor on Earth, and it was realized that many more of them were coming from the Sun. Every second, 200 trillion trillion trillion, 2×10^{38}, neutrinos stream out in all directions from the core of the Sun, moving at the velocity of light through nearly everything in their path, including our "solid" Earth. A few of the elusive solar neutrinos have been indirectly observed, using massive detectors placed deep underground so that only neutrinos can reach them. By finding solar neutrinos in roughly the predicted numbers, four pioneering neutrino detectors have demonstrated that the Sun is energized by proton fusion. However, these experiments always come up short, detecting only one-third to one-half of the neutrinos that accepted theory predicts. This discrepancy is known as the solar neutrino problem. There are two possible explanations: either we don't know exactly how nuclear fusion energizes the Sun; or we have an incomplete knowledge of neutrinos. Astronomers have used sound waves to probe the internal constitution of the Sun with results that verify their models, so they are convinced that the problem is related to our understanding of neutrinos. The solar neutrino problem might be explained if the electron neutrinos produced inside the Sun transform, or oscillate, into neutrinos of a different type while traveling to Earth, thereby escaping detection by instruments that can only respond to electron neutrinos. Measurements of a deficit of muon neutrinos, produced by cosmic rays in the Earth's atmosphere, indicate that neutrinos can undergo such a metamorphosis, and that they have the slight mass needed to do so. Underground instruments currently in operation will be able to measure the different types of neutrinos, and determine if neutrino oscillation is the correct explanation to the solar neutrino problem.

Neutrinos from the Sun

Neutrinos are very close to being nothing at all. They move at or very near the velocity of light, have no electric charge, and have so little mass that until very recently scientists were not sure if neutrinos have any mass at all. Lacking any bonds to matter, neutrinos are extraordinarily antisocial. They just don't like to interact with anything in the material world.

Almost nothing gets in the way of these unstoppable particles. Neutrinos are so insubstantial, and interact so weakly with matter, that they travel almost unimpeded through the Sun, the Earth and nearly any amount of matter, like ghosts that move through walls. Neutrinos are the true ghost riders of the Universe; to them almost all of the cosmos is transparent.

As John Updike put it in his poem "Cosmic Gall":

> Neutrinos, they are very small.
> They have no charge and have no mass
> And do not interact at all.
> The Earth is just a silly ball
> To them, through which they simply pass,
> Like dust maids down a drafty hall.

Neutrinos were postulated to explain a kind of radioactivity, called beta decay, in which a neutron in the nucleus of a radioactive atom spontaneously turns into a proton and emits an electron to balance the charge. The disturbing thing was that the energy remained unbalanced. The electron emerged from the atomic nucleus with all kinds of energy, rather than a fixed amount, and the varying amount of energy coming out was less than the amount lost by the nucleus. Yet, according to a fundamental principle of physics, called the conservation of energy, the total energy of a system must remain unchanged.

In an informal letter, written in 1930, the Austrian physicist Wolfgang Pauli proposed a "desperate way out" of the beta-decay problem, arguing that a low-mass, uncharged and unseen particle was emitted at the same time as the electron. The mysterious, unknown, sub-atomic particle escaped from the nucleus in every decay event, carrying away the missing energy. It was a desperate solution, since the particle was unobservable by the technology of the day and might never be detected.

In 1934 Enrico Fermi developed Pauli's idea into a successful theory for beta decay, and gave the putative particle its name neutrino, Italian for little neutral one. When a neutron decays, its emits both an electron and a neutrino and becomes a proton. A similar inverse beta decay occurs when a proton is turned into a neutron during fusion reactions in the Sun (Section 3.3). The neutrinos were nevertheless dismissed as mathematical figments for more than two decades, just something to make the equations balance.

The elusive neutrino was not discovered until 1956, when two Americans, Frederick Reines and Clyde Cowan, recorded flashes of light induced in a 10-ton (10 000-liter) tank of water by the impacts of neutrinos coming from a nearby nuclear reactor. The neutrinos were not themselves observed, and they never have been. Their presence was inferred by an exceedingly rare interaction. One out of every billion billion (10^{18}) neutrinos that passed through the water tank hit a proton, producing the tell-tale burst of radiation. The poltergeist, as Reines called it, had finally been detected. In 1995 he was awarded the Nobel Prize in Physics for this accomplishment, but his colleague Cowan just did not live long enough to share it.

Radiation is not the only thing coming from the Sun. Two neutrinos are produced each time a helium nucleus is made in the Sun's energy-generating core, as a by-product of neutralizing two of the fusing protons into neutrons. So much helium is made during the nuclear fusion that energizes the Sun, that trillions and trillions and trillions of neutrinos are produced every second (Focus 3.6).

To neutrinos, the Sun is transparent, and large amounts of them move out into all directions of space at nearly the speed of light. About three million billion (3×10^{15}) solar

Focus 3.6 Trillions upon trillions of neutrinos

The aggregate number of solar neutrinos is staggering. Every time the proton–proton chain creates one helium nucleus, it releases an energy, $\Delta E = (\Delta m)c^2 = 0.007(4m_p) \, c^2 = 4.2 \times 10^{-12}$ J, where Δm is the mass difference between the helium nucleus and the four protons that went into making it. The mass of the proton is $m_p = 1.6726 \times 10^{-27}$ kg, and the velocity of light $c = 2.9979 \times 10^8$ m s^{-1}. The energy released by the Sun's ongoing nuclear fusion reactions works its way out of the Sun to provide its present luminosity $L_\odot = 3.85 \times 10^{26}$ J s^{-1}. The total number of helium-producing proton-proton chains required to fuel the Sun's energy every second is $L_\odot / \Delta E$, and since there are two neutrinos emitted every time one helium nucleus is made, we conclude that:

the Sun emits $\dfrac{2L_\odot}{\Delta E} = 2 \times 10^{38}$ neutrinos every second.

Every second, 200 trillion trillion trillion neutrinos are generated inside the Sun, and move out in all directions from it at the velocity of light. The number of neutrinos passing through the Earth each second will be less than those emitted by the Sun, diminished by the ratio of the Earth's cross-sectional area to the area of a sphere with a radius equal to the mean distance from the Sun to Earth. Thus, we have:

number of neutrinos passing through Earth per second =
$\left(\dfrac{R_E}{AU}\right)^2 \left(\dfrac{2L_\odot}{\Delta E}\right) = 4 \times 10^{29}$,

where the radius of the Earth is $R_E = 6.378 \times 10^6$ m and the mean distance between the Earth and the Sun is 1 AU = 1.496×10^{11} m.

So, there are 0.4 million trillion trillion neutrinos passing through the Earth every second. To obtain the number of neutrinos passing through every square meter of the side of the Earth facing the Sun, just divide by the Earth's area πR_E^2 to get 3×10^{15}, or 3 million billion, neutrinos per square meter.

neutrinos enter every square meter of the Earth's surface facing the Sun every second, and pass out through the opposite surface unimpeded. Each second there are about 100 billion ghostly solar neutrinos passing through the tip of your finger, and every other square centimeter of your body, whether you are indoors or outdoors, or whether it is day or night, and without your body noticing them, or them noticing your body. At night they go through the entire Earth before reaching you.

The ubiquitous solar neutrinos convey a unique signal from the center of the Sun, telling us about nuclear processes that are otherwise unobservable. A by-product of fusion, the neutrinos are like the exhaust from an automobile, that can tell us how the nuclear engine is running.

The Sun produces neutrinos with a range of energies, and both the amount and energy of solar neutrinos depend on the particular reactions that produced them. Their expected flux at the Earth is calculated using large computers that produce theoretical models, culminating in the Standard Solar Model that best describes the Sun's luminous output, size and mass at its present age (Focus 3.2). The results of these calculations (Fig. 3.3) indicate that the great majority of solar neutrinos have the lowest energy, and that they are generated during the nuclear fusion of two protons, the reaction that initiates the proton–proton chain. Smaller amounts of high-energy neutrinos are produced from the decay of boron-8 during a rare termination of the proton–proton chain (Focus 3.4). Different neutrino detectors are sensitive to different energy ranges (Fig. 3.3), and so tell us about different nuclear reactions in the Sun.

Detecting neutrinos

The only way to detect solar neutrinos is through their exceptionally rare collisions with ordinary matter. Although the vast majority of neutrinos pass right through matter more easily than light through a window pane, there is a finite chance that a neutrino will interact with a sub-atomic particle. When this slight chance is multiplied by the enormous quantities of neutrinos flowing from the Sun, we conclude that once in a great while a solar neutrino will score a direct hit, and the resulting blast of nuclear debris can signal the existence of the otherwise invisible neutrino.

Fortunately for science, a miniscule proportion of the Sun's neutrinos do collide with more palpable sub-atomic particles, and when such a collision occurs inside a massive detector its effects reveal the neutrino's presence. The neutrino detectors contain large amounts of material, literally tons of it, to measure even a few of the solar neutrinos. The massive tanks must also be placed deep underground, beneath a mountain or inside a mine, so that only neutrinos can reach them (Fig. 3.4). The thick layers of intervening rock are transparent to neutrinos, but they filter out other energetic particles, generated by cosmic rays, and shield the detectors from their confusing signals. So, all of the solar neutrino hunters work deep underground where the Sun never shines.

Massive, subterranean detectors have been snagging just a handful of elusive neutrinos for more than a quarter-century. The first such experiment, code-named HOMESTAKE, and constructed by Raymond Davis Jr in 1967, is a 615-ton tank containing 378 thousand liters of cleaning fluid, technically called perchloroethylene or "perc". Each molecule of the fluid, denoted by C_2Cl_4, contains two carbon atoms, C, and four chlorine atoms, Cl. Most of the solar neutrinos pass through the ground and the tank without being noticed. Nevertheless, once every 2.5 days a neutrino scores a direct hit with the stable isotope of chlorine, ^{37}Cl, transforming it into a radioactive isotope of argon, ^{37}Ar. The radioactive argon is collected by radiochemical means, and yields a direct estimate of the high-energy solar neutrino flux.

The nuclear reaction is denoted by an arrow $\rightarrow$, with the particles on the left side of the arrow interacting to produce those on the right side. So, we have:

$$\nu_e + {}^{37}Cl \rightarrow e^- + {}^{37}Ar \ ,$$

where ν_e denotes an electron neutrino and e^- is an electron. The neutrino turns one of the neutrons in the chlorine nucleus into a proton, emitting an electron to conserve charge. Due to its high energy threshold, the chlorine experiment only detects relatively rare, energetic neutrinos produced during a rare side reaction involving boron-8.

This pioneering chlorine detector measured solar neutrinos for more than a quarter-century, always detecting about one third of the expected amount (Fig. 3.5). It measures an average neutrino flux at the Earth of 2.55 ± 0.25 SNU. In contrast, the Standard Solar Model predicts that it should observe a flux of about 8.0 ± 1.0 SNU. This discrepancy between the number of detected neutrinos and the number predicted is known as the solar neutrino problem.

The neutrino reaction rate with even massive underground detectors is so slow that a special unit has been invented to specify the flux. This solar neutrino unit, or SNU, is equal to one neutrino interaction per second for every 10^{36} atoms, or 1 SNU $= 10^{-36}$ neutrino absorptions per target atom per second. The number that follows the $\pm$ sign after a measurement denotes the statistical uncertainty, at one standard deviation, in the preceding number; a definite detection has to be above three standard deviations and preferably above five of them.

A second experiment, that began operating in 1987, was

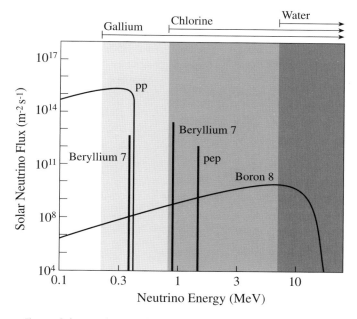

Fig. 3.3 **Solar neutrino energies** Neutrinos are produced inside the Sun as a byproduct of nuclear-fusion reactions in its core, but both the amounts and energies of the neutrinos depend on the element fused and the detailed model of the solar interior. Here we show the neutrino flux predicted by the Standard Solar Model. The largest flux of solar neutrinos is found at low energies; they are produced by the main proton–proton, pp, reaction in the Sun's core. Less-abundant, high-energy neutrinos are produced by a rare side reaction involving boron-8. The shading denotes the detection range for gallium, chlorine and water experiments (see text), with detection thresholds marked by the vertical lines. The gallium experiment can detect the low-energy pp neutrinos, as well as those of higher energy; the chlorine and water detectors are both sensitive to the high-energy boron-8 neutrinos. Neutrinos should also be generated at two specific energies when beryllium–7 captures an electron, and also during a relatively rare proton–electron– proton (pep) reaction; the fluxes are given in number $m^{-1}s^{-1}$. The neutrino fluxes for the pp and boron-8 reactions are in units of number $m^{-1} s^{-1} MeV^{-1}$.

Fig. 3.4 **Underground neutrino detector** The original solar neutrino detector located 1500 m underground in the Homestake Gold Mine, near Lead, South Dakota, to filter out strong signals from energetic cosmic particles. The huge cylindrical tank was filled with 378 thousand liters of cleaning fluid. When a high-energy solar neutrino interacted with the nucleus of a chlorine atom in the fluid, radioactive argon was produced, which was extracted to count the solar neutrinos. This experiment operated for more than 25 years, always finding fewer neutrinos than expected from the Standard Solar Model – see Fig.3.5. (Courtesy of Brookhaven National Laboratory.)

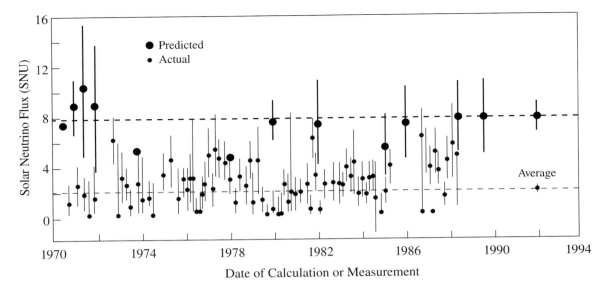

Fig. 3.5 **Solar neutrino fluxes** Calculated and measured solar neutrino fluxes have consistently disagreed over the past two decades. The fluxes are measured in solar neutrino units, or SNU, defined as one neutrino interaction per 10^{36} atoms per second. Measurements from the HOMESTAKE chlorine neutrino detector (*small filled circles*) give an average solar neutrino flux of 2.55 ± 0.25 SNU (*lower broken line*). The most recent theoretical calculations (*large filled circles*) predict a flux of 8.0 ± 1.0 SNU (*upper broken line*) for the Standard Solar Model. Other experiments have also observed a deficit of solar neutrinos, suggesting that either some process prevents neutrinos from being detected or the method by which the Sun shines differs from that predicted by current theoretical models.

located 1000 meters down in the Kamioka zinc mine under Mt Ikena in the Japanese Alps. It was also limited to the rare, high-energy solar neutrinos, but it could uniquely tell where the neutrinos come from. The detector consisted of a 3000-ton (3-million-liter) tank of pure water. When a neutrino beats the odds by striking an electron orbiting an atomic nucleus within a water molecule, the electron recoils and generates a flash of light. There are no new particles detected, just an exchange of energy between the neutrino and the electron, with the scattering reaction:

$$\nu_e + e^- \rightarrow \nu_e{'} + e^{-'},$$

where ν_e denotes an electron neutrino, e^- is an electron, and the unprimed and primed sides respectively denote the incident and scattered neutrinos and electrons. The electron moves through the water faster than light travels in water, generating the optical equivalent of a sonic boom, where an electromagnetic shock wave is emitted as the electron propagates through the water and slows down. A conical flash of blue light, called Cherenkov radiation, is spawned by the recoiling electron.

Thousands of light sensors, called photo-multiplier tubes, lined the inner walls of the water tank, and monitored the light cone. The intensity of the light and the tightness of the light cone measured the energy of the incoming neutrino, and the axis of the light cone pinpointed where the neutrino came from. The observed signals not only confirmed the shortfall in solar neutrinos, detecting slightly less than one-half the expected amounts (Table 3.1), but also showed that the neutrinos are coming from the direction of the Sun and are presumably made inside it.

The fundamental reaction in the solar energy-generating process is the interaction of two protons that begins the proton–proton chain, but these neutrinos have energies

Target	Experiment	Threshold energy (MeV)	Measured neutrino flux (SNU)[a]	Predicted neutrino flux (SNU)	Ratio: measured/predicted
Chlorine 37	HOMESTAKE	0.814	$2.56 \pm 0.16 \pm 0.14$	$9.5^{+1.2}_{-1.4}$	0.27 ± 0.022
Water	KAMIOKANDE[b]	7.5	$2.80 \pm 0.19 \pm 0.35$	$6.62^{+0.93}_{1.12-}$	0.42 ± 0.060
Gallium 71	GALLEX	0.2	$69.7 \pm 6.67^{+3.9}_{-4.5}$	136.8^{+8}_{-7}	0.51 ± 0.058
Gallium 71	SAGE	0.2	72^{+12+5}_{-10-7}	136.8^{+8}_{-7}	0.53 ± 0.095

[a] Here the uncertainties are one standard deviation, and the first and second measurement uncertainties correspond, respectively, to the statistical and systematic uncertainties.

[b] The units of the measured and predicted values for the KAMIOKANDE experiment are 10^{10} m^{-2} s^{-1}, while the solar neutrino unit, or SNU, is used for the other three experiments.

Table 3.1 **Solar neutrino experiments**

below the detection thresholds of the chlorine and water experiments. The numerous low energy neutrinos as well as a few high-energy neutrinos are detected in two gallium experiments using the reaction:

$$\nu_e + {}^{71}Ga \rightarrow e^- + {}^{71}Ge,$$

where an electron neutrino, ν_e, strikes a stable isotope of gallium, ${}^{71}Ga$, producing an electron, e^-, and a radioactive isotope of germanium, ${}^{71}Ge$, that can be collected by radiochemical means. In 1990, the Soviet-American Gallium Experiment, or SAGE, began operation in a long tunnel, some 2000 meters below the summit of Mt Andyrchi in the northern Caucasus. A second multi-national experiment, dubbed GALLEX for GALLium EXperiment, started operating in 1991, located in the Gran Sasso Underground Laboratory some 1400 meters below a peak in the Appenine Mountains of Italy.

The GALLEX and SAGE experiments measure about half the predicted flux of neutrinos (Table 3.1). Moreover, if you believe the HOMESTAKE and KAMIOKANDE results, then the gallium experiments have apparently detected the low-energy neutrinos, confirming that the Sun shines by the nuclear fusion of protons.

By finding solar neutrinos in roughly the predicted numbers, those four pioneering neutrino detectors have now demonstrated that the Sun is indeed energized by hydrogen fusion. However their apparent inability to detect closer to the predicted value further confirmed the solar neutrino problem.

The KAMIOKANDE experiment was updated in 1996 to a high-tech, $100 million SUPER KAMIOKANDE status, also located deep underground in the Kamioka zinc mine. The new detector is a stainless-steel cylinder, roughly 40 meters in diameter and height, that contains 50 000 tons (50 million liters) of ultra-pure water. About 13 000 light sensors are uniformly arrayed on all the inner walls of the cistern (Fig. 3.6). These photo-multiplier tubes are so sensitive that they can detect a single photon of light – a light level approximately the same as light visible on Earth from a candle on the Moon. Since there are practically no impurities in the very clean water, light can travel for almost 100 meters without being noticeably attenuated; for ordinary water it is less than 3 meters. This means that the light sensors can monitor the entire water volume for the bluish Cherenkov light generated by an electron recoiling from a direct hit by a neutrino.

Like its prototype KAMIOKANDE, the SUPER KAMIOKANDE experiment provides information about the direction and energy of neutrinos, confirming that electron neutrinos come from the direction of the Sun in less than predicted amounts. But there is something else streaking through the water. SUPER KAMIOKANDE detects two kinds of neutrinos, and measures a deficit in both of them.

During the past few decades, scientists have learned that there are three separate types, or flavors, of neutrinos, each

Fig. 3.6 **SUPER KAMIOKANDE** This neutrino detector has been built a kilometer underground in a Japanese zinc mine. The huge stainless-steel vessel, 40-m tall and 40-m wide, has been filled with 50 000 tons (50 million liters) of highly purified water. Its walls are lined with 13 000 light sensors, called photo-multiplier tubes, that pick up a flash of light generated by electrons recoiling from neutrino collisions in the water. (Courtesy of Yoji Totsuka, Institute for Cosmic Ray Research, University of Tokyo.)

named after the fundamental, sub-atomic particle with which it is most likely to interact. All of the neutrinos generated inside the Sun are electron neutrinos, ν_e; this is the kind that interacts with electrons, e^-. The other two flavors, the muon neutrino, ν_μ, and the tau neutrino, ν_τ, interact with muons, μ, and tau particles, τ, respectively. The muon and tau particles are unfamiliar to most of us because they die shortly after birth. The muon decays into an electron, a muon neutrino and an electron anti-neutrino in just 2×10^{-6} s, and the tau particle disappears just 3×10^{-13} s after it is made.

The muon particle is a kind of fat electron, and the tau particle is an even fatter relative of the electron. Yet, electrons, muons and tau particles are all significantly less massive than other elementary particles, and hence collectively known as *leptons*, the Greek word for "slender". The muon was discovered in 1936 as part of the particle fall out generated when a high-energy, cosmic-ray particle from deep space slams into the Earth's atmosphere. The tau particle is a newcomer; it was unexpectedly discovered by Martin Perl and his colleagues in

Fig. 3.7 **Sudbury Neutrino Observatory (SNO)** The central spherical task of this neutrino observatory is 12 m in diameter and is surrounded by a geodesic array of thousands of light sensors to detect the flash of light from the interaction of a neutrino with heavy water in the central tank - see also Fig. 3.8. (Courtesy of Kevin Lesko, Lawrence Berkeley National Laboratory.)

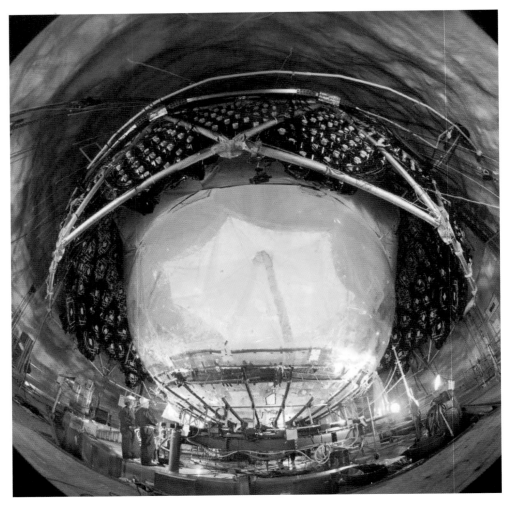

1975–76 using the Stanford linear particle accelerator.

SUPER KAMIOKANDE indirectly detects both electron neutrinos, created by nuclear fusion in the Sun, and muon neutrinos, created in the Earth's atmosphere when fast-moving protons from outer space collide with air molecules. Far more solar electron neutrinos than atmospheric muon neutrinos pass through the detector. The solar electron neutrinos are also distinguished by their relatively low energy, near the 5 MeV threshold of the detector, while a high energy of 1000 MeV is typical of the atmospheric muon neutrino.

Both types of neutrinos can bang into an electron orbiting an atom in the detector's water. The electron recoils, giving off a cone of light, and the higher the energy of the incoming neutrino the tighter the cone. The observed distribution of light can therefore be used to identity the type of neutrino. The electron neutrino makes a fuzzy, blurred and ragged light pattern, while the muon neutrino produces a neat, sharp-edged ring of light.

The important thing is that the atmospheric muon neutrinos also fail to turn up in predicted numbers. After monitoring light patterns for more than 500 days, scientists reported in 1998 that roughly twice as many muon neutrinos were detected coming from the atmosphere directly above SUPER KAMIOKANDE than were detected coming though from the other side of the Earth. The muon neutrinos are produced in the atmosphere above every place on our planet, but some of them apparently disappeared while traveling through the Earth to arrive at the detector from below. As we shall see, it is not that the neutrinos are not there, but they are instead traveling in disguise. But first we discuss another relatively new detector that is sensitive to all three kinds of neutrinos.

The Sudbury Neutrino Observatory, or SNO pronounced "snow", is located 2000 meters underground in a working nickel mine near Sudbury, Ontario, Canada. Like the previous water detectors, it will see only boron-8 solar neutrinos with energies above 5 MeV. But unlike SUPER KAMIOKANDE, the heart of the SNO detector contains heavy water. One thousand tons (one million liters) of heavy water, with a value of $300 million, has been placed in a central spherical cistern with transparent acrylic walls (Figs. 3.7, 3.8). Since the scientists cannot afford its cost, the heavy water has been borrowed from Atomic Energy of Canada Limited, who have stockpiled it for use in its nuclear power reactors – the heavy water moderates neutrons created by uranium fission in the reactors.

A geodesic array of 9500 light sensors surrounds the

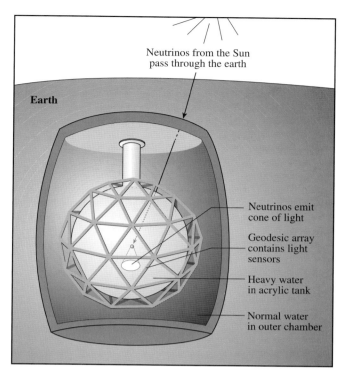

Fig. 3.8 **How SNO works** Neutrinos from the Sun travel through more than 2000 m of rock, entering an acrylic tank containing 1000 tons (1 million liters) of heavy water. When one of these neutrinos interacts with a water molecule, it produces a flash of light that is detected by a geodesic array of light sensors. Some 7800 tons (7.8 million liters) of ordinary water surrounding the acrylic tank blocks radiation from the rock, while the overlying rock blocks energetic particles generated by cosmic rays in our atmosphere. The heavy water is sensitive to all three types of neutrinos.

vessel to detect the flash of light given off by heavy water when it is hit by a neutrino. Both the light sensors and the central tank are enveloped by a 7800-ton (7.8-million-liter) outer chamber jacket of ordinary water (Fig. 3.8), to shield the heavy water from weak natural radiation, gamma rays and neutrons, in the underground rocks. As with the other neutrino detectors, the overlying rock blocks energetic particles generated by cosmic rays. If the detector was put on the surface of the Earth, the high-energy, cosmic-ray particles would make the detector glow like a giant neon sign.

Heavy water is made of molecules consisting of oxygen atoms, denoted O, and the heavy isotope of hydrogen, called deuterium, ^{2}H. So the hydrogen atoms, H, in regular water H_2O, have been replaced with the deuterium atoms to make ^2H_2O, which is about 10 percent heavier than H_2O.

The atomic nucleus of a deuterium atom consists of a proton and a neutron; a nucleus of the normal hydrogen atom contains a single proton and no neutron. The neutron in the nucleus of heavy-water atoms doesn't make the water's appearance, chemistry or taste any different from regular water used in other neutrino detectors. In fact, heavy water exits naturally as a constituent of ordinary tap or lake water in a ratio of about one part in seven thousand, and it can be separated by expensive chemical and physical processes. The

neutron in the nucleus of deuterium does, however, change the water's nuclear structure enough to make SNO sensitive to not just one but to all three known varieties of neutrinos.

A rare direct hit by a solar electron neutrino can change the neutron in deuterium to a proton, and the neutrino into an electron, by the "charged current" reaction:

$$\nu_e + D \rightarrow {}^1H + {}^1H + e^-,$$

where ν_e is the electron neutrino, D is a deuteron, the nucleus of a deuterium atom, and ^{1}H and e^- respectively denote a proton and an electron. The electron will get most of the neutrino energy since it has a smaller mass than the proton, just as when a gun is fired, the bullet, being lighter, gets most of the energy. The electron will be so energetic that it will be ejected from the deuteron at the speed of light, producing a flash of blue Cherenkov light. This occurs about 10 times more often than neutrino scattering off electrons in an equal volume of either heavy or regular water, so the SNO is expected to detect about as many electron neutrinos as SUPER KAMIOKANDE, with much less water.

The nucleus of heavy water is also a target for all three flavors of neutrinos. When a neutrino of type x, denoted ν_x, interacts with a deuteron, D, it can break the nucleus into its sub-atomic constituents, the proton, ^{1}H, and a neutron, denoted by n, by the "neutral current" reaction:

$$\nu_x + D \rightarrow {}^1H + n + \nu_x.$$

The neutron that is knocked loose by the impacting neutrino can be readily detected by the gamma rays which are emitted when the neutron is captured by another nucleus; the gamma rays will scatter electrons which produce detectable Cherenkov light.

These two reactions between neutrinos and the nucleus of deuterium can tell scientists if solar electron neutrinos are changing form *en route* to the Earth. The first reaction can only be induced by electron-type neutrinos, whereas the second reaction is equally sensitive to neutrinos of all three types. If more neutrinos are detected via the first reaction than the second one, that would be direct evidence that some electron-type neutrinos have changed, or oscillated, into neutrinos of some other type. As we shall next see, such a finding would help resolve the solar neutrino problem.

Solving the solar neutrino problem

Massive subterranean instruments always come up short, detecting only one-third to one-half the number of solar electron neutrinos that the Standard Solar Model says they should detect. This discord between measurements and theoretical predictions is known as the solar neutrino problem, and there are two possible explanations for it. Either we don't know exactly how nuclear fusion reactions generate energy in the Sun and other stars, or we have an incomplete knowledge of neutrinos.

One solution would be to alter our model of the Sun's

Focus 3.7 Neutrino oscillations

In the bizarre world of quantum mechanics, a neutrino is not born as a single particle. Instead, it seems to exist in a confused schizophrenic state that is a combination of different, co-existing types of neutrinos, sometimes called flavors. The three types are the electron, muon and tau neutrino.

A neutrino can oscillate back and forth between flavors, mixing between different mass states as it travels through space and time. The probability that a neutrino starting out as type 1 will switch into another kind of type 2 depends on the difference in the square of the masses, Δm^2, between the two flavors, and the distance, D, traveled, or:

probability of neutrino changing =

$$\sin^2 2\theta_V \sin^2\left(\frac{\pi D}{L_V}\right) = \sin^2 2\theta_V \sin^2\left(\frac{\Delta m^2 D}{4E}\right).$$

where the neutrino energy before transformation is E, the vacuum oscillation length $L_V = 4\pi E/\Delta m^2$, and the difference in energy, between the two kinds of neutrinos, is given by:

neutrino energy difference $= \Delta E = \frac{\Delta m^2}{2E}$,

where the mass difference, Δm, is specified in units of energy.

The probability of neutrino change depends on the strength of the mixing, specified by the vacuum mixing angle, θ_V; the $\sin^2 2\theta_V$ term lies between 0 and 1 depending on θ_V. The second sinusoidal term shows that the probability of transformation oscillates as the neutrino moves through greater distances in space, changing back and forth as first one type and then another predominates. Neutrinos traveling a greater distance will generally exhibit greater depletion from oscillation.

Neutrino oscillations in the vacuum of space, first suggested by Bruno Pontecorvo in the late-1950s, provide an important theoretical background, but they are an unlikely solution to the solar neutrino problem. This is because the amount of mixing required is large, and because the maximum reduction achievable is only $\frac{1}{2}$ with two neutrino flavors. Physicists now believe that neutrinos might change flavor in either a vacuum or while passing through some substance, but the oscillation rates under these two conditions might differ. Neutrino oscillations in matter can, for example, cause the almost complete conversion of solar electron neutrinos to neutrinos of a different flavor on their way out of the Sun, thereby resolving the solar neutrino problem.

The transformation of neutrinos in passing through matter, rather than a vacuum, is known as the MSW effect after the first letters of the last names of the scientists who developed the theory – Mikheyev, Smirnov and Wolfenstein. They also showed that a non-trivial suppression of the Sun's neutrino flux can occur for a wide range of mass differences $\Delta m^2 = 10^{-4} - 10^{-8}$ $(eV)^2$ and mixing angles $\sin^2\theta m > 10^{-4}$.

There are no vacuum or matter oscillations if the masses of the two neutrino types are not different. The key parameter is the mass difference squared, Δm^2. The smaller the difference to be measured, the greater is the required distance between the neutrino source and the detector. Because of the very large distance from the Earth to the Sun, the solar neutrino experiments are uniquely sensitive to very small neutrino mass differences. The probability of observing solar electron neutrinos at Earth can be significantly altered for mass differences greater than:

minimum solar mass differences squared =

$$\Delta m^2_{solar} = 1.6 \times 10^{-12} \left(\frac{E}{1 MeV}\right) eV^2.$$

Here a typical solar neutrino energy, E, is in the MeV, or millions of electron volts, range, and the mass difference, Δm, is expressed in energy units of electron volts, abbreviated eV.

An energy of 1 eV $= 1.602 \times 10^{-19}$ J. When mass is expressed in units of energy, the mass can be obtained by dividing by c^2, where the velocity of light $c = 2.9979 \times 10^8$ m s^{-1}. An energy of 1 eV $= 0.178 \times 10^{-35}$ kg.

The distance to the Sun varies by about 5 percent as the Earth moves along its annual, egg-shaped, elliptical orbit, and this variation may lead to a detectable change in neutrino transformation during the year. A reduction in the amount of solar electron neutrinos might also be observed for those that pass through the Earth at night.

Neutrino oscillations have recently been observed for atmospheric muon neutrinos of much greater energy, E = 1 GeV = 1000 MeV, than solar electron neutrinos, and over much shorter distances comparable to the size of the Earth. The characteristic vacuum oscillation length, L_V, in kilometers is $E/(1.27 \Delta m^2)$ if the neutrino energy is in GeV and Δm is in eV. The relevant distance is the Earth's mean diameter D = 12 742 km, and a measurable neutrino depletion only occurs for much greater mass differences than the solar case. Recent measurements with the SUPER KAMIOKANDE detector indicate that the muon neutrinos, produced when energetic cosmic-ray protons hit the Earth's atmosphere, are changed into tau neutrinos when traveling through the Earth. The measured mixing strength $\sin^2 2\theta_V$ exceeds 0.82, and the mass difference squared, Δm^2, lies somewhere between 0.0005 and 0.006 eV^2.

If one neutrino is much lighter than the other, then one gets something like 0.05 eV for the mass of the heavier neutrino – small but definitely not zero. That amounts to 8.9×10^{-38} kg, which is 10 million times less massive than the electron and five-billionths the mass of a proton.

Intense neutrino beams generated by particle accelerators are being sent hundreds of kilometers through the Earth, from accelerator to detector, providing sensitive tests of this type of neutrino flavor oscillation and possible measurements of the neutrino mass.

The Universe is teeming with neutrinos formed in the big bang that created the expanding Universe, outnumbering the electrons and protons by about a billion to one. If these neutrinos have a mass of just 0.1 eV, then their combined mass could be comparable to the total mass of all visible stars and galaxies. However, a much larger mass (energy) of about 30 eV is required to stop the expansion of the Universe in the future, so neutrinos are not likely to be the dominant force in the evolution and eventual fate of the Universe.

By way of contrast, the solar neutrino problem could be resolved with a neutrino mass as little as 0.003 eV or less. We should nevertheless remember that the atmospheric result applies to the muon and tau neutrinos, and has no direct bearing on the solar neutrino problem since it is electron neutrinos that are made in the Sun's energy-generating core.

interior so that it produces the observed number of neutrinos. If the Sun's core is, for example, slightly cooler than previously thought, fewer neutrinos would be produced. However, this would be inconsistent with other observations. The predicted rate of neutrino production depends on specific assumptions of temperature and composition, which also dictate the Sun's internal sound velocity. Observations of these sound waves, discussed later in Section 4.3, indicate that the measured and predicted sound velocities do not differ from each other by more than 0.2 percent down to very near the Sun's center. This pins down the central temperature very close to the calculated value of 15.6 million degrees kelvin, and indicates that the solar neutrino problem cannot be solved with plausible variations in the Standard Solar Model. Most scientists therefore agree that the problem lies in their understanding of neutrinos.

In one attractive solution to the solar neutrino problem, the neutrinos are produced at the Sun's center in the quantity predicted by the Standard Solar Model, but the neutrinos change form and switch identity as they propagate out from the Sun. The neutrinos have an identity crisis on their way to us from the center of the Sun, transforming themselves into a form that we have not yet detected from the Sun and a flavor that we have not yet tasted.

As previously mentioned, neutrinos exist in three types, or flavors, the electron, muon and tau neutrinos. Nuclear reactions in the Sun only produce the electron neutrinos, and it is the only kind of solar neutrino that our detectors have responded to, so far. Could some of the electron neutrinos be switching to another type during their 8.3-minute journey from the center of the Sun to Earth, thereby escaping detection? This could neatly account for why we find only a fraction of the expected number.

The effect is called neutrino oscillation, since the probability of metamorphosis between neutrino types has a sinusoidal, in and out, or up and down, oscillating dependence on path length (Focus 3.7). The chameleon-like change in identity is not one-way, for a neutrino can change into another kind and back again as it moves along. Like the Cheshire cat, the elusive neutrino can appear and disappear; only an as yet undetectable grin remains on the vanished cat.

In one theoretical model, the switchover is modulated by the Sun's matter as the neutrinos pass through it. This transformation is known as the MSW effect after Lincoln Wolfenstein who originated the theory in the late-1970s, and S. P. Mikheyev and Alexis Y. Smirnov who further developed it about a decade later. The MSW effect was suggested as a possible explanation for the solar neutrino problem by Hans Bethe in 1986, and by Bethe and John Bahcall a few years later. That is the same Hans Bethe who discovered the proton–proton chain more than half-a-century before his discussions of solar neutrinos. According to this explanation, some of the Sun's electron neutrinos undergo metamorphosis into another type of neutrino, disappearing somewhere on their way out of the Sun. That

doesn't mean that the solar neutrinos aren't there; they have just changed into a flavor that couldn't be seen with the pioneering terrestrial detectors.

Neutrino oscillations were next used to explain the disappearing act of muon neutrinos, created when cosmic rays strike the Earth's atmosphere. Muon neutrinos arriving at SUPER KAMIOKANDE came in the expected number from above, but too few arrived from below. The further a neutrino travels, the more time it has to oscillate (Focus 3.7), and that would account for why there are fewer muon neutrinos arriving from the back side of Earth, some 12 million meters away, than from 40 thousand meters up in the atmosphere directly above the detector. Some of the muon neutrinos generated in the atmosphere on the Earth's far side had apparently changed or oscillated into undetected tau neutrinos somewhere along their way through the Earth.

According to the theory of quantum mechanics, any particle capable of transforming itself in mid-flight must have some substance, or mass, to begin with. The observed neutrino oscillations do not tell us the exact mass, only the very small difference in mass between two neutrino types (Focus 3.7). However, since there is a mass difference, at least one type of neutrino, and most likely all of them, must have a non-zero mass.

If neutrinos have even the slightest mass, that is no small matter. As presently understood, all neutrinos are assumed to be completely without mass. So the SUPER KAMIOKANDE result will most likely require the development of new particle physics that explains why neutrinos have mass, and why all sub-atomic particles have the mass that they do.

This powerful new evidence for mass and oscillation of the atmospheric muon neutrinos has no direct, immediate bearing on the solar neutrino problem. The neutrinos involved in the atmospheric anomaly do not come from the Sun, and have nothing to do with electron neutrinos, the only kind made by nuclear reactions in the Sun. Yet, it suggests that the solar electron neutrinos may be capable of transforming into another type that underground experiments have not yet detected, and if this metamorphosis is observed it could explain the solar neutrino shortfall.

The resolution of the solar neutrino problem rests with the refinement of our detection techniques. Experiments currently under way should reveal new secrets of the Sun's core, and settle the question of whether solar neutrinos switch identities while traveling to Earth. They will also tell us about neutrinos themselves, providing stringent limits to, or measurements of, the neutrino mass. This may yield an improved estimate of the mass of the Universe.

The combined mass of cosmic neutrinos is very large indeed. The big bang that started the expanding Universe created several billion neutrinos for every proton, and these invisible relic neutrinos still outnumber all other sub-atomic particles in the Universe by this amount. Yet, they are unlikely to affect the fate of the expanding universe (Focus 3.7).

3.5 The Sun's remote past and distant future

Nothing in the cosmos is fixed and unchanging, and nothing escapes the ravages of time. Everything moves and evolves, and that includes the seemingly constant and unchanging star of our solar system. The Sun and planets in our solar system formed 4.6 billion years ago when a spinning cloud of dust and gas in space fell in on itself. The center got denser and denser, until it became so packed, so tight and so hot that protons came together and fused into helium, making the Sun glow. Ever since then, the Sun has slowly grown in luminous intensity with age, a steady, inexorable brightening that is a consequence of the increasing amount of helium accumulating in the Sun's core.

As the hydrogen in the Sun's core is slowly depleted over time, the core must keep producing enough pressure to keep the Sun from collapsing in on itself. The pressure of a gas depends upon the temperature and the number of particles per unit volume, but not upon the kind of particle. Because four protons are used to make each helium nucleus, the number density of sub-atomic particles in the core gradually decreases. The only way to maintain the pressure and keep on supporting the weight of overlying material is to increase the central temperature. As a result of the rise in temperature, the rate of nuclear fusion increases and the star slowly brightens.

The luminosity, effective temperature and radius of the Sun have all slowly increased with time (Fig. 3.9). The Sun is now 30 percent brighter than it was 4.6 billion years ago. The brightening is enough to make the visible solar disk 300 degrees kelvin hotter and its radius 6 percent greater than when the Sun first shone. The luminosity has only increased by a miniscule 0.000 0023 (2.3×10^{-6}) percent during the past 350 years, and there is no way that this small change will ever be directly measured. Yet, it has profound implications over cosmic periods of time.

If the Sun was significantly dimmer in the remote past, then the Earth should have been noticeably colder then, but this does not agree with geological evidence. Assuming an unchanging atmosphere, with the same composition and reflecting properties as today, the decreased solar luminosity would have caused the Earth's global surface temperature to drop below the freezing point of water about 2 billion years ago. The oceans would have been frozen solid, there would be no liquid water, and the entire planet would have been locked into a global ice age something like Mars is now.

Yet, sedimentary rocks, which must have been deposited in liquid water, date from 3.8 billion years ago. There is fossil evidence in those rocks for living things at about that time. Thus, for billions of years the Earth's surface temperature was not very different from today, and conditions have remained hospitable for life on Earth throughout most of the planet's history.

The discrepancy between the Earth's warm climatic record and an initially dimmer Sun has come to be known as the faint-young-Sun paradox. It can be resolved if the Earth's primitive atmosphere contained about a thousand times more carbon dioxide than it does now. Greater amounts of carbon dioxide would enable the early atmosphere to trap greater amounts of heat near the Earth's surface, warming it by the greenhouse effect (Section 8.5).

Over time the Sun grew brighter and hotter. The Earth could only maintain a temperate climate by turning down its greenhouse effect as the Sun turned up the heat. Our planet's atmosphere, rocks, oceans, and perhaps life itself, apparently combined to decrease the amount of carbon dioxide over time.

In the end, the prospects for life as we know it are not all that great anyway. The Earth's self-regulating thermostat will

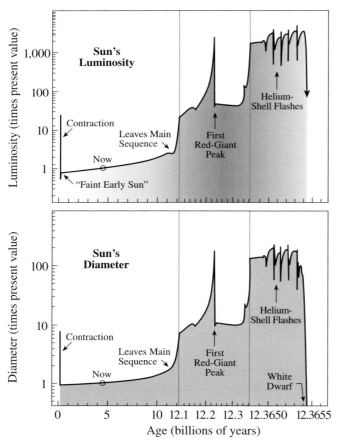

Fig. 3.9 **The Sun's fate** In about eight billion years the Sun will become much brighter (*top*) and larger (*bottom*). The time-scale has been expanded near the end of the Sun's life to show relatively rapid changes. (Courtesy of I-Juliana Sackmann and Arnold I. Boothroyd).

eventually go out of control, and the Sun will eventually consume the life it once nurtured. We can anticipate an additional seven billion years of slow luminosity increase (Fig. 3.9), but terrestrial life will be wiped out well before then. Any Earthly life is doomed to fry in about three billion years from now. The Sun will then be hot enough to boil the Earth's oceans away, leaving the planet a burned-out cinder, a dead and sterile place.

The Sun cannot shine forever, because it will eventually use up the hydrogen fuel in its core. Although it has converted only a trivial part of its original mass into energy, the Sun has processed a substantial 37 percent of its core hydrogen into helium during the past 4.6 billion years. The Sun will have used up all its available core hydrogen in about seven billion years, and will then balloon into a red-giant star with a dramatic increase in size and a powerful rise in luminosity (Fig. 3.9).

When the core hydrogen is exhausted, the central part of the Sun will be undergoing a slow collapse, and the gradually increasing core temperature will cause the outer layers of the Sun to expand. Mercury will become little more than a memory, being pulled in and swallowed by the swollen Sun. Eventually the star will become 170 times larger than it is now; the Earth and Venus will probably escape wholesale consumption, but their prospects are dim. The orbits of Venus, Earth and other planets will become slightly larger as the result of the Sun's lower gravitational pull.

The giant Sun will be 2300 times brighter than it is now, resulting in a substantial rise in temperatures throughout the solar system. It will become hot enough to melt the Earth's surface. The frozen moons of the outer planets might then spring to life as the Sun melts their ice into seas of liquid water. The only imaginable escape would then be interplanetary migration to distant moons or planets with a warm, pleasant climate.

Meanwhile, during the apocalyptic period of planetary destruction, the core of the Sun will continue to contract until the central temperature is hot enough to burn helium – at about a hundred million degrees kelvin. But this conversion of helium into carbon does not last very long, compared to the Sun's 12 billion years of hydrogen burning. In about 35 million years, the core helium will have been used up, and there will be no heat left to hold up the Sun. In a last desperate gasp of activity, the Sun will shed the outer layers of gas to produce an expanding "planetary" nebula around the star, while the core collapses into a white-dwarf star (Section 2.3).

By this time the intense winds will have stripped the Sun down to about half its original mass, and gravitational collapse will squeeze that into an insignificant cinder about the size of the present-day Earth. Nuclear reactions will then be a thing of the past, and there will be nothing left to warm the Sun or planets. The former Sun will just gradually cool down and fade away into old age, plunging all of the planets

Focus 3.8 Lifetime on the main sequence

How long can a star continue to shine? A star on the main sequence of the Hertzsprung–Russell diagram shines by converting the nuclei of hydrogen atoms, or protons, into helium nuclei. These nuclear fusion reactions are limited to the hot, dense stellar core. Outside of the core, where the overlying weight and compression are less, the gas is cooler and thinner and nuclear fusion cannot exit. When all of the hydrogen within the core has been converted into helium, the star has exhausted its fuel supply and can no longer reside on the main sequence.

In 1942 Mario Schönberg and Subrahmanyan Chandrasekhar considered stellar models in which hydrogen is burned inside a star's core, or in a thin shell between the exhausted core and the overlying material. They found that it was impossible to construct models in which more than 12 percent of the mass of the star is included in the exhausted core. This meant that the lifetime of a star on the main sequence is limited to the time it takes to convert 12 percent of its hydrogen into helium. After the star has reached this Schönberg–Chandrasekhar limit, the core contracts under its own gravity. The heating up of the core makes the outer gases expand; the star leaves the main sequence and evolves into a giant star.

The lifetime on the main sequence is therefore given by:

main-sequence lifetime =

$$0.12 \left(\frac{m_{He}}{4m_H}\right) \frac{Mc^2}{L} = 3.90 \times 10^{17} \left(\frac{M}{M_\odot}\right) \left(\frac{L_\odot}{L}\right) \text{ s}$$

where the mass of the helium nucleus, m_{He}, is 0.7 percent less than the mass of the four protons, $4m_H$, that went into making it, or $m_{He}/(4m_H) = 0.007$. The energy is equal to the product of the mass and the square of the velocity of light $c = 2.9979 \times 10^8 \text{ m s}^{-1}$. In the numerical expression, the star's mass, M, and absolute luminosity, L, are given in solar units, where the mass of the Sun is $M_\odot = 1.989 \times 10^{30}$ kg and the Sun's luminosity is $L_\odot = 3.85 \times 10^{26}$ W.

Since there are $3.155\ 69 \times 10^7$ seconds in a year, the main-sequence lifetime of the Sun is 1.2×10^{10}, or 12 billion, years. The Sun has been shining for 4.6 billion years, so it is expected to evolve into a giant star in about 7.4 billion years from now.

The absolute luminosity, L, of main-sequence stars, increases with the fourth power of the mass, M, or L = constant $\times M^4$ (Section 2.3), so we also have:

main-sequence lifetime = $12 \left(\frac{M_\odot}{M}\right)^3$ billion years.

A main-sequence star that is ten times more massive than the Sun consumes its nuclear fuel a thousand times faster, and lives a correspondingly shorter time of about 12 million years.

into a deep freeze.

Such events are in the very distant future, of course. For now, the Sun provides us with an up-close laboratory of how a star behaves, from deep inside to its expanding atmosphere.

CHAPTER FOUR

Inside the Sun

Tones of the oscillating Sun Sound waves resonate deep within the Sun, enabling scientists to determine its internal constitution. Only waves with specific combinations of period (*left axis*) and horizontal wavelength (*bottom axis*) resonate within the Sun, producing the fine-tuned "ridges" in the oscillation power depicted in this *SOHO* image. (Courtesy of the *SOHO*/MDI consortium. *SOHO* is a project of international collaboration between ESA and NASA.)

4.1 How the energy gets out

All of the Sun's energy is produced by nuclear fusion reactions deep down inside a dense, high-temperature core, which extends from the Sun's center to about one-quarter of its radius, or 1.74×10^8 m out. It thus accounts for only 1.6 percent of the Sun's volume – but about one-half its mass since the gas is so compacted down there.

Nuclear reactions cannot occur in the cooler, less-dense regions outside the Sun's core, so no energy is produced there. Energy therefore has to be transported from the core through the rest of the solar interior to reach the Sun's visible disk and make it shine. The mechanisms that transport the energy out define the structure and behavior of the layers inside the Sun.

Energy moves from the core to the rest of the Sun through two spherical shells that surround the core like nested Russian dolls (Fig. 4.1). The inner shell is called the radiative zone, and the outer one is called the convective zone. Radiation and convection are the two ways that energy can travel from one place to another inside a star.

Radiation involves the movement of energy, but not the movement of material. The gas within the inner radiative zone is therefore very calm and still. It reaches from the core boundary out to 71.3 percent of the Sun's radius, or to 4.96×10^8 m. Above this lies the convective zone, within which

churning, wheeling, turbulent motions of matter transport energy out.

The Sun's interior cools with increasing distance from the core, as heat and radiation spread out into an ever-larger volume. The temperature always decreases outward, so the Sun is hottest at the center and coolest at the photosphere. Because the flux of radiation depends on temperature, there is slightly more outbound radiation than there is inbound. The energy is therefore always flowing outward, and the steeper the temperature gradient, the greater the flow. This gradient is controlled by the gas' opacity, its ability to absorb the radiation.

Although light is the fastest thing around, radiation does not move quickly through the radiative zone. Instead, it diffuses slowly outward in a haphazard zig-zag pattern, called a random walk, that resembles a drunk staggering through a crowd of people. An insulating shroud of charged particles in the radiative zone controls the flow, reducing the energy content of the radiation by repeatedly absorbing, re-radiating and deflecting it. Each time the energy is re-emitted, it comes on the average from a layer at a slightly lower temperature, so the energy of the radiation is degraded as it works its way out.

The energetic gamma rays created by nuclear fusion in the Sun's core cannot move more than 0.0009 meters, or 0.09

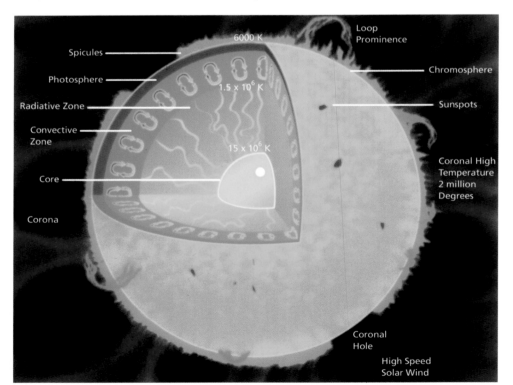

Fig. 4.1 **Cutaway drawing of Sun's interior** The Sun is powered by the nuclear fusion of hydrogen in its core, Energy produced by fusion reactions is transported outward, first by countless absorptions and emissions within the radiative zone, and then by convection. The temperature has dropped to about 1.5 million degrees kelvin at the bottom of the convective zone. The visible disk of the Sun, called the photosphere, contains dark, magnetic sunspots. It is enveloped by a million-degree-kelvin corona. A high-speed solar wind gushes out of coronal holes.

centimeters, before colliding with one of the many charged particles, producing shorter X-rays with slightly less energy. At each encounter with the plasma in the radiative zone, the solar radiation downshifts to lower energy and its wavelength becomes longer.

Because of this continuing ricocheting in the radiative zone, it takes about 170 thousand years, for radiation to work its way out from the Sun's core to the bottom of the convective zone. That amounts to an average speed of about 10^{-4} m s^{-1}. In contrast, sunlight moves through interplanetary space at the velocity of light, 2.9979×10^8 m s^{-1} taking only eight minutes to travel from the Sun to the Earth.

At the bottom of the convective zone, the temperature has become cool enough, about two million degrees kelvin, to allow some heavy nuclei to capture electrons; because of their lighter weight and greater speed, the abundant hydrogen and helium remain fully ionized. The heavy particles absorb light and block the flow of heat more efficiently than do bare nuclei or electrons. The increased light-absorbing ability, or opacity, of the solar gas obstructs the outflowing radiation, like dirt on a window. The radiative energy streaming out of the interior is therefore blocked, and the pent-up energy must be released in some other way. Material convection currents take over the transport of energy in the outer 28.7 percent of the solar interior.

The cool, opaque material at the bottom of the convective zone absorbs great quantities of radiation without re-emitting it. This causes the material to become hotter than it

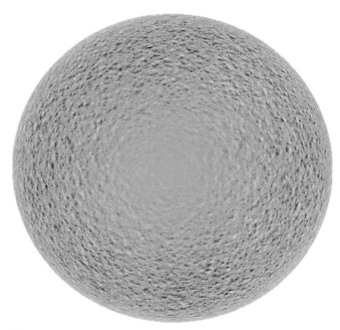

Fig. 4.3 **Supergranulation** This velocity image, or Dopplergram, of the Sun's photosphere, with rotational motion removed, shows motion toward and away from the observer, along the line of sight, as light and dark patches, respectively. It was obtained using the Doppler effect of a single spectral line with the MDI instrument on *SOHO*. The velocity field is made up of about 2500 supergranules, each about 35 million meters across (the Sun's radius is about 695.5 million meters). Near the disk center , where the Doppler effect detects radial motion, the supergranulation is hardly visible at all, thus indicating that the velocities are predominantly horizontal. (Courtesy of the *SOHO* MDI/SOI consortium. *SOHO* is a project of international cooperation between ESA and NASA.)

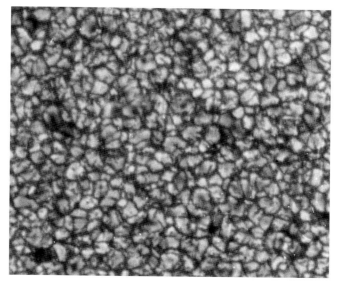

Fig. 4.2 **Double, double, toil and trouble** At high magnification, the photosphere appears completely covered with an irregular, strongly-textured pattern of cells called granules. The hot granules, each 1.4 million meters or smaller across, rise at speeds of about 1000 m s^{-1}. The rising granules radiate their energy, and cool material then sinks down along the dark, intergranular lanes. This photograph has an exceptional angular resolution of 0.2 seconds of arc or 1.5×10^5 m at the Sun. (Courtesy of Richard Muller and Thierry Roudier, Observatoire du Pic-du-midi et de Toulouse.)

would otherwise be. The heated material expands and becomes less dense than the gas in the overlying layers. Due to its low density, the hot gas rises. On Earth you can see hot air rising when watching the smoke above a fire or hawks riding on upward currents of heated air.

The heated material carries energy through the convective zone, from bottom to top, in about 10 days. The hot material then cools by radiating sunlight into space, and sinks back down to become reheated and rise again. Such wheeling convective motions occur in a kettle of boiling water, with hot rising bubbles and cooler sinking material. They produce the plopping sounds of a simmering pot of oatmeal or the more violent sounds of Yellowstone's mud geysers. Similar convective flow also produces towering thunder clouds, the great ocean currents and the shimmering of air currents over a paved road on a sunny day.

High-resolution images of the Sun taken in white light, or in all the visible colors of the Sun combined, show a granular pattern that marks the top of the convective zone (Fig. 4.2). This solar granulation exhibits a non-stationary, overturning motion, a visible manifestation of convection. The bright center of each granule, or convection cell, is the highest point of a rising column of hot gas. The dark edges of each grain are the cooled gas beginning its descent to be heated once again.

The mean angular distance between the bright centers of adjacent granules is about 1.9 seconds of arcs, corresponding to 1.4 million meters at the Sun. The largest granules have a size comparable to this separation, but there are more smaller granules than larger ones. At least a million granules cover the visible solar disk at any instant. They constantly evolve and change on time-scales of minutes, like the seething surface of an immense boiling caldron, with a typical granule lifetime of only ten minutes.

A larger cellular pattern, called the supergranulation, has typical sizes of about 35 million meters (Fig. 4.3). About 2500 supergranules are seen on the visible solar disk, each persisting for one or two days. Like the ordinary granulation, the supergranulation is generally accepted as convection. The material in these large-scale convection cells rises in the center at 50 to 100 m s^{-1}, moves away from the centers with velocities of about 200 to 400 m s^{-1}, and sinks down again at about 200 m s^{-1}.

As the supergranulation flows across the photosphere, it carries the magnetic field along with it, sweeping the magnetism to the cell edges where the field collects and strengthens. In the center of the supergranulation cells, where the material is rising, only weak magnetic fields are present, and the magnetic flux is concentrated into the downflowing boundaries of the supergranules. There is a strong spatial correspondence between the supergranule boundaries and the bright emission features above them, indicating that the overlying gases are heated at places where the magnetism is concentrated.

The convective zone is capped by a thin veneer of solar radiation, called the photosphere (from *photos*, the Greek word for light), at a temperature of 5780 K (Focus 1.2). About 500 km thick, the photosphere is a zone where the gaseous material changes from being completely opaque to radiation to being transparent. It is the layer from which the light we actually see is emitted and where most of the Sun's energy finally escapes into space.

Our model of the Sun's interior is a simple one that provides a good description of the Sun's global properties. However, it omits some of the details by assuming complete spherical symmetry and ignoring internal rotation and possible mixing of core material to surrounding gas. This model can be improved by using sound waves to open a window into the Sun's interior. They show how the Sun's gases rotate and flow inside, and pinpoint where its magnetism is probably generated.

4.2 Taking the Sun's pulse

We can look inside the Sun by observing the slow, rhythmic, in-and-out motions of the photosphere. These widespread throbbing oscillations are caused by internal sound waves. When the sounds strike the photosphere and rebound back down, they disturb the gases there, causing them to move in and out with a period of about five minutes.

Sound waves, whether on Earth or in the Sun, are waves carried by matter. They travel by compressing material in their path. Because they rely on matter, sound waves cannot travel through a vacuum, or an area in which no matter is present. Air carries most of the sound we hear on Earth. The hot plasma of the Sun carries sound waves within the Sun.

The solar oscillations are excited by vigorous turbulent motion in the convective zone, somewhat like the deafening roar of a jet aircraft, but much, much louder. The violent convective motions occur continuously, so the visible solar disk is always oscillating in tune with its internal sounds. Although the sounds are produced within the convective zone, they echo and resonate throughout the Sun. Scientists can therefore decipher this endless melody to reveal the Sun's internal constitution, in much the same way that an ultrasonic scanner can peer inside a mother's womb and map out the shape of an unborn infant.

The sound waves are trapped inside the Sun and cannot travel through the near-vacuum of space. Even if they could reach the Earth, the sounds are too low-pitched for the human ear to hear. A period of five minutes corresponds to 0.003 vibrations per second. This is 12.5 octaves below the lowest note audible to humans, and many of the Sun's notes are even lower. Sound waves that are heard by even a sensitive ear have a frequency of 25 to 20000 vibrations a second; the latter is about equivalent to a bad telephone connection.

Nevertheless, when these sounds propagate outward to the photosphere they disturb the gases there and cause them to rise and fall, producing widespread throbbing motions (Fig. 4.4). These vertical oscillations can be tens of thousands of meters high and travel a few hundred meters per second. Such movements are imperceptible to the eye, but sensitive instruments on the ground and in space routinely pick them out. They are detected as tiny, periodic changes in the

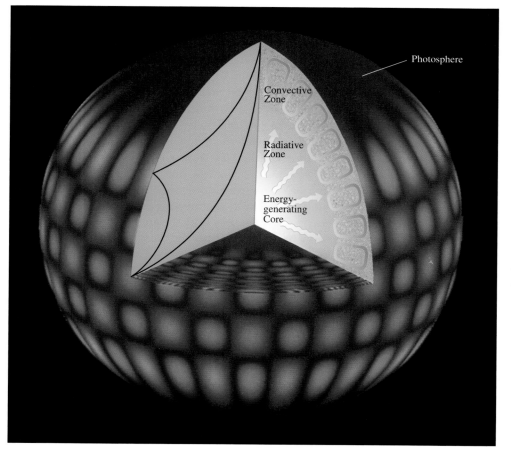

Fig. 4.4 **The pulsating Sun**
Sound waves inside the Sun cause the visible solar disk to move in and out. This heaving motion can be described as the superposition of literally millions of oscillations, including the one shown in this illustration for regions pulsing inward (*red regions*) and outward (*blue regions*). The sound waves, whose paths are represented here by black lines inside the cutaway section, resonate through the Sun. They are produced by hot gas churning in the convective zone, which lies above the radiative zone and the Sun's core. (Courtesy of John W. Harvey, National Optical Astronomy Observatories, except cutaway.)

Photosphere

Convective Zone

Radiative Zone

Energy-generating Core

wavelength of a well-defined spectral line, or as miniscule variations in the Sun's total light output.

Local areas of the photosphere swell and contract, slowly and rhythmically at five-minute intervals, moving in and out like the tides in a bay or a beating heart. Each five-minute period is the time it takes for the localized motion to go from its highest point to its lowest point and back again. These oscillations are superficially similar to ripples formed when a stone hits a pond or when a floating log has been pushed below a pond's surface and let go. However, unlike a water ripple, the photosphere's heaving motions never stop, and the entire Sun can move in and out.

When discovered by Robert B. Leighton in 1960, the five-minute oscillations of the photosphere seemed to be a chaotic, short-lived and purely local effect, with each small region moving independently of nearby ones. They seemed to be caused by the local convective rising of gases to the granulation in the photosphere, something like the waves generated in a swimming pool by the head of a diver rising to the surface. However, subsequent observations and theoretical considerations, in the early-1970s, demonstrated that all the local motions are driven by sound waves that echo and resonate through the solar interior. Every point of the photosphere is being pushed in and out by numerous sound waves arriving from every direction inside the Sun. Eventually observers showed that the entire Sun is vibrating with ponderous rhythms that extend to its very core.

A major obstacle to studies of the pulsating Sun is the Earth's rotation, which keeps us from observing the Sun around the clock. Nightly gaps in the data create background noise that hides all but the strongest oscillations. The low frequencies that probe the solar interior to the greatest depths are especially difficult to detect. Fortunately, continuous uninterrupted observation is possible from space. The *SOlar and Heliospheric Observatory* (SOHO) is located at a place where the Earth's gravitational pull just equals that of the Sun, and SOHO therefore orbits the Sun along with the Earth, and does not move around our planet (Section 9.3). Previous spacecraft observing the Sun orbited the Earth, which would regularly obstruct their view.

Our atmosphere makes some kinds of solar oscillations difficult to record, especially very low or exceptionally high frequencies. For these sounds, ground-based observations are something like trying to listen to a Mozart piano concerto when your son is blasting rap music on his boom box. Out in quiet, peaceful and tranquil space, unperturbed by terrestrial interference, SOHO has a long, clear undistorted view. It obtains recordings of both large-scale global oscillations and small-scale ones with unprecedented quality.

Two of the instruments aboard SOHO, the Michelson Doppler Imager (MDI) and the Global Oscillations at Low Frequency (GOLF), record small periodic changes in the wavelength of light, measuring the oscillation speeds using the Doppler effect (Focus 2.2). This effect occurs whenever a source of light or sound moves with respect to the observer,

and results in a change in the wavelength and frequency of the waves.

A similar thing happens to sunlight in localized regions of the photosphere (Fig. 4.4). When oscillations move part of the photosphere toward Earth, the wavelength of light emitted from that region becomes shorter, the wave fronts or crests appear closer together, and the light therefore becomes bluer. This blueshift occurs because each successive wave has a shorter distance to travel than the one before it did in order to reach Earth, so the distance between waves, the wavelength, becomes shorter. When the oscillations carry localized regions away from Earth, the wavelength becomes longer with a redshift. Each wave has farther to travel than the one before it did. The magnitude of the wavelength change, in either direction, establishes the velocity of motion along the line of sight, which is called the radial velocity.

A single, very narrow spectral feature, called a line, is observed, which is sharply defined in wavelength. MDI watches the motions of nickel atoms, in their red spectral line (Ni I at 676.8 nm), as they bob like buoys on the undulating Sun. The nickel line is one of the dark absorption lines formed by the relatively cool gas of the photosphere. Such spectral lines appear when the intensity of sunlight is displayed as a function of wavelength like a rainbow (Section 1.5).

The photosphere motions can be inferred by taking solar images at different wavelengths on both the long-wavelength and short-wavelength sides of the absorption line, and fitting a model to them in order to determine the Doppler shift. Sequences of such measurements, taken at regular, successive intervals of time, can be used to determine the periodicity of the motions.

A single measurement determines the velocity with an accuracy of about 5 m s^{-1}. When a lot of MDI's measurements are averaged over adjacent places or long times, the velocity can be determined with a remarkable precision of better than one millimeter per second – when the Sun's surface is heaving and dropping about a million times faster.

Moreover, the MDI telescope detects these motions at a million points across the Sun every minute! Because the instrument is positioned well above Earth's obscuring atmosphere, MDI continuously resolves fine detail that cannot always be seen from the ground. The blurring effects of our air, cloudy weather and the day – night cycle prohibit long, continual, spatially-resolved velocity measurements, such as those made with MDI for oscillations of relatively small scales.

A third device tracks another change caused by the sound waves. As these vibrations heat and cool the gases in light-emitting regions of the Sun, the entire orb regularly flickers. When the sound waves alternately compress and relax the photosphere, they alter the brightness of the whole Sun, over periods of about five minutes, by a tiny fraction of a few millionths, or a few times 10^{-6}, of the Sun's average brightness. The minuscule but regular changes in the Sun's total output for each oscillation mode correspond to a change in temperature of just 0.005 K. The net brightness variation caused by all of the five-minute oscillations is about 20 parts per million, corresponding to an average temperature change of 0.03 K.

SOHO's Variability of solar IRradiance and Gravity Oscillations (VIRGO) instrument records these intensity changes, precisely determining variations in the Sun's total irradiance of Earth (Section 8.4). The amplitude of the solar intensity oscillation is about 1 percent, and the brightest emission occurs when the solar gases are most compressed downward. VIRGO's telescope gathers and integrates light over the entire disk of the Sun, resolving it coarsely into 12 elements. It therefore tends to average out the peaks and troughs of the smaller localized undulations that occur randomly at many different places and times within the field of view. Yet, VIRGO provides a sensitive record of global, long-period oscillations, and refines our knowledge of the physical and dynamical properties of the central regions of the Sun.

Ground-based observatories are also making significant contributions to studies of solar pulsation. The Sun is being observed around the clock by a world-wide network of imaging observatories, known as the Global Oscillations Network Group (GONG). It consists of six identical instruments distributed in longitude around the world, producing a map of photospheric velocities every minute that the Sun is visible at each site. These electronically-linked sentinels follow the Sun as the Earth rotates. In effect, the Sun never sets on the GONG telescopes.

Other imaged and non-imaged network observations of solar oscillations include those of BiSON (Birmingham Solar Oscillation Network), operating continuously since 1981, the IRIS (International Research on the Interior of the Sun) project, operating since 1989, and more recently the HDHN (High Degree Helioseismology Network), the RODOMA (ROma DOppler and MAgnetic) network, and the ECHO (Experiment for Coordinated Helioseismic Observations) network.

4.3 Sounds inside the Sun

Astronomers peel back the outer layers of the Sun, and glimpse inside it by examining sounds with different paths within the Sun. They use the term helioseismology to describe such investigations of the solar interior. It is a hybrid name combining the Greek word *helios* for Sun or light, the Greek word *seismos* meaning quake or tremor, and *logos* for reasoning or discourse. So, literally translated, helioseismology is the logical study of solar tremors. Geophysicists similarly unravel the internal structure of the Earth by recording earthquakes, or seismic waves, that travel to different depths; this type of investigation is called seismology. The techniques resemble the way that Computed Axial Tomography (CAT) scans derive views inside our bodies from numerous readings of X-rays that cross them from different directions.

Sound waves, produced by hot gas churning in the convective zone, circle within resonant cavities inside the Sun. Starting near the photosphere, a sound wave moves into the Sun toward the center, traveling at the local speed of sound. Since the speed of sound increases with temperature (Focus 4.1), which in the Sun increases with depth, the wavefront's deeper, inner edge travels faster than its shallower outer edge, so the inner edge pulls ahead. Gradually an advancing wavefront is refracted, or bent, until the wave is once again headed toward the Sun's surface. At the same time, the enormous drop in density near the photosphere reflects waves traveling outward back in. Thus, the sounds move within solar cavities. They resemble an airplane's noise that gets trapped near the ground on a hot day, or the shouts of a child that can travel far across a cloud-covered lake on a summer night.

The photosphere's oscillations are the combined effect of about ten million separate notes. Each of these notes travels along a unique path inside the Sun, and sounds of different frequency or pitch descend to different depths (Fig. 4.5). Some of them stay within the convective zone, while others travel to the very center of the Sun. To trace our star's physical landscape all the way through – from its churning convective zone down into its radiative zone and core – we must determine the precise pitch (frequency) of all the notes.

Sound waves that travel far into the Sun move nearly perpendicular to the photosphere when they hit it, and cause the entire star to ring like a bell. Those with shorter trajectories strike the photosphere at a glancing angle, and travel through shallower and cooler layers. Sounds with both deep and shallow turning points go around and around within cavities inside the Sun, like hamsters caught in an exercise wheel.

The combined sound of all the notes reverberating inside the Sun has been compared to a resonating gong in a sandstorm, being repeatedly struck with tiny particles and randomly ringing with an incredible din. The Sun produces order out of this chaos by reinforcing certain notes that resonate within it, like the plucked strings of a guitar. They are called standing waves.

The oscillations seen at a single point on the Sun are the incoherent addition of millions of oscillations, each with velocity amplitudes of no more than 0.1 m s^{-1}. When superimposed, they produce oscillations that move with thousands of times this speed. They are strongest at the center

---

Focus 4.1. The speed of sound

Sound waves are produced by perturbations in an otherwise undisturbed gas. They can be described as a propagating change in the gas mass density, ρ, which is itself related to the pressure, P_g, and temperature, T, by the ideal gas law:

$$\text{gas pressure} = P_g = NkT = \frac{\rho kT}{m_u \mu},$$

where N is the number of particles per unit volume, Boltzmann's constant $k = 1.38066 \times 10^{-23}$ J K^{-1}, and the atomic mass unit $m_u = 1.66054 \times 10^{-27}$ kg. The mean molecular weight, μ, is given by $\mu = 2/(1 + 3X + 0.5Y)$ where X, Y, and Z are the concentration by mass of hydrogen, helium and heavier elements, respectively, and $X + Y + Z = 1$. For a fully ionized hydrogen gas, $\mu = 1/2$, and for a fully ionized helium gas, $\mu = 4/3$.

The velocity of sound, s, is given by:

$$\text{velocity of sound} = s = \left(\frac{\partial P}{\partial \rho}\right)^{1/2} = \left(\frac{\gamma kT}{m_u \mu}\right)^{1/2} = \text{constant} \times \left(\frac{T}{\mu}\right)^{1/2},$$

where the symbol ∂ denotes a differentiation under conditions of constant entropy and γ is the adiabatic index. The wavelength, λ, and frequency, ν, of the sound waves are related by $\lambda \nu = s$.

Helioseismologists measure the sound speed, and this is affected by both temperature and composition. Without additional information they cannot be distinguished as causes of discrepancies from standard models.

For an adiabatic process, the pressure, P_g, and mass density, ρ, are related by $P_g = K\rho^\gamma$ where K is a constant. For adiabatic perturbations of a monatomic gas the index $\gamma = 5/3$ and for isothermal perturbations $\gamma = 1$. For both adiabatic and isothermal perturbations the sound speed is on the order of the mean thermal speed of the ions of the gas or, numerically:

$$\text{approximate velocity of sound} = s \approx 10^4 \left(\frac{T}{10^4}\right)^{1/2} \text{m s}^{-1}$$

where T is the temperature in degrees kelvin.

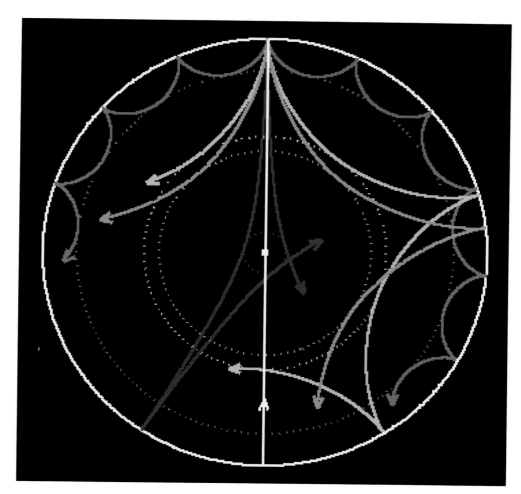

Fig. 4.5 **Sound paths** The trajectories of sound waves are bent inside the Sun, like light within the lens of an eye, and circle the solar interior in spherical shells or resonant cavities. Each shell is bounded at the top by a large density drop near the photosphere and bounded at the bottom by an increase in sound speed with depth that refracts a downward-propagating wave back toward the photosphere. The bottom turning points occur along the dotted circles shown here. How deep a wave penetrates and how far around the Sun it goes before it hits the photosphere depends on the harmonic degree, l. The white curve is for $l = 0$, blue for $l = 2$, green for $l = 20$, yellow for $l = 25$ and red for $l = 75$. (Courtesy of Jørgen Christensen-Dalsgaard and Philip H. Scherrer.)

of the solar disk, with a velocity amplitude of about 300 m s^{-1}. Since the motions are nearly radial, their observed strength decreases systematically with increasing distance away from disk center. When averaged over the entire solar disk, the amplitude of the five-minute oscillation is reduced to only 0.7 m s^{-1}.

At any given point on the photosphere, the five-minute oscillations build up in amplitude and then, within a few periods, decay, and the phase changes from one burst of oscillations to the next. The low-amplitude components temporarily reinforce each other, producing the well-known, five-minute oscillations that grow and decay as numerous vibrations go in and out of phase to combine and disperse and then combine again, somewhat like groups of birds or schools of fish that gather together, move apart and congregate again. A drum or horn similarly produces a definite musical note as the result of the constructive interference of sound waves within a cavity of some definite size and shape. Only certain standing waves can fit exactly within the cavity without any overlap.

Scientists have examined the power in the various oscillations, or how often each and every note is played, confirming that the power is concentrated into such resonant standing waves. Instead of meaningless, random fluctuations, orderly motions are detected with specific

combinations of size and period, or wavelength and frequency (Fig. 4.6). Destructive interference filters out all but the resonant waves that combine and reinforce each other. Yet, there are still millions of such notes resonating in the Sun, so prolonged observations with high spatial resolution and detailed computer analysis are required to sort them all out.

The spatial patterns of oscillations detected at the photosphere can be separated into their component standing waves, each characterized by nodes at which the motion vanishes. For example, if you shake one end of a rope that is fixed at the opposite end, a standing wave can be created. The nodal points are the places on the rope that are standing still while the rest of the rope moves up and down. In two dimensions, such as a drum skin or the solar photosphere, the nodal points are arranged in nodal lines.

Since the oscillations are due to sound waves that reverberate in three dimensions within the Sun, three whole numbers are needed to specify each standing wave. For a nearly perfect sphere such as the Sun, spherical harmonic functions are used to represent the trapped modes, and a single mode is designated by a set of three integers l, m and n. The spherical harmonic degree, l, indicates the total number of nodal lines on the solar disk, ranging from zero to thousands. It is a measure of the spatial scale of the mode,

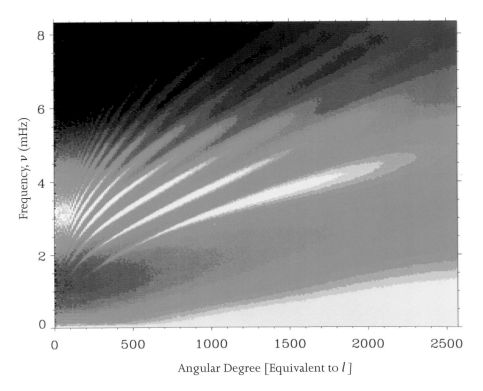

Fig. 4.6 **An l–nu diagram** The frequency, nu, of sound waves is plotted as a function of the spherical harmonic degree, l, expressed as the angular degree, for just eight hours of high-resolution data taken with the *SOHO* MDI instrument. A frequency of 3 mHz, or 0.003 cycles per second, corresponds to a wave period of 5 minutes. The degree, l, is the inverse of the spatial wavelength, or surface size; an l of 400 corresponds to wavelengths on the order of 10^7 m in size. The oscillation power is contained within specific combinations of frequency and degree, demonstrating that the oscillations detected in the photosphere are due to internal standing waves confined within resonant cavities. (Courtesy of the *SOHO* MDI/SOI consortium. *SOHO* is a project of international cooperation between ESA and NASA.)

which extends from the size of small granules to the entire Sun, or from a thousand meters to hundreds of millions of them. For $l = 0$, the whole solar globe resonates across its radius of 696 million meters. The lowest-degree modes are expected to last without significant change for several years, while the high-degree ones reflect the varying conditions in the turbulent convective zone.

The azimuthal order, m, describes the number of nodal lines that cross the equator, and ranges from zero to plus and minus the degree number 1. The m represents the concentration of the mode near the equator. Oscillations with m = 1 are located close to the equator, while oscillations with low m reach high latitudes.

The radial order n is the number of nodal surfaces in the radial direction inside the Sun, on a line from the center of the Sun to the photosphere, and has been observed to range from zero to nearly 50.

The Sun's notes can be portrayed in a two-dimensional power plot, called the $l–\nu$ or l–nu diagram, that displays how much sound energy there is at each frequency, ν, for every one of the different spatial scales or l values (Fig. 4.6). This spectrum removes the effects of solar rotation and then averages the results over m, so it represents the spectrum that the Sun would produce if it were not rotating.

The frequency, ν, is the number of wave crests passing a stationary observer each second, and it therefore tells us how fast the sound wave moves up and down in a given cavity. Most of the power is concentrated near a frequency of 0.003 Hz, or 3 mHz. That corresponds to a period of 5 minutes, the time interval between wave peaks. The highest frequency that is expected is 0.0056 Hz, with a 180-second period. Sound waves have been observed with periods as long as 20 minutes.

At constant frequency, ν, there is a relationship between the degree, l, and the depth of propagation in the sense that sound waves with low degree travel to the solar core, while high-degree sound waves are confined closer to the photosphere. For $l = 0$ the waves actually go to the solar center; such waves correspond to purely radial oscillations in which the Sun expands and contracts but keeps its spherical shape. The $l = 0$ oscillations are "breathing modes" where the whole Sun moves in and out at the same time. All the other higher-order values of l correspond to non-radial oscillations that deform the surface into non-spherical shapes. The higher the degree the smaller the spatial scale of the distortion.

The deepest parts of the Sun are probed by the oscillations with the largest sizes and the lowest spherical harmonic degree ls, thus requiring the least surface resolution. Full-Sun, integrated-light observations provide this information. The SOHO instruments GOLF and VIRGO respectively observe $l = 0$ to 3 and $l = 0$ to 7, that are able to penetrate the deep solar interior.

In contrast, a large number of shallower depths are accessible to MDI. It observes sound waves with very high values of l up to $l = 4500$ that can dive just a short distance below the surface and return quickly. They correspond to a horizontal scale as small as $1/4500$ of the solar circumference or about a million meters.

The dominant factor affecting each sound is its speed, which in turn depends on the temperature and composition of the solar regions through which it passes (Focus 4.1). The sound waves move faster through higher-temperature gas, and their speed increases in gases with lower-than-average molecular weight. So, by investigating many different waves we can build up a very detailed three-dimensional picture of the physical conditions inside the Sun, including the temperature and chemical composition, from just below the surface down to the very core. The observed frequencies are integral measures of the speed along the path of the sound wave; helioseismologists have to use complex mathematical techniques and powerful computers to invert this measured data and get the sound speed.

Scientists have used inversion methods to determine the difference between the observed sound speed and that of a numerical model. These differences are obtained from discrepancies between the detected and calculated sound-wave frequencies. Relatively small differences between the computer calculations and the observed sound speed are used to fine-tune the model and establish the Sun's radial variations in temperature, density and composition. Early comparisons of observations with theoretical expectations led to improvements in our understanding of the internal opacity that blocks the flow of radiation and governs the Sun's temperature structure. More recently, a small but definite change in sound speed has been detected at the lower boundary of the convective zone, pinpointing it at a radius of 71.3 percent of the radius of the visible solar disk.

The sound-speed measurements have important implications for the solar neutrino problem (Section 3.4). The difficulty is that the massive, underground neutrino detectors observe fewer neutrinos than expected. The discrepancy might be resolved if the Sun's central temperature is lower than current models predict, for the number of neutrinos produced by nuclear reactions is a sensitive, increasing function of the temperature. However, the measured and predicted sound velocities do not now differ from each other by more than 0.2 percent from 0.95 solar radii down to 0.05 solar radii from the center.

By measuring the speed of the sound waves that course through the Sun, we can accurately determine its internal temperature. The sound waves act as a sort of internal thermometer. A similar connection between sound and temperature can be heard when morning cold makes a piano sound out of tune, or noting a change in the frequency of a cricket's chirp when the weather is warm or chilly.

Since the square of the sound speed is proportional to the temperature (Focus 4.1), even tiny fractional errors in the model values of temperature would produce measurable discrepancies in the precisely determined sound speed. This novel thermometer has placed the central temperature of the Sun very close to the 15.6 million degrees kelvin predicted by theory. Since nuclear fusion reactions occur at a rate that increases with temperature, this measurement removes uncertainty in predictions of the amount of neutrinos coming from the Sun. It confirms that the solar neutrino problem results from an incomplete understanding of

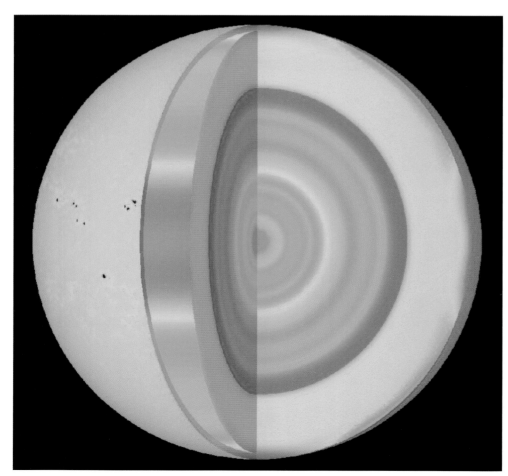

Fig. 4.7 **Radial variations of sound speed.** Red and blue correspond to faster and slower sound speeds, respectively, relative to the Standard Solar Model (*yellow*). When the sound travels faster than predicted by theory, the temperature is higher than expected; slow sound waves imply temperatures that are colder than expected. The conspicuous red layer, about a third of the way down, shows unexpected high temperatures at the boundary between the turbulent outer region (convective zone) and the more stable region inside it (radiative zone). Latitudinal variations in temperature are seen near the photosphere. These sound-speed measurements were made by the MDI/SOI and VIRGO instruments aboard *SOHO*. (Courtesy of Alexander G. Kosovichev, the *SOHO* MDI/SOI consortium, and the *SOHO* VIRGO consortium. *SOHO* is a project of international cooperation between ESA and NASA.)

neutrinos. They seem to have an identity crisis on their way out of the Sun, transforming into a form that was not detected by pioneering neutrino detectors (Section 3.4).

Nevertheless, very small sound-speed discrepancies between measurements and theory are significant (Fig. 4.7). Just below the convective zone, there is an increase in the observed sound speed compared to the model, suggesting that turbulent material is mixing in and out within this base layer. Since the speed of sound depends on temperature as well as composition, the temperature might also increase in this place. Without additional information, scientists cannot distinguish between temperature and compositional mixing as causes of discrepancies from standard models. There is a sharp decrease of the observed speed relative to theoretical expectations at the boundary of the energy-generating core, hinting that either cooler material or turbulent churning motions might occur there. Scientists speculate that they could be due to unstable nuclear-burning processes. If substantiated by further studies, this could be very important for studies of stellar evolution; they usually assume that nuclear reactions proceed without any mixing of fresh material into the core.

Scientists are also using measurements of sound-wave frequencies to infer rotational and other motions inside the Sun. The moving material produces slight changes in the frequency of a sound wave that passes through it, as expected from the well-known Doppler effect.

4.4 Internal motions

Helioseismologists have shown how the Sun rotates inside, using the Doppler effect in which motion changes the pitch of sound waves. Regions near the Sun's poles rotate with exceptionally slow speeds, while the equatorial regions spin rapidly. This differential rotation persists to about a third of the way inside the Sun; just below this the rotation becomes uniform from pole to pole. The Sun's magnetism is probably generated at the place where the differential rotation changes to uniform rotation. Internal flows have also been discovered by helioseismologists. White-hot currents of gas move beneath the Sun we see with our eyes, streaming at a leisurely pace when compared to the rotation. They circulate near the equator, and between the equator and poles, describing a sort of solar meteorology. Internal tremors, or sunquakes, have also been detected, generated by flaring explosions in the solar atmosphere.

Rotational motions

Like Earth, the Sun rotates, or spins, around an imaginary line that runs through its center. This line is called the Sun's rotational axis, and the top and bottom of this line mark the Sun's north and south poles, in the same way that the Earth's axis marks the North Pole and South Pole on Earth. The Sun and all the planets in the solar system lie on one plane, and the Sun's north pole and Earth's North Pole are oriented in the same direction relative to the plane. The Sun's equator, like Earth's, is an imaginary line halfway between the north and south poles, that runs east and west. Like Earth, the Sun rotates from west to east when viewed from above the north pole, but unlike Earth, different parts of the Sun rotate at different rates. Also, because the Earth orbits the Sun, we observe a rotation period that is about 2 days longer than the true value. The synodic rotation period of the visible solar equator, as observed from Earth, is about 27.6 days, while the equatorial photosphere is intrinsically spinning about the Sun's axis once every 25.67 days.

For at least a century, astronomers have known from watching sunspots that the photosphere rotates faster at the equator than it does at higher latitudes, decreasing in speed evenly toward each pole. The photosphere spins about the Sun's rotation axis with a sidereal rotation period, from east to west against the stars, of 25.67 days at the solar equator; its rotation period reaches 33.4 days at 75 degrees latitude north or south. These rotation periods can be converted into velocities – just divide the circumference at the latitude by its period (Table 4.1). Regions near the poles rotate with very slow speeds, in part because the rate of rotation is

Solar latitude (degrees)	Rotation period (days)	Rotation speed (km h^{-1})	Rotation speed (m s^{-1})
0 (Equator)	25.67	7097	1971
15	25.88	6807	1891
30	26.64	5922	1645
45	28.26	4544	1262
60	30.76	2961	823
75	33.40	1416	393

[a] Data from the MDI instrument aboard SOHO.

Table 4.1 **Differential rotation at the photosphere**[a]

smaller, but also because the surface near the poles is closer to the Sun's axis and the distance around the Sun is shorter.

The varying rotation of the photosphere is known as differential rotation, because it differs with latitude. In contrast, every point on the surface of the Earth rotates at the same speed, so a day is 24 hours everywhere on the Earth. Such a uniform spin is called solid-body rotation. Only a gaseous or liquid body can undergo differential rotation; it would tear a solid body into pieces.

Although radial convection churns the Sun's outermost layers, rotation produces much faster, circumferential motions. So, rotation dominates the motion of the Sun's deep interior. This rotation can be specified as a function of heliographic latitude, the angular distance north or south of the solar equator in a spherical coordinate system.

Helioseismologists can use sound waves to measure its internal spin. Sound waves moving with the direction of rotation appear to move faster, like birds flying with the wind, so their measured periods are shorter. Waves propagating against the rotation will be slowed and show longer periods. These opposite effects split a given oscillation into a pair of closely spaced frequencies, like two notes that are not quite at the same pitch.

The size of the expected frequency splitting can be estimated from the visible rotation. The solar oscillations have a period of about five minutes and the photosphere's equatorial rotation period is about 25 days. So the rotational splitting is roughly five minutes divided by 25 days, or almost one part in ten thousand. The oscillation frequencies have to be measured ten or a hundred times more accurately to determine subtle various in the Sun's rotation, or as precisely as one part in a million.

Armed with this sensitive technique, scientists have found that the latitude-dependent rates exhibited by the

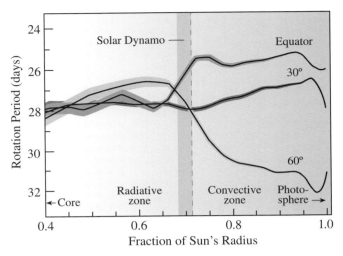

Fig. 4.8 **Internal rotation of the Sun**. The rotation rate inside the Sun at latitudes of zero (solar equator), 30 and 60 degrees has been inferred using data from the Michelson Doppler Imager (MDI) aboard *SOHO*. Just below the convective zone, the rotational speed changes markedly. The outer parts of the gaseous Sun rotate far faster at the equator than at the poles; this differential rotation persists to the bottom of the convective zone at 28.7 percent of the way down. Below that, uniform rotation appears to be the norm. Shearing motions along this interface may be the dynamo source of the Sun's magnetism. We don't know much about the rotation of the central parts of the Sun, within its energy-generating core, since most of the observed sound waves do not reach that far. (Courtesy of Alexander G. Kosovichev and the *SOHO* MDI/SOI consortium. *SOHO* is a project of international cooperation between ESA and NASA.)

photosphere persist throughout the convective zone (Fig. 4.8). However, the rotation speed becomes uniform from pole to pole nearly one-third of the way to the core, 220 million meters beneath the photosphere. Lower down the rotation rate remains independent of latitude, acting as if the Sun were a solid body.

Thus, the rotation velocity changes sharply at the top of the radiative zone. There the outer parts of the radiative interior, which rotates at one speed, meet the overlying convective zone, which spins faster in its equatorial middle. Scientists suspect that the forces generated by the two zones moving against each other may create the Sun's magnetic field.

The roughly 20-million-meter-wide layer where these very different zones meet and shear against one another, called the tacholine, is the likely site of the solar dynamo, the source of the Sun's magnetism. The dynamo amplifies and regenerates the Sun's magnetic field within the solar interior. The hot circulating gases, which are good conductors of electricity, generate electrical currents that create magnetic fields; these fields in turn sustain the generation of electricity, just as in a power-plant dynamo. The process of field amplification is cumulative, so an intense magnetic field can be generated from an initially weak one. These strong fields eventually rise through the convective zone and emerge at the photosphere.

Below the boundary layer, or tacholine, the Sun rotates like a solid object, with too little variation in spin to drive a solar dynamo. Closer to the surface, the rotation rates at different latitudes diverge over broad areas that are not focused enough to play much of a role in the dynamo. Moreover, the MDI team found that sound waves speed up more than expected in this shear layer, indicating that

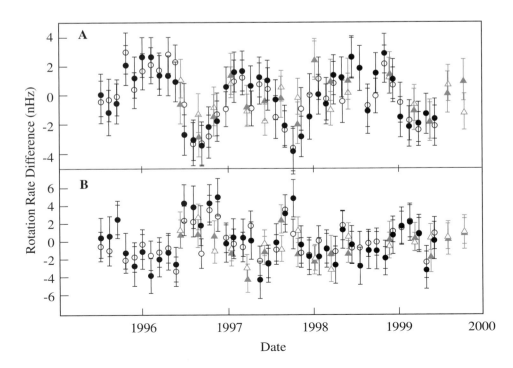

Fig. 4.9 **Pulse of the solar dynamo**. Temporal variations in the rotation of the Sun near the base of the convective zone and close to the probable site of the solar dynamo. In equatorial regions, the rotation speeds up, slows down, and speeds up again with a period of about 16 months, or 1.3 years. Here we display differences of the rotation rate from the temporal mean at $0.72\,R_\odot$ (**A**) and $0.63\,R_\odot$ (**B**). When the lower gas speeds up, the upper gas slows down, and *vice versa*. The solar radius $R_\odot = 6.955 \times 10^8$ m, and the base of the convective zone is located at a radius of $0.713\,R_\odot$ This data was obtained from both the ground (GONG) and from space (*SOHO* MDI). (Courtesy Rachel Howe and colleagues, the GONG consortium, the *SOHO* MDI/SOI consortium, NASA, NSF, NSO, NOAO, and AURA.)

turbulence and mixing associated with a dynamo are most likely present.

Scientists have recently used observations from both the ground and space to examine rotation rates at the bottom edge of the convective zone, providing important clues to how the solar dynamo works. The Sun does not rotate at a fixed rate down there. The rotation speed varies periodically, spinning fast and slow and then fast again (Fig. 4.9). These alterations in rotation speed have a period of 1.3 years, or 16 months, in equatorial regions. There is a more complicated variation with a dominant 1.0-year period at higher latitudes.

The periodic variations at different depths are out of phase with each other, and the contrast in speed above and below the dynamo region can change as much as 20 percent in six months. When the lower gas speeds up the upper gas slows down, and *vice versa* (Fig. 4.9). These pulsating relative motions between neighboring layers of electrified gas probably drive the dynamo that amplifies and generates the Sun's magnetic field.

A more esoteric implication of the rotation results involves tests of Einstein's theory of gravity (Focus 4.2).

Solar meteorology

Our star's interior flows in ways other than rotation. Ponderous, slow rivers of gas circulate beneath the visible disk of the Sun. They have been detected by measuring travel-times and distances for sound waves that probe regions just below the photosphere.

The method is quite straightforward: *SOHO*'s MDI instrument records small periodic changes in the wavelength of light emitted from a million points across the Sun every minute. By keeping track of them, it is possible to determine how long it takes for sound waves to travel from one point on the solar photosphere, through the Sun's interior, to another point on the photosphere. The time taken to skim through the Sun's outer layers tells of both the temperature and gas flows along the internal path connecting the two points. If the local temperature is high, sound waves move more quickly – as they do if they travel with the flow of gas. This data is then inverted in a computer to chart the three-dimensional internal structure and dynamics of the Sun, including the sound speed, flow speed, and direction of motion.

The MDI has provided travel-times for sounds crossing thousands of paths, linking myriad surface points. Researchers have fed one year of nearly continuous observations into a supercomputer, using it to work out temperatures and flows along these intersecting paths. After weeks of number crunching, the *SOHO* scientists have identified vast new currents coursing through the Sun and clarified the form of previously discovered ones (Fig. 4.10). These flows are not the dominant, global rotational motions, but rather the ones found when rotation is removed from the data. Their speeds only reach tens of meters per second.

These white-hot rivers of plasma describe a solar

Focus 4.2 Solar oblateness and tests of Einstein's General Theory of Relativity

According to Einstein's General Theory of Relativity, space is distorted or curved in the neighborhood of matter, and this distortion is the cause of gravity. The result is a gravitational effect that departs slightly from Newton's expression, and the planetary orbits are not exactly elliptical. Instead of returning to its starting point to form a closed ellipse in one orbital period, a planet moves slightly ahead in a winding path that can be described as a rotating ellipse. Einstein first used his theory to describe such a previously unexplained twisting of Mercury's orbit.

Mercury's anomalous orbital shift, of only 43 seconds of arc per century, is in almost exact agreement with Einstein's prediction, but this accord depends on the assumption that the Sun is a nearly perfect sphere. If the interior of the Sun is rotating very fast, it will push the equator out further than the poles, so its shape ought to be somewhat oblate rather than perfectly spherical. The gravitational influence of the outward bulge, called a quadrupole moment, will provide an added twist to Mercury's orbital motion, shifting its orbit around the Sun by an additional amount and lessening the agreement with Einstein's theory of gravity.

Fortunately, helioseismology indicates that most of the inside of the Sun does not rotate significantly faster than the outside, at least down to the energy-generating core. The internal rotation is not enough to produce a substantial asymmetry in the Sun's shape, even if its core is spinning rapidly. So, we may safely conclude that measurements of Mercury's orbit confirm the predictions of The General Theory of Relativity. In fact, the small quadrupole moment inferred from the oscillation data, about one-ten-millionth rather than exactly zero as Einstein assumed, is consistent with a very small improvement in the accuracy of the measurement of Mercury's orbit in recent times. So, the Sun does have an extremely small, middle-aged bulge after all.

---

meteorology that resembles weather patterns on Earth, but on a much larger scale with hotter temperatures and no rain in sight. Broad zonal bands sweep around the equatorial regions at different speeds, reminding us of the trade winds in the Earth's tropics. The Sun's "trade winds" are detected to a considerable depth below the photosphere. The banding is apparently symmetric about the solar equator, with at least two zones of faster rotation and two zones of slower rotation in each hemisphere of the Sun.

The velocity of the faster zonal flows is about 5 m s^{-1} higher than gases to either side. This is substantially slower than the mean velocity of rotation, which is about $2\,000$ m s^{-1}, so the fast zones glide along in the spinning gas, like a wide, lazy river of fire. Helioseismologists speculate that sunspots might originate at the boundaries of zones moving at different speeds, where the shearing force and turbulence might twist the magnetic fields and intensify magnetic activity, like two nearby speed boats moving at different velocities and churning the water between them.

The existence of bands of fast and slow rotation was first

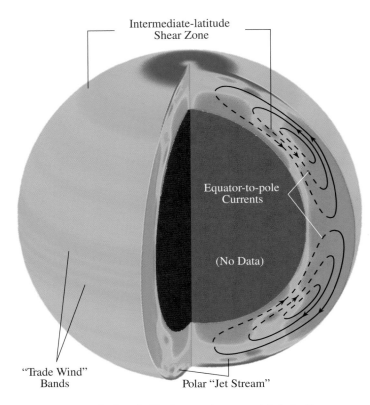

Fig. 4.10 **Interior flows** In this picture, red corresponds to faster-than-average flows, yellow to slower than average, and blue to slower yet. On the left side, deeply rooted zones (*yellow bands*), analogous to the Earth's trade winds, travel slightly faster than their surroundings (*blue regions*). The right-hand cutaway reveals a slow movement poleward from the equator shown by the streamlines; the return flow below it is inferred. This image is the result of computations using one year of continuous observation, from May 1996 to May 1997, with the Michelson Doppler Imager (MDI) instrument aboard *SOHO*. (Courtesy of Philip H. Scherrer and the *SOHO* MDI/SOI consortium. *SOHO* is a project of international cooperation between ESA and NASA.)

noticed in 1980, as the result of photosphere observations from the Mount Wilson Observatory. However, only with the new techniques of helioseismology could we realize that the zonal flow bands are not a superficial phenomenon, but instead extend deeply into the Sun. A single zonal flow band is broad and deep, more than 65 million meters wide and penetrating as far as 20 million meters below the surface. The full extent of this component of the Sun's stormy weather could never have been seen looking at the visible layer of the solar atmosphere.

Helioseismologists have shown that the deep zonal bands of slightly faster and slower rotation gradually move from high latitudes toward the equator as the years go by (Fig. 4.11). The observations were made over a 4.5-year period at a depth of 1 percent of the Sun's radius, or at about 7 million meters down. There is evidence that these bands may extend deeper; down to the base of the convective zone. Their slow motion toward the equator is consistent with measurements of the bands in the photosphere, as well as with the similar drift of the location of sunspots as the 11-year solar cycle of magnetic activity (Section 5.1) winds down to its minimum and approaches the next maximum.

Both the sunspots and the zonal bands are moving against another flow, from equator to pole, like a boat moving up river against a strong current of water. Helioseismologists have demonstrated that this so-called meridional flow penetrates deeply. The entire outer layer of the Sun, to a depth of at least 25 million meters, is slowly but steadily flowing from the equator to the poles with a speed of about 20 m s^{-1}. At this rate, an object would be transported from the equator to a pole in a little more than one year. Of course, the Sun rotates at more than 100 times this rate, completing one revolution at the equator in about 25 days. Thus, as the

Fig. 4.11 **Migrating zonal bands inside the Sun** The variation of internal rotation rate with latitude and time, when the temporal average has been subtracted. The residual rotation rate is shown for a depth of 7×10^6 m below the photosphere. Deeply rooted zonal bands of faster and slower rotation converge toward the equator over a 4.5-year period. This figure was derived from the Global Oscillation Network Group (GONG) data. (Courtesy of Rachel Howe, Rudi Komm, Frank Hill, the GONG consortium, NSO, AURA, NOAO and NSF.)

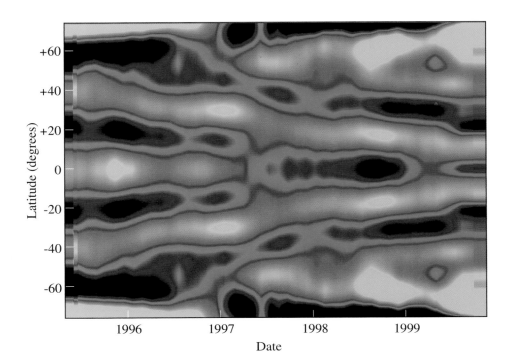

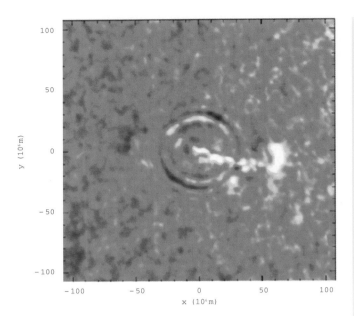

Fig. 4.12 Sunquakes Massive sunquakes were produced when a powerful explosion, or flare, sent shock waves into the underlying gases. The explosion, on 9 July 1996, produced concentric rings of a sunquake that spread away from the flare, traveling across a distance equal to 10 Earth diameters in just one hour. This solar flare left the Sun quaking with about 40 thousand times the energy released in the great earthquake that devastated San Francisco in 1906. (Courtesy of Alexander G. Kosovichev and Valentina V. Zharkova, and the *SOHO* MDI/SOI consortium. *SOHO* is project of international cooperation between ESA and NASA.)

Focus 4.3 Sunquakes

Rare, powerful explosions in the solar atmosphere can make the Sun vibrate inside. These awesome explosions, called solar flares, can release stored magnetic energy equivalent to billions of terrestrial nuclear bombs (Section 6.1). Scientists have speculated that powerful shocks are produced when a flare-associated beam of electrons slams down into the dense solar atmosphere. Now SOHO's Michelson Doppler Imager, or MDI, has detected a powerful sunquake caused by an exploding solar flare higher up, and observed circular flare-generated seismic waves moving across the photosphere like ripples spreading out in all directions from a pebble thrown into a pond.

The explosion's particle beams and shocks came down at the Sun with such force that they caused waves three thousand meters high on the Sun's surface. These waves moved out to a distance of at least 120 million meters, equal to ten Earth diameters, before fading into the background of the Sun's photosphere (Fig. 4.12). Unlike water ripples, that travel outward at constant velocity, the solar waves accelerated from about 30 thousand meters per second to approximately 100 thousand meters per second before disappearing.

Seismic waves produced by sunquakes can shake the Sun to its very center, just as earthquakes can cause our entire planet to vibrate and ring like a bell. However, unlike their terrestrial counterparts, sunquakes involve unheard-of amounts of energy. The observed sunquake, which was produced by a perfectly "ordinary" solar flare, was equivalent to an earthquake of magnitude 11.3 on the Richter scale. That is 40 thousand times the energy released in the great earthquake that devastated San Francisco in 1906, at a magnitude of 8 on the same scale. Scientists remain somewhat perplexed about how such a relatively modest flare could have the downward thrust and power to generate such powerful seismic waves.

material in the outer layers of the Sun rotates rapidly, it also flows slowly from equator to poles in about a year. Researchers suspect that an equatorward return flow exists, but they have not yet observed detailed motions down there.

A new technique uses observations of the oscillations to create a sort of mathematical lens that focuses to different depths. This method is called helioseismic holography because of its similarity to laser holography. It has been used to look through the Sun to its back side. The images show the structure, evolution and eruption of solar active regions on the far side of the Sun, many days before they rotate onto the side facing Earth (see Section 6.1).

Detailed helioseismology should tell us more about how the churning motions of hot solar gas interact with concentrations of the Sun's magnetism and give rise to its explosive activity. These incredible explosions threaten astronauts and satellites, and can affect Earth with power and communications disruptions (Section 8.4). They can also move down into the Sun, producing seismic waves that propagate deep within it (Focus 4.3).

CHAPTER FIVE

The magnetic solar atmosphere

The entire Sun is a giant mass of incandescent gas, unlike anything we know on Earth. The Sun has no surface; its gas just becomes more tenuous the farther out you go. Although we cannot see it with our eyes, very diffuse solar gas extends all the way to the Earth and beyond.

The diaphanous solar atmosphere includes, from its deepest part outward, the photosphere, chromosphere and corona (see Fig. 4.1). The Sun's magnetism plays an important role in molding, shaping and heating the coronal gas.

The visible photosphere, or "sphere of light", is the part of the Sun we can watch each day. It is the level of the solar atmosphere from which we get our light and heat. Thephotosphere contains sunspots, thousands of times more magnetic than the Earth, and the number and position of sunspots varies over an 11-year cycle of solar magnetic activity.

The visible sharp edge of the photosphere is something of an illusion. It is merely the level beyond which the gas in the solar atmosphere becomes thin enough to be transparent. The chromosphere and corona are so rarefied that we look right through them, just as we see through the Earth's clear air.

The chromosphere is very thin, but the Sun does not stop there. Its atmosphere extends way out in the corona, to the edge of the solar system. The corona's temperature is a searing million degrees kelvin, so hot that the corona is forever expanding into space.

The entire solar atmosphere is permeated by magnetic fields generated inside the Sun, rooted in the photosphere, and extending into the chromosphere and corona.

Magnetic fields near and far In the low solar corona, strong magnetic fields are tied to the Sun at both ends, trapping hot, dense electrified gas within magnetized loops. Far from the Sun, the magnetic fields are too weak to constrain the outward pressure of the hot gas, and the loops break open to allow electrically-charged particles to escape, forming the solar wind and carrying magnetic fields away. (Courtesy of Newton Magazine, the Kyoikusha Company.)

5.1 The photosphere and its magnetism

The photosphere is the lowest, densest level of the solar atmosphere. It is the place where most of the Sun's energy escapes into space. The photosphere is the source of our visible sunlight, so it is appropriately named from the Greek word *photos* meaning "light". It is the only part of the Sun that our eyes can see. Traveling at the speed of light, it takes only about 8.3 minutes for sunlight to get from the photosphere to the Earth.

All that sunlight comes from a thin, bright shell, only a few hundred thousand meters thick, or less than 0.05 percent (0.0005) of the Sun's radius. At the distance of the Sun, that thickness corresponds to an angular width of a fraction of a second of arc, too small to be resolved with any telescope on Earth. All stars have photospheres, although they are not all so thin as that of the Sun.

The photosphere has an effective temperature of 5780 K (Focus 1.2), far above the boiling point of water at 373 K, but its gases are so rarefied that we would call them a

vacuum here on Earth. Even though the photosphere is the densest part of the solar atmosphere, its pressure is only one-ten-thousandth (0.0001) of the pressure of the Earth's atmosphere at sea level.

Yet, the very tenuous, gaseous photosphere looks like a solid ball with a sharp edge. This apparently hard "surface" of the Sun is an illusion caused by gases of extremely high opacity. There is no surface on the Sun, for its atmosphere extends to the Earth and beyond.

Why don't we see through the tenuous photosphere? It is cool enough for electrons to join protons to make hydrogen atoms. In a rare collision, a hydrogen atom in the photosphere can briefly attach a free electron to itself and temporarily become an ion with a negative electrical charge. These negative hydrogen ions absorb radiation coming from the interior and re-emit visible sunlight. Even though their concentration is only one-millionth that of normal hydrogen atoms, the negative hydrogen ions make the photosphere as opaque as a brick wall.

Since you cannot see beneath the photosphere, the solar interior remains hidden. And you cannot use your eyes to see anything in front of the photosphere because we look through the overlying gas. So, the photosphere looks like a solid "surface", but appearances are deceiving.

People sometimes say that the familiar photosphere, or visible solar disk, is the "surface" of the Sun, but the photosphere is not really a surface and we do not call it that in this book. Being entirely gaseous, the Sun has no solid surface and no permanent features. There are seething gases everywhere above, in and below the photosphere. Moreover, under careful inspection all kinds of varying features can be detected in the photosphere.

Granulation

When telescopes zoom in to take a close look at the photosphere in the white light of the Sun, or in all of the colors of sunlight combined, it breaks into a million tiny, varying spots of light, called granules (Fig. 5.1, see also Fig. 4.2). This distinctly textured pattern is created and driven by convective motions beneath the photosphere. Each bright granule is a packet of hot gas rising at speeds of 1000 m s^{-1}, like supersonic bubbles in an immense boiling cauldron. Upon reaching the photosphere, the hot bubble releases its energy by radiation and then cools, darkens and sinks downward along the dark intergranular lanes.

The average diameter of the largest granules is about 1.4 million meters – roughly the same as the distance between

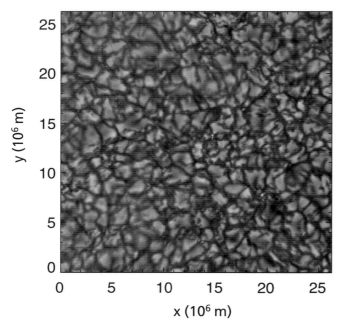

Fig. 5.1 **The solar granulation** Underlying convection shapes the photosphere, producing tiny, varying regions called granules. They are places where hot, and therefore bright, material reaches the visible solar disk. The largest granules are about 1.4 million meters across. Notice that they are not circular, but angular in shape. This honeycomb pattern of rising (*bright*) and falling (*dark*) gas is in constant turmoil, completely changing on time-scales of minutes and never exactly repeating itself. This photograph was taken with exceptional angular resolution of 0.2 seconds of arc or 150 thousand meters at the Sun, using the National Solar Observatory's Vacuum Tower Telescope at the Sacramento Peak Observatory. (Courtesy of Thomas R. Rimmele, AURA, NOAO and NSF.)

Boston and Chicago, or between Paris and Madrid. Larger and deeper convection cells, called supergranules, are about 35 million meters across; but they are not visible in white light. The supergranulation is observed in Dopplergrams that detect motion (see Section 4.1, Fig. 4.3).

The solar granulation marks the top of a seething mass of rising and falling gas, presenting an ever-changing picture. Each bright cell lasts only about 10 minutes before it is replaced by another. So, the honeycomb pattern captured in a single photograph can be misleading; it provides only a frozen snapshot of shapes and forms that are never repeated with exact precision (Fig. 5.2, Focus 5.1).

The most familiar part of the Sun – its photosphere – is in constant turmoil and change. Larger, darker places of colossal magnetism also come and go, spotting the Sun's varying face.

Sunspots and faculae

To most of us, the Sun looks like a perfect, white-hot globe, smooth and without a blemish. However, detailed scrutiny indicates that the photosphere is often pitted with dark, ephemeral spots, called sunspots (Fig. 5.3). The largest spots

Focus 5.1 Order out of chaos

Convection currents are formed by hot gas rising and cool gas sinking. It occurs when a gas or fluid is heated from below. Examples include a boiling pot of water, currents in air heated by the Sun-baked ground, and the churning, visible disk of the Sun.

Such motions are described by nonlinear equations that relate several parameters, each changing with time. These variable quantities include the velocity, pressure, and density of the moving fluid or gas. The relevant equations connecting them are called the Navier–Stokes equations after the two mathematicians who discovered them in the mid-19th century.

In an important early study, Edward Lorenz solved a simple version of the convection equations using a computer, and examined the time variation of three variables. If just one parameter changed with time, you would portray it in a one-dimensional plot of its value as a function of time. To illustrate the variation of all three parameters, Lorenz used three axes, creating a diagram that resembles the wings of a butterfly and is now known as the Lorenz attractor (Fig. 5.2). At any instant of time, the three variables describe a point in the diagram. As time goes on, the points describe a trajectory that loops around forever, never intersecting itself.

The large-scale structure of the Lorenz attractor nevertheless describes the overall ordered pattern of convection. For example, the crossover from one wing of the butterfly to the other corresponds to a reversal in the direction of convection – hot gas rising to cool gas sinking or vice versa. Convection also always stays within the bounds of the diagram. The fine structure of the never-ending loops tells us a different story, displaying an infinite complexity that never exactly repeats itself.

can be many times bigger than the Earth. They can be seen with the unaided eye through fog, haze or cloud, when the Sun's usual brightness is heavily dimmed. (You cannot look directly at the Sun without severely damaging your eyes.) Ancient Chinese records indicate that such rare, large sunspots were seen at least 2000 years ago.

In Western culture, sunspots were discovered in 1610–11, when the newly-invented telescope was turned to the Sun. It enabled several persons to discern smaller sunspots that appear more frequently than the exceptional, largest ones that can be seen without the aid of a telescope. Johannes Fabricus, of Germany, the English mathematician and philosopher, Thomas Harriot, the Italian scientist, Galileo Galilei, and the Jesuit priest, Christoph Scheiner, also of Germany, all claimed to observe sunspots through telescopes sometime in late-1610 or in 1611. All four of them apparently made drawings of the changing size and shape of sunspots, and watched them move slowly across the visible face of the Sun as they rotate with it.

Sunspots appear dark because they are much cooler than their bright surroundings. The temperature at the center of a sunspot is about 2000 degrees kelvin lower than the surrounding gas in the photosphere, at 5780 K. While sunspots are dark, they still radiate light. A typical sunspot is about ten times brighter than the full Moon. So the appearance of sunspots is as deceiving as the seemingly "solid" photosphere.

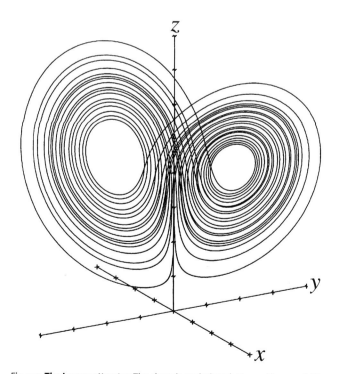

Fig. 5.2 **The Lorenz attractor** The changing relations between three variables are shown in this curve, with each point plotted in three dimensions using the three axes x, y and z. The trajectory loops about continuously and never intersects itself. It reveals the fine structure of convection that never exactly reproduces itself.

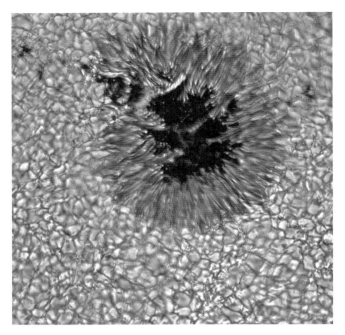

Fig. 5.3 **Spot on the Sun** This sunspot consists of a dark umbra inside a light, filamentary penumbra. It is about twice as large as the Earth, or about 24 million meters in diameter. Outside the penumbra the solar granulation is visible. This photograph was taken with exceptional angular resolution of 0.2 seconds of arc or 150 thousand meters at the Sun, using the National Solar Observatory's Vacuum Tower Telescope at the Sacramento Peak Observatory. (Courtesy of Thomas R. Rimmele, AURA, NOAO and NSF.)

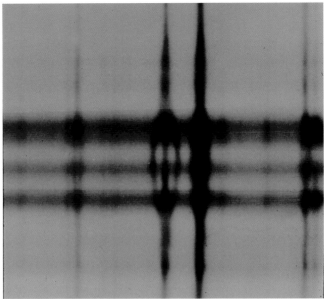

Fig. 5.4 **The Zeeman effect** In a sunspot, the spectral lines that are normally at a single wavelength become split into two or three components in the presence of a magnetic field, depending on the orientation of the field with respect to the line of sight. The separation of the outermost components is proportional to the strength of the magnetic field, in this sunspot about 0.4 T, or 4000 G. The components also have a circular or plane polarization, and the circular polarization, or orientation, indicates the direction, or polarity of the longitudinal magnetic field. (Courtesy of NOAO.)

Exceptionally bright patches are also found in the white-light photosphere. Called *faculae* – Latin for "little torches", they are associated with strong magnetic fields, both in the vicinity of sunspots and in smaller structures that are found all over the photosphere. Faculae are nevertheless usually only visible near the limb, or apparent edge, of the Sun.

Exceptionally long-lived equatorial sunspots move from east to west against the stars with a rotation period of about 25 days. Because our planet orbits the Sun in the same direction that the Sun rotates, the rotation rate observed from the Earth, called the synodic rotation period, is about two days longer than this sidereal rotation period. For at least a century, astronomers have known that sunspots at higher latitudes move at somewhat slower rates (see Section 4.4, Table 4.1). Helioseismologists have recently shown that this differential rotation persists through the convective zone (Section 4.4).

The sunspots appear and disappear, rising from inside the Sun and moving back into it. Most sunspots remain visible for only a few days, and some last just a few hours; others persist for weeks and even months. As sunspots form, they can also coalesce, and move past or even through each other. Sunspots have therefore long indicated that the apparently serene Sun is a dynamic place in a continual state of change.

Modern telescopes have revealed the detailed features of sunspots (Fig. 5.3). A simple sunspot has a dark center, called the umbra, surrounded by a lighter penumbra. The umbral

area is a constant fraction (0.17 ± 0.03) of the total area of the spot. A fully-developed sunspot has a typical penumbral diameter of between 20 and 60 million meters, which can be compared with the Earth's mean diameter of 12.7 million meters.

A large sunspot can have several umbrae within a single penumbra, and fibrils in the filamentary penumbra can extend nearly to the center of an umbra. The penumbra consists of radial bright and dark filaments that arch and splay out from the umbra, somewhat like a sea anemone.

Matter flows outward along these filaments, moving across the photosphere at velocities of a few thousand meters per second. This outflow is known as the Evershed effect, after John Evershed who first observed it in 1908 from the Kodaikanal Observatory in India. The material flow is inward and downward at higher levels just above the penumbra, and a predominantly downward flow is observed above the umbra.

If a sunspot is near the center of the solar disk, the penumbra has about the same width in all directions (Fig. 5.3). When the spot is near the edge, or limb, of the Sun, the portion of the penumbra closer to the disk center appears narrower than the region closer to the solar limb. This asymmetric distortion of sunspots observed near the solar limb is named the Wilson effect, after the Scottish astronomer Alexander Wilson who was among one of the first to observe it, in the 18th century. (The effect was independently noted at about the same time by the Würtemberg priest Ludwig Christoph Schülen.) It can be explained if the sunspot umbra is apparently depressed 0.4 to 0.6 million meters below the surrounding photosphere.

Small spots, called pores, contain a single dark umbra, and lack the surrounding penumbra. Pores are typically smaller than 2.5 million meters in diameter, and often persist for only 10 or 15 minutes before fading away inside the Sun. Neighboring pores can also merge to form a larger spot.

We can describe the complicated structures of sunspots, but that does not mean that they are fully understood. Scientists cannot describe exactly how sunspots form, remain cool and stable, and then disappear. All parts of a sunspot may be composed of fine-scale, transient features that have not yet been resolved. A complete understanding of sunspots will also involve descriptions of the way in which magnetic fields are spawned deep inside the Sun, breaking out though the photosphere into higher levels of the solar atmosphere.

Islands of magnetism

When placed in a strong external magnetic field, the electrons in an atom interact with the magnetic field and adjust their energy. This energy change produces a shift in the wavelength of a spectral line emitted during the electron's orbital transition. In effect, the orbiting electron gives rise to a tiny electric current with its own weak magnetism that points in one direction like a tiny compass. The interaction of the tiny internal magnetic field with the stronger external one results in an increase or decrease in the electron's energy, depending on the varying orientation of the two fields.

As a result, a spectral line is split into two components displaced to either side of the normal line wavelength seen without an external magnetic field (Fig. 5.4). This so-called Zeeman effect was predicted by Hendrick Lorentz, and measured in a terrestrial laboratory by Pieter Zeeman; the two Dutch physicists shared the 1902 Nobel Prize in Physics for these investigations of the influence of magnetism on radiation.

The Zeeman effect can be used to measure the magnetic field strength in the solar photosphere. The size of an atom's internal adjustments to an external magnetic field, and the extent of its spectral division, increase with the strength of the magnetism. So, the size of the line-splitting can be used to measure the magnetic field strength (Focus 5.2).

In 1908 the American astronomer George Ellery Hale used the Zeeman effect to show that sunspots contain magnetic fields that are thousands of times stronger than the Earth's magnetic field. The sunspot magnetic field strengths are as large as 0.3 T, or 3000 G; in comparison, the Earth's magnetic field which orients our compasses is about 0.00003 T, or 0.3 G, at the equator. Such an intense sunspot magnetic field, sometimes encompassing an area larger than our planet,

---

Focus 5.2 The Zeeman effect

When placed in an external magnetic field, a spectral line is split into components whose total separation increases with the strength of the magnetic field. We can understand this by considering the motion of a free electron in the presence of a magnetic field. The electron will circle or gyrate about the magnetic field with a gyroradius, R_g, and with a period, P, given by (see Section 7.3, Focus 7.4):

$$P = \frac{2\pi R_g}{V_\perp} = \frac{2\pi m_e}{eB},$$

where e and m_e respectively denote the charge and mass of the electron. At the velocity $V_\perp$, the electron goes once around the circumference $2\pi R_g$ in the period P. The frequency, ν_g, of this motion is

$$\nu_g = \frac{1}{P} = \frac{eB}{2\pi m_e} = 2.8 \times 10^{10} B \text{ Hz},$$

where B is the magnetic field strength in tesla.

When an atom is placed in a magnetic field, a very similar thing happens to its electrons and the spectral lines they emit. A line that radiates at a wavelength λ_0 without a magnetic field, becomes split into two or three components depending on the orientation of the magnetic field. The shift, $\Delta\lambda$, in wavelength of the two outer components from the central wavelength is given by:

$$\Delta\lambda = \frac{e}{4\pi c m_e} \lambda_0^2 B = 46.7 \lambda_0^2 B \text{ m},$$

where λ_0 is in meters, B is in tesla, and the total frequency shift, $\Delta\nu$, is given by:

$$\Delta\nu = \frac{2(\Delta\lambda)c}{\lambda^2} = \nu_g = 2.8 \times 10^{10} B \text{ Hz},$$

where the velocity of light is $c = 2.9974 \times 10^8 \text{ m s}^{-1}$.

If the magnetic field lines run across the line of sight, each line is split into three components. One of them, called the π component, is at the normal line wavelength seen when no magnetic field is present, and the other two, called the σ components, are displaced by an equal amount to either side of the π one, with a total separation given by $\Delta\lambda$. If the magnetic field lines are directed toward or away from the observer, only the two σ components are observed, with the separation $\Delta\lambda$, and the longitudinal magnetic field strength is measured.

The π component is linear, or plane, polarized, while the two σ components are circularly polarized with a sense that depends on the direction of the magnetic field. Longitudinal, or line-of-sight, magnetic fields that point out of the Sun have positive magnetic polarity, and the split σ components have right-hand circular polarization; to the observer the plane of polarization rotates in a clockwise sense. When the longitudinal magnetic field points into the Sun, with negative magnetic polarity, left-hand, counterclockwise circular polarization is observed.

---

requires an electrical current flow of a million million (10^{12}) amperes.

The magnetism is strongest at the center of a sunspot, within the umbra, where the magnetic field is nearly radial and pokes straight out of, or into, the Sun. The magnetic field weakens and is strongly inclined in the penumbra, where the magnetism spreads out in umbrella fashion. The penumbral filaments are aligned along the nearly horizontal magnetic fields in the penumbra. In the outer penumbra the magnetic field strength is about 0.1 T, and it steps down to 0.01 T at the sharp penumbral edge.

So, it is powerful magnetic fields that protrude to darken the skin of the Sun, forming dark, Earth-sized sunspots. The intense sunspot magnetism acts as a filter or valve, choking off the upward convection and outward flow of heat and energy from the solar interior. This keeps a sunspot thousands of degrees colder than the surrounding gas in the photosphere, like a giant refrigerator. Sunspots can also temporarily reduce the Sun's total radiation of the Earth by as much as 0.1 percent for minutes or days. [On longer times of months or years, the brightness increase due to faculae dominates the dimming in the solar irradiance of the Earth caused by sunspots (Section 8.4).]

Scientists were perplexed for decades over what holds sunspots together. The outward pressure of the strong magnetism ought to make sunspots expand and disperse, but they are held together by material that flows beneath them. Scientists have shown that the internal gas forces the magnetic fields together, concentrating and holding them within sunspots.

Magnetic fields are defined by lines of force, like those joining the north and south poles of a bar magnet or the Earth. These magnetic field lines point out of a south pole and into a north pole. When the lines are close together, the force of the field is strong; when they are far apart the force is weak.

The direction of the magnetic lines of force on the Sun can be inferred from the polarization of the Zeeman line-splitting. Normal sunlight is not polarized and has no preferred direction, but the components of the Zeeman effect radiate either about a circle or along a preferred plane. The two components displaced on either side of the normal line wavelength, called the σ components, are circularly polarized, while the third undisplaced π component is linear, or plane, polarized. When viewed along the magnetic field lines, that run parallel to the line of sight, only the σ components are visible. Astronomers then measure the longitudinal magnetic field strength from the size of the splitting, and the direction of circular polarization establishes if the magnetism is pointing directly into or out of the Sun. When viewed perpendicular to the magnetic field lines, that are running across the line of sight, all three components are visible and one can study the transverse field.

Nowadays, astronomers use magnetographs to portray the Sun's magnetism. These instruments consist of an array of tiny detectors that measure the Zeeman effect at different locations across the photosphere. Two images are usually produced, one in each circular polarization, and the difference of these images produces a magnetogram (Fig. 5.5). Strong magnetic fields show up as bright or dark regions, depending on their polarity; weaker ones are less

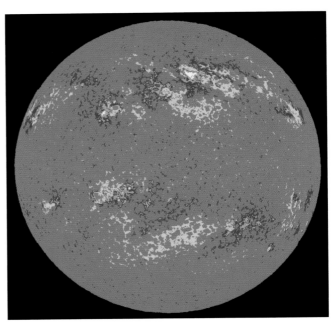

Fig. 5.5 **Magnetogram** This magnetogram was taken on 12 February 1989, close to sunspot maximum. Yellow represents positive or north polarity pointing out of the Sun, with red the strongest fields which are around sunspots; blue is negative or south polarity that points into the Sun, with green the strongest. In the northern hemisphere (*top half*) positive fields lead; in the southern hemisphere (*bottom half*) the polarities are exactly reversed and the negative fields lead. The Sun rotates east to west so that leading parts of active regions are to the right. (Courtesy of William C. Livingston, NSO and NOAO.)

bright or dark. These magnetograms display the longitudinal component of the magnetic field in the photosphere, or the component which is directed toward or away from us. They chart the magnetic fields running in and out of the photosphere.

The magnetograms show that sunspots normally occur in pairs of opposite magnetic polarity, with one directed out of the Sun and the other pointing into it. Thus, each pair of sunspots is a magnetic dipole, consisting of two magnetic poles, like the north and south poles of a bar magnet. The magnetic field lines emerge from a sunspot of one polarity, loop through the low solar atmosphere above it, and enter a neighboring sunspot of opposite polarity. These magnetic loops constrain the hot gas above pairs or groups of sunspots, and are among the brightest features seen in ultraviolet and X-ray images of the Sun (Section 5.3).

In addition to the longitudinal components, some modern instruments, called vector magnetographs, also measure the component of the magnetic field directed across our line of sight, the transverse component of the magnetic field. They provide a vector measurement of both the strength and direction of the magnetic field in the photosphere.

A vector magnetograph works by measuring the sense and degree of circular polarization in a spectral line and the direction and degree of the linear polarization in the same spectral line, often the magnetically sensitive line at a

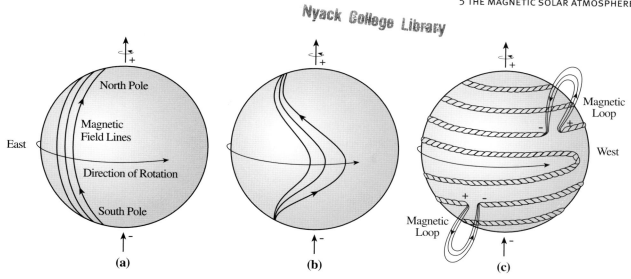

(a)   **(b)**   **(c)**

Fig. 5.6 **Winding up the field** A model for generating the orientation and polarity of the sunspot magnetic fields. At the beginning of the 11-year cycle of magnetic activity, when the number of sunspots is at a minimum, the magnetic field is the dipolar (poloidal) field seen at the poles of the Sun (a). The internal magnetic fields then run just below the photosphere from the south to north poles. As time proceeds, the highly-conductive, rotating material inside the Sun carries the magnetic field along and winds it up. Because the equatorial regions rotate at a faster rate than the polar ones, the internal magnetic fields are stretched out and wrapped around the Sun's center, becoming deformed into a partly toroidal field (b) and (c). At the time of activity maximum (c), active regions are formed in two belts, in the northern and southern hemispheres, respectively, and bipolar sunspot groups are created when the magnetic loops break through the photosphere.

wavelength of 525.02 nm. The circular polarization gives a measure of the longitudinal magnetic field, the strength of the field directed toward and away from the instrument. The linear polarization provides information on the strength and direction of the magnetic field transverse to, or across, the line of sight. Thus, the vector magnetograph measures the full vector and so returns three quantities: the size of the line-of-sight component, the size of the transverse or horizontal component, and the direction of the transverse component. This three-dimensional description of the magnetic field in the photosphere can be extrapolated into the overlying corona, and compared to radio measurements of the magnetism there.

Solar magnetism does not consist of just one simple dipole or even just a few of them. The magnetograms show that the Sun is spotted all over. There are big and small spots; strong and weak ones. But this magnetism is not distributed in a haphazard, chaotic way; it is highly organized.

Most of the sunspots occur in pairs or groups of opposite magnetic polarity, and they are usually oriented roughly parallel to the Sun's equator, in the east–west direction of the Sun's rotation. Moreover, all of the sunspot pairs in either the northern or southern hemisphere have the same orientation and polarity alignment, with an exactly opposite arrangement in the two hemispheres (Fig. 5.5).

The Sun has a general, weak dipole magnetic field with a north and south magnetic pole of perhaps 0.001 T in strength. If the north pole of the Sun has a positive magnetic polarity that points out of the Sun, then the westernmost leader spot of the pair in that hemisphere – the one that is ahead in the direction of rotation – will always be positive, the follower negative (Fig. 5.5). In the southern hemisphere, the polarities will be reversed. This orderly arrangement, known as Hale's law of polarity, is described by a simple model in which the magnetic field gets amplified, coiled, and wrapped around inside the Sun, looping through the photosphere to make the bipolar sunspot pairs (Fig. 5.6). Such a magnetic field is probably generated by a solar dynamo located at the base of the convective zone (Section 4.4).

The Sun's magnetic-activity cycle

The highly magnetized realm in, around and above bipolar sunspot pairs or groups is a disturbed area called an active region. Neighboring sunspots of opposite polarity are joined by magnetic loops that rise into the overlying atmosphere, so an active region consists mainly of sunspots and the magnetic loops that connect them.

Each magnetic loop in a solar active region acts as a barrier to electrically-charged particles. They move back and forth along the magnetic field lines but cannot move across them. Hot, electrified gas is therefore slaved to the magnetism, and becomes trapped within the closed magnetic loops where the gas emits bright ultraviolet and X-ray radiation (Section 5.3). Indeed, the magnetic loops in active regions usually dominate the Sun's X-ray, ultraviolet and radio radiation.

Active regions are never still, but instead continually alter their magnetic shape. The magnetic loops can rise up out of the solar interior, and submerge back within it in hours or days. But the active regions in which they develop may last for many weeks. They are the seats of profound change and unrest. The stressed magnetic fields build up magnetic energy that is waiting to be released, and the ongoing magnetic interaction can trigger sudden and catastrophic explosions, such as powerful solar flares (Section 6.1).

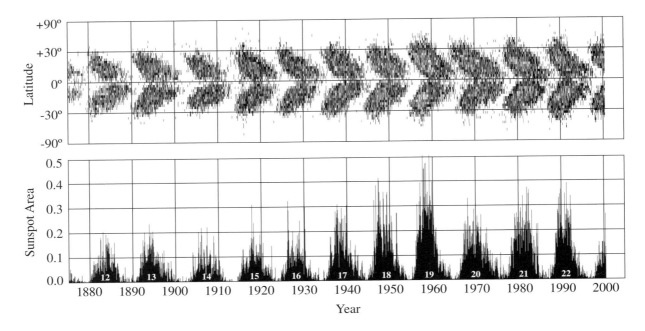

Fig. 5.7 **Sunspot cycle** The location of sunspots (*upper panel*) and their total area (*bottom panel*) have varied in an 11-year cycle for the past 120 years, but this activity cycle varies both in cycle length and maximum amplitude. As shown in the upper panel, the sunspots form at about 30 degrees latitude at the beginning of the cycle and then migrate to near the Sun's equator at the end of the cycle. Such an illustration is sometimes called a "butterfly diagram" because of its resemblance to the wings of a butterfly. The total area of the sunspots (*bottom panel*), given here as a percent of the visible hemisphere, follows a similar 11-year cycle. (Courtesy of David Hathaway, NASA/MSFC.)

All of this activity varies in step with the periodic 11-year change in the total number of sunspots. The existence of this sunspot cycle was first suggested in the early-1840s by Samuel Heinrich Schwabe, a pharmacist and amateur astronomer of Dessau, Germany, who diligently and meticulously observed the Sun for more than forty years, noting a decade-long variation in the total number of sunspots.

In 1848 Rudolf Wolf, a Swiss astronomer from Zurich, Switzerland, introduced a relative sunspot number, R, that could be compared for all observers, and initiated an international collaboration that has fully confirmed the sunspot cycle. The periodic variation in both the number and position of sunspots has now been carefully observed for more than a century (Fig. 5.7). Moreover, Wolf's reconstruction of the historical evidence indicated that the 11-year cycle in the number of sunspots has been present since 1700.

The Wolf sunspot number, R, given by $R = k (10g + s)$, helps to distinguish between the number, g, of sunspot groups counted on a specific day, and the number, s, of individual sunspots counted in all the groups on the same day. The k is a factor based on the estimated efficiency of observer and telescope.

The number of sunspots visible on the Sun varies from a maximum to a minimum and back to a maximum in a cycle that lasts about 11 years, but can vary between 10 and 12 years. At the maximum in the cycle, we may find 100 or more spots on the visible disk of the Sun at one time; at sunspot minimum very few of them are seen, and for periods as long as a month none can be found. The maximum number of sunspots in the cycle has also fluctuated by as much as a factor of two over the past century (Fig. 5.7), and the spots practically disappeared from the face of the Sun for a 70-year interval ending in 1715 (Section 8.5).

Since most forms of solar activity are magnetic in origin, they also follow an 11-year cycle. Thus, the sunspot cycle is also known as the solar cycle of magnetic activity. The number of active regions, with their energetic radiation and magnetized loops also varies from a maximum to a minimum and back to a maximum in about 11 years (Table 5.1).

The number and frequency of solar explosions, including powerful flares, are greatest at the maximum of the cycle; but they are not caused by sunspots. The explosive flares are powered by magnetic energy stored high in active regions above the sunspots (Section 6.3).

The positions of sunspots and their associated active regions also vary during the cycle (Fig. 5.7). Active regions form in two belts of activity, one north and one south of the solar equator. At the beginning of the sunspot cycle, when

Cycle number	Date of minimum	Date of maximum
20	1964.9	1968.9
21	1976.5	1979.9
22	1986.8	1989.6
23	1996.9	2000.5

[a] Courtesy of David H. Hathaway.

Table 5.1 **Dates of the minimum and maximum in solar activity since 1960**[a]

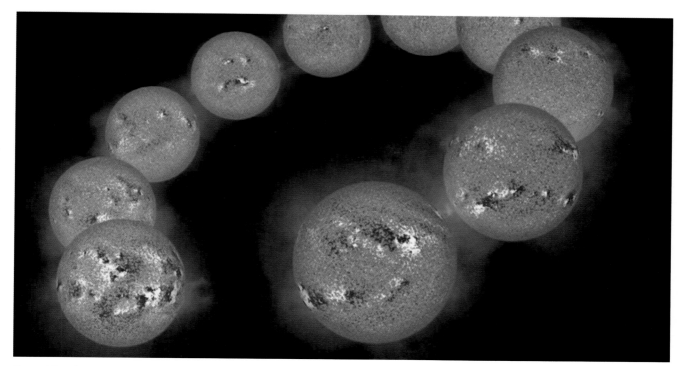

Fig. 5.8 **The solar magnetic field in time** These magnetograms portray the polarity and surface distribution of the magnetism in the solar photosphere. They were made with the Vacuum Tower Telescope of the National Solar Observatory at Kitt Peak from 8 January 1992, during the last maximum in the sunspot cycle (*lower left*) to 25 July 1999, well into the next maximum (*lower left*). They show opposite polarities as darker and brighter than the average tint. When the Sun is most active, the number of sunspots is at a maximum, and solar magnetism is dominated by large bipolar sunspots oriented in the east–west (*left–right*) direction within two parallel bands. At times of low activity (*top middle*), there are no large sunspots and tiny magnetic fields of different magnetic polarity can be observed all over the photosphere. The haze around the images is the inner solar corona. (Courtesy of Karel J. Schrijver, NSO, NOAO and NSF.)

solar activity rises to its maximum, the belts of activity appear about one-third of the way toward each pole. The active region belts gradually drift from these mid-latitudes to lower ones as the cycle progresses, reaching the Sun's equatorial regions at activity minimum. The active regions fizzle out and gradually disappear at sunspot minimum, just before coming together at the equator. The cycle then renews itself once more, and active regions emerge again about one-third of the way toward the poles.

As old spots linger near the equator, new ones break out at mid-latitudes, but the magnetic polarities of the new spots are reversed with north becoming south and *vice versa* – as if the Sun had turned itself inside out. During one 11-year cycle, the magnetic polarity of all the leading (westernmost) spots in the northern solar hemisphere is the same, and is opposite to that of leading spots in the southern hemisphere. The magnetic polarity of the leading spots reverses in each hemisphere at the beginning of the next 11-year cycle. The dipolar magnetic field at the solar poles also reverses at the beginning of the cycle, so the magnetic north pole switches to

a magnetic south pole and *vice versa* near the cycle maximum. Then, for the next 11 years, in the new cycle, all the polarities will be exchanged, including those of all the sunspots and that of the general polar magnetic field. Thus, the full magnetic cycle on the Sun takes an average of 22 years.

Magnetograms indicate that there is still plenty of magnetism at the minimum of the solar cycle of magnetic activity, when there are no large sunspots present. The magnetism then resides in a large number of very small regions spread all over the Sun (Fig. 5.8). The magnetic field averaged over vast areas of the Sun at activity minimum is only a few ten-thousandths of a tesla, or a few gauss, but the averaging process conceals a host of fields of small size and large strength. When the telescope resolution is increased, we find evidence of finer and finer magnetic fields with higher and higher field strengths.

The principal difference between magnetic fields in a solar active region and in the non-active quiet Sun is not the strength of the magnetic fields, but rather the size of the localized magnetic concentrations. When the activity cycle is at a minimum, the magnetism is concentrated in numerous small but intense regions, with magnetic field strengths comparable to those found in the much larger sunspots. These small, focused pockets of magnetism are constantly rising out of the photosphere, disappearing within it, and are then regenerated by internal motions and currents.

So, the magnetic fields are never smoothly distributed across the photosphere, either during activity minimum or maximum. Instead, they are highly concentrated, inhomogeneous, and everywhere clumped together into intense bundles that cover only a few percent of the photosphere's surface area.

5.2 The solar chromosphere

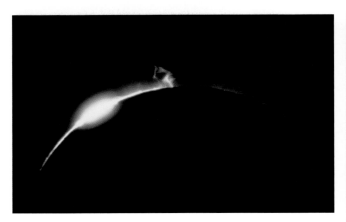

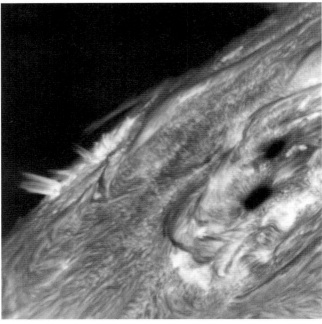

Fig. 5.9 **Chromosphere during eclipse** A reddened protuberance, called a prominence, provides a brief glimpse of the chromosphere during a total eclipse of the Sun on 11 July 1991. The red light is emitted by the abundant hydrogen atoms in the chromosphere at the hydrogen-alpha electron transition with a wavelength of 656.3 nm. (Courtesy of Dennis di Cicco, *Sky and Telescope*.)

Fig. 5.10 **Hydrogen-alpha chromosphere** This photograph of the Sun's chromosphere was made by tuning into the red line of atomic hydrogen, at the hydrogen-alpha line with a wavelength of 656.3 nm. Two round, dark sunspots, each about the size of the Earth, are present in the right half of this image, together with bright plages that mark highly-magnetized regions. (Courtesy of Victor Gaizauskas, obtained at the Ottawa River Solar Observatory, a facility operated by the Herzberg Institute of Astrophysics for the National Research Council of Canada.)

Just above the photosphere lies a relatively thin layer, about 2.5 million meters thick, called the chromosphere, from *chromos*, the Greek word for "color". The chromosphere is so faint, and the underlying photosphere so bright, that the chromosphere was first observed during a total eclipse of the Sun. It became visible a few seconds before and after the eclipse totality, creating a narrow, rose-colored band at the limb of the Sun, punctuated by extended curls of red (Fig. 5.9).

The Sun's temperature rises to about ten thousand degrees kelvin in the chromosphere, but the gas density in the chromosphere drops to roughly a million times less than that of the photosphere. Because of its very low density, the chromosphere does not create absorption lines. Instead of absorbing radiation, the tenuous gas is heated to incandescence and emits spectral lines. Whenever the light from the chromosphere is isolated, during a total solar eclipse or by other means, we see bright emission lines shining at precisely the same wavelengths as many dark absorption lines in the photosphere's light.

Studies of the Fraunhofer absorption lines in the spectrum of normal sunlight showed that hydrogen is by far the most-abundant element in the photosphere (Section 1.5). So, it is not surprising that the brightest emission line in the chromosphere is due to hydrogen. Indeed, the red color of the chromosphere is supplied by hydrogen atoms emitting at a single red wavelength called hydrogen alpha, at 656.3 nm. This is Fraunhofer's C line seen in emission rather than

absorption. The brightness of the chromosphere's hydrogen line confirms the great abundance of hydrogen in the solar atmosphere, for greater amounts of a substance tend to produce a brighter spectral line.

The second most-abundant element in the Sun is helium. This element was discovered during the solar eclipse on 18 August 1868, as an intense yellow emission line at a wavelength of 587.6 nm. It was probably not until the following year that the English solar physicist Norman Lockyer convinced himself that the yellow line could not be identified with any known terrestrial element, and coined the term "helium" after the Greek Sun god *Helios*. Helium was not found on the Earth until 1895 when William Ramsey detected the line in gases given off by a terrestrial mineral. Ramsey received the Nobel Prize in Chemistry in 1904 for his discovery of inert gaseous elements in the air, and his determination of their place in the periodic system.

In normal white light, the chromosphere is completely invisible against the blinding glare of the photosphere. By tuning into the red emission of hydrogen, and rejecting all

Fig. 5.11 **Spicules** Rows of dark spicules, or little spikes, are seen in the red hydrogen-alpha light of the solar chromosphere. A spicule is a short-lived (minutes) narrow jet of gas spurting out of the solar chromosphere at supersonic speeds to heights as great as 15 million meters. Spicules are a thousand times denser than the surrounding gas, so they are seen in absorption against the bright chromospheric background. They appear to cluster in "hedgerows" or "tufts". (Courtesy of NSO, NOAO and NSF.)

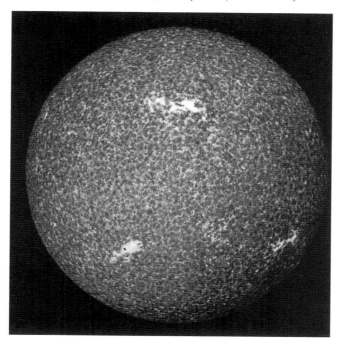

Fig. 5.12 **The Sun in the light of calcium ions** This global spectroheliogram of the Sun was taken in the light of singly-ionized calcium, or Ca II, at the core of the violet K line with a wavelength of 393.4 nm, on 02 June 1997. The emission outlines the chromospheric network where magnetic fields are concentrated, like the edges of the tiles in a mosaic. The brightest extended regions are called plages; they are dense places in the chromosphere found above sunspots or other active areas of the photosphere in regions of enhanced magnetic field. (Courtesy of the NSO, NOAO and NSF.)

the other wavelengths or colors of sunlight, astronomers can separate the light of the chromosphere. This makes it visible at all daylight times rather than during the few fleeting moments of infrequent total solar eclipses. This was initially realized in 1868 when both Lockyer and Jules Janssen, Director of the Meudon Observatory near Paris, independently showed that a spectroscope could be used to observe the red rim at the Sun's edge outside of an eclipse.

Then, building upon this result, the American astronomer George Ellery Hale invented, in 1890–92, an instrument named the spectroheliograph that can photograph the entire solar disk in just one color or wavelength. By isolating the bright red emission line of hydrogen atoms, Hale provided a new perspective in which the chromosphere can be observed across the face of the Sun on any clear day. Solar astronomers have continued to monitor the chromosphere in this way for more than a century, studying the fascinating features that come into view.

Dark regions and bright plages

Sunspots extend from the photosphere into the chromosphere, creating dark regions in hydrogen-alpha photographs (Fig. 5.10). Bright regions, called *plages* from the French word for "beaches", glow in hydrogen light; they are often located near sunspots in places with intense magnetism. The plages are a chromospheric phenomena detected in monochromatic hydrogen-alpha light; they are associated with, and often confused with, bright patches in the photosphere, called faculae, that are seen near the solar limb in white light.

Spicules

The chromosphere is jagged, irregular and by no means smooth or homogeneous. When observed in hydrogen alpha, numerous thin, luminous extensions, dubbed spicules by Father Angelo Secchi, may be seen (Fig. 5.11). They rise and fall like chopping waves on the sea or a prairie fire of burning, wind-blown grass. The needle-shaped spicules are about two thousand meters in width, and shoot up to heights of 15 million meters at speeds of about 20 thousand meters per second. Individual spicules persist for only five or ten minutes, but new ones continuously arise as old ones fade away. Approximately half-a-million of the evanescent, flame-like spicules are dancing in the chromosphere at any given moment.

Calcium network

We can also study the chromosphere across the visible solar disk in the violet light of calcium ions. Designated Ca II, the calcium is singly ionized, so the calcium atoms are missing one electron. They emit radiation at wavelengths of 393.4 and 396.8 nm, and are often called the calcium H and K lines

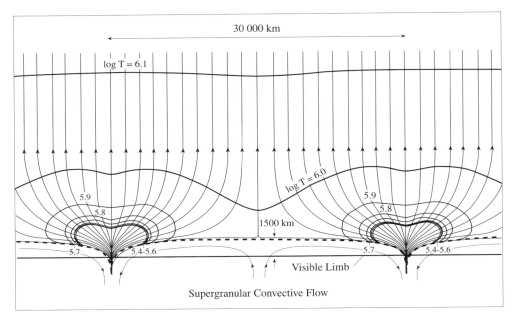

Fig. 5.13 **Magnetic concentration** A two-dimensional, radial cross-section of the magnetic-network model of the solar transition region. The motion of supergranular convective cells (*bottom*) concentrates magnetic fields at their boundaries in the photosphere. The magnetic fields (*arrowed lines*) are pushed together and amplified up to 0.1 tesla at the cell edges. Heating in the chromosphere above this magnetic network produces bright calcium emission (Fig. 5.12). The concentrated magnetic fields expand and flare out with height in the overlying corona. Temperature contours between log T = 6.1 (*corona*) and log T = 5.4 (*upper transition region*) are marked.

after Fraunhofer's designation of the corresponding absorption lines in the underlying photosphere.

We obtain a completely different view when the chromosphere is pictured in a calcium emission line (Fig. 5.12). Bright regions of calcium light correspond to places where there are strong magnetic fields, both above sunspots and all over the Sun in a network of magnetism.

The calcium, or magnetic, network, coincides with the pattern formed by large-scale convective cells, known as the supergranulation, each about 35 million meters in diameter or 2.5 times the diameter of the Earth (Section 4.1). The giant cells move across the photosphere, carrying the magnetic fields with them. Each supergranulation cell sweeps the magnetic fields to its outer edges, where the field collects and strengthens (Fig. 5.13). Chromospheric heating is produced above these field concentrations, resulting in the bright calcium emission that outlines the magnetic network (Fig. 5.12).

Filaments and prominences

The hydrogen-alpha photographs of the chromosphere reveal massive loops of cool dense gas that arch up over the photosphere. Seen from above, they are elongated, dark features, called filaments, that stretch up to halfway across the face of the Sun (Fig. 5.14). The cool gas looks dark against the brightness of the hot Sun beneath it. When seen from the side at the edge of the solar disk, where the chromosphere extends beyond the lowest layers of the Sun's atmosphere, these same features light up as bright loops, called prominences, against the dark background. They can be detected during a solar eclipse as large curling pink protuberances that extend beyond the Moon's edge, hence the name *prominence* from the French word for "protuberance".

So, filaments and prominences are two words that describe different perspectives of the same thing. They are dark sinuous filaments that snake across the solar disk, and bright arching or looping prominences at the solar rim. There are two main types of prominences: quiescent and active.

Fig. 5.14 **The Sun in the light of hydrogen atoms** This global image of the Sun was taken in the light of hydrogen atoms, at the red hydrogen-alpha transition with a wavelength of 656.3 nm, on 11 August 1980. Small, dark regions above sunspots, bright active regions and plages, and dark string-like filaments are seen on the visible disk; bright prominences extend outward over the disk edge or limb. (Courtesy of the Space Environment Center, National Oceanic and Atmospheric Administration (NOAA) under partial support from the United States Air Force.)

Fig. 5.15 **Active prominence** When seen at the edge of the Sun, a filament takes the name prominence. This bright loop prominence outlines magnetic fields above sunspots; the loop is big enough to swallow the Earth. The photograph was taken in the green line of ionized iron, designated Fe XIV. (Courtesy of NSO, NOAO, and NSF.)

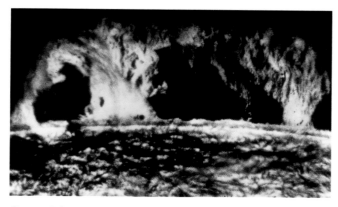

Fig. 5.16 **Quiescent prominence** Dense, cool gas, seen in the red light of hydrogen alpha at the rim of the Sun, outlines magnetic arches that are silhouetted against the dark background. The prominence material, appearing as a flaming curtain up to 65 million meters above the photosphere, is probably injected into the base of the magnetic loops in the chromosphere. (Courtesy of the Big Bear Solar Observatory.)

Active prominences lie along the polarity inversion line of strong magnetic fields connected to sunspots within active regions. Active prominences are dynamic structures with violent motions; they have lifetimes of only minutes or hours. There are various types, such as surges, sprays and loop prominences (Fig. 5.15).

Quiescent prominences are long, thin, vertical sheets of dense plasma, with a characteristic width of about 5 million meters and length of 100 million meters. They can extend tens and even hundreds of millions of meters above the edge of the Sun (Fig. 5.16). Some of them are big enough to girdle the Earth or even to stretch from the Earth to the Moon. Quiescent prominences are exceedingly stable structures that can last for many months. They lie along the magnetic neutral lines of weak bipolar magnetic regions of the solar photosphere, and can form within the cavity below a coronal streamer (Section 5.3).

The quiescent prominences are hundreds of times cooler and denser than the surrounding material, with temperatures between 5000 and 15 000 K and densities of 10^{16} to 10^{17} hydrogen atoms per cubic meter. The mass contained within a filament, or prominence, is billions of tons, or trillions of kilograms, comparable to that in some terrestrial mountains. All of this enormous mass is supported against the downward pull of gravity by powerful magnetic fields.

A quiescent filament, or prominence, can hang and seemingly float almost motionless above the photosphere for weeks or months at a time, never falling down; but sometimes the magnetism disconnects and the material flows out into space (Section 6.2). Moreover, it is all held together, suspended, and embedded within a tenuous, hot, million-degree gas called the corona.

5.3 The solar corona – loops, holes and unexpected heat

Total eclipse of the Sun

Still higher, above the chromosphere, is the *corona*, from the Latin word for "crown". The corona becomes momentarily visible to the unaided eye when the Sun's bright disk is blocked out, or eclipsed, by the Moon and it becomes dark during the day. During such a total solar eclipse, the corona is seen at the limb, or apparent edge, of the Sun, against the blackened sky as a faint, shimmering halo of pearl-white light (Fig. 5.17). But be careful if you go watch an eclipse, for the light of the corona is still very hazardous to human eyes and should not be viewed directly.

A total eclipse of the Sun occurs when the Moon passes between the Earth and the Sun, and the Moon's shadow falls on the Earth. In an incredible cosmic coincidence, the Moon is just the right size and distance to blot out the bright photosphere when properly aligned and viewed from the Earth. In other words, the apparent angular diameters of the

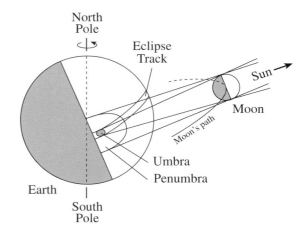

Fig. 5.18 **Solar eclipse** During a total solar eclipse, the Moon casts its shadow upon the Earth. No portion of the Sun's photosphere can be seen from the umbral region of the Moon's shadow (*small gray spot*), but the Sun's light is only partially blocked in the penumbral region (*larger half-circle*). A total solar eclipse, observable only from the umbral region, traces a narrow path across the Earth's surface.

Moon and the visible solar disk are almost exactly the same, about 30 minutes of arc, so that under favorable circumstances the Moon's shadow can reach the Earth and cut off the light of the photosphere.

A total eclipse does not happen very often. Since the Moon and the Earth move along different orbits whose planes are inclined to each other, the Moon only passes directly between the Earth and the Sun about three times every decade on average. Even then, a total eclipse occurs along a relatively narrow region of the Earth's surface, where the tip of the Moon's shadow touches the Earth (Fig. 5.18). At other nearby places on the Earth, the Sun will be partially eclipsed, and at more remote locations you cannot see any eclipse.

The Moon's orbital motion carries its shadow rapidly eastward across the ground at about 1.6 million meters per hour. As a result, the longest total eclipse observed at a fixed point on the ground lasts just under eight minutes. The dates, duration of totality, and path of totality for the total eclipses of the Sun during the first half of the 21st century are given in Table 5.2.

If the Moon is at a distant part of its orbit at the time of eclipse, the Moon appears smaller than the Sun, and the tip of the Moon's shadow does not quite reach the Earth. The bright ring of the Sun's disk is then seen around the edge of the Moon. This is an annular eclipse and it has none of the darkness and excitement of a total eclipse of the Sun. An annular eclipse of the Sun is similar to a partial eclipse,

Fig. 5.17 **Gossamer corona** The Sun's corona as photographed during the total solar eclipse of 26 February 1998, observed from Oranjestad, Aruba. To extract this much coronal detail, several individual images, made with different exposure times, were combined and processed electronically in a computer. The resultant composite image shows the solar corona approximately as it appears to the human eye during totality. Note the fine rays and helmet streamers that extend far from the Sun and correspond to a wide range of brightness. (Courtesy of Fred Espenak.)

Date	Maximum duration (minutes)	Path of totality
21 June 2001	4.95	Atlantic Ocean, Angola, Zambia, Zimbabwe, Mozambique, Madagascar
04 December 2002	2.07	Angola, Botswana, Zambia, Zimbabwe, South Africa, Mozambique, south Indian Ocean, southern Australia
23 November 2003	1.99	Antarctica
08 April 2005	0.70	Eastern Pacific Ocean
29 March 2006	4.12	Eastern Brazil, Atlantic Ocean, Ghana, Togo, Benin, Nigeria, Niger, Chad, Libya, Egypt, Turkey, Russia
01 August 2008	2.45	Northern Canada, Greenland, Arctic Ocean, Russia, Mongolia, China
22 July 2009	6.65	India, Nepal, Bhutan, Burma, China, Pacific Ocean
11 July 2010	5.33	South Pacific Ocean, Easter Island, Chile, Argentina
13 November 2012	4.03	Northern Australia, south Pacific Ocean
03 November 2013	1.67	Atlantic Ocean, Gabon, Congo, Zaire, Uganda, Kenya
20 March 2015	2.78	North Atlantic Ocean, Faeroe Islands, Arctic Oceans, Svalbard
09 March 2016	4.17	Indonesia (Sumatra, Borneo, Sulawesi, Halmahera), Pacific Ocean
21 August 2017	2.67	Pacific Ocean, United States (Oregon, Idaho, Wyoming, Nebraska, Missouri, Illinois, Kentucky, Tennessee, North Carolina, South Carolina), Atlantic Ocean
02 July 2019	4.55	South Pacific Ocean, Chile, Argentina
14 December 2020	2.17	Pacific Ocean, Chile, Argentina, south Atlantic Ocean
04 December 2021	1.92	Antarctica
20 April 2023	1.27	South Indian Ocean, Western Australia, Indonesia, Pacific Ocean
08 April 2024	4.47	Pacific Ocean, Mexico, United States (Texas, Oklahoma, Arkansas, Missouri, Kentucky, Illinois, Indiana, Ohio, Pennsylvania, New York, Vermont, New Hampshire, Maine), southeastern Canada, Atlantic Ocean
12 August 2026	2.32	Greenland, Iceland, Spain
02 August 2027	6.38	Atlantic Ocean, Morocco, Spain, Algeria, Libya, Egypt, Saudi Arabia, Yemen, Somalia
22 July 2028	5.17	South Indian Ocean, Australia, New Zealand
25 November 2030	3.73	South West Africa, Botswana, South Africa, south Indian Ocean, southeastern Australia
14 November 2031	1.13	Central Pacific Ocean
30 March 2033	2.62	Alaska, Arctic Ocean
20 March 2034	4.17	Atlantic Ocean, Nigeria, Cameroon, Chad, Sudan, Egypt, Saudi Arabia, Iran, Afghanistan, Pakistan, India, China
02 September 2035	2.90	China, Korea, Japan, Pacific Ocean
13 July 2037	3.98	Australia, New Zealand, south Pacific Ocean
25-26 December 2038	2.30	Indian Ocean, western and southern Australia, New Zealand, Pacific Ocean
15 December 2039	1.85	Antarctica
30 April 2041	1.85	South Atlantic Ocean, Angola, Congo, Uganda, Kenya, Somalia
20 April 2042	4.85	Indonesia, Malaysia, Philippines, Pacific Ocean
23 August 2044	2.08	Greenland, Canada, United States (Montana, North Dakota)
12 August 2045	6.01	United States (California, Nevada, Utah, Colorado, Kansas, Oklahoma, Arkansas, Mississippi, Alabama, Georgia, Florida), Haiti, Dominican Republic, Venezuela, Guyana, Suriname, French Guyana, Brazil
02 August 2046	4.85	Eastern Brazil, Atlantic Ocean, Namibia, Botswana, South Africa, south Indian Ocean
05 December 2048	3.47	South Pacific Ocean, Chile, Argentina, south Atlantic Ocean, Namibia, Botswana
25 November 2049	0.63	Indian Ocean, Indonesia
20 May 2050	0.37	South Pacific Ocean
30 March 2052	4.13	Pacific Ocean, Mexico, United States (Louisiana, Alabama, Florida, Georgia, South Carolina), Atlantic Ocean

[a] Adapted from *Totality Eclipses of the Sun*, second edition, by M. Littmann, K. Wilcox and F. Espenak, New York: Oxford University Press, 1999.

Table 5.2 **Total eclipses of the Sun in the first half of the 21st century**[a]

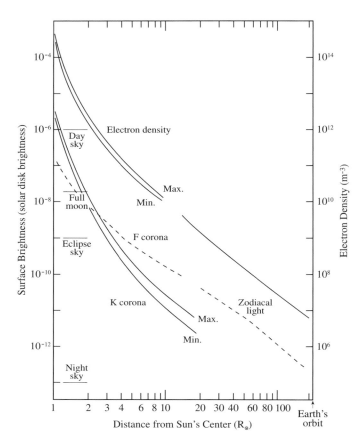

Fig. 5.19 **Corona brightness and electron density** Variation of coronal surface brightness (*left scale*) and electron density (*right scale*) with distance from the Sun's center, in units of the Sun's radius, $R_\odot = 6.96 \times 10^8$ m. The maximum and minimum of the solar magnetic-activity cycle are designated by Max. and Min., respectively. The F corona values (*dashed line*) are continuous with the zodiacal light. For comparison, the surface brightness for the full Moon, clear sky for day and night, and the clear sky during a total solar eclipse are indicated.

because the Sun is still very bright. In contrast, during a total eclipse the main body of the Sun is hidden from view and we can glimpse the ghostly light from the outer layers of the Sun.

The corons is very faint. Its light is only about one-millionth as intense as the light from the photosphere. Thus it is no surprise that the corona is only visible to the unaided eye during a total solar eclipse.

The low corona, that is close to the photosphere, shines by visible sunlight scattered by electrons there. This electron-scattered component of the corona's white light has been named the K corona. It emits a continuous spectrum without absorption lines, and the K comes from the German *Kontinuum*, or *Kontinuierlich*.

The amount of observed coronal light is proportional to the electron density integrated along the line of sight, so we can use observations of the K corona to infer the density of electrons there (Fig. 5.19). At the base of the corona there are almost a million billion (10^{15}) electrons per cubic meter. These coronal electrons are so tenuous and rarefied that a million billion cubic meters would only weigh one kilogram. Since protons are 1836 times more massive than

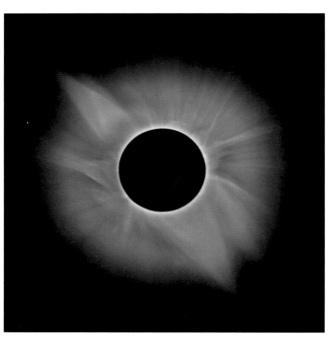

Fig. 5.20 **Helmet streamers** The Sun's corona photographed in white light during the solar eclipse on 11 July 1991; it extended several solar radii and had numerous fine rays as well as larger helmet streamers. (Courtesy of Shigemi Numazawa, Niigata, Japan.)

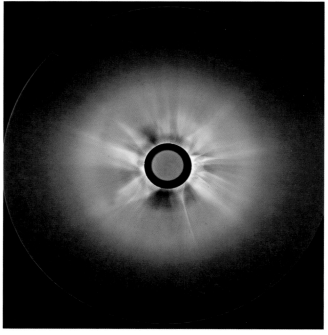

Fig. 5.21 **Corona at sunspot maximum** The dim, ghostly light of the Sun's outer atmosphere, the corona, during a total eclipse on 16 February 1980, observed from Yellapur, India. This was a time near sunspot maximum, when the sunspots are most numerous. The bright helmet streamers were then distributed about the entire solar limb, resembling the petals on a flower. The stalks of the streamers stretch out four billion meters, or about six solar radii from the center of the Sun in this photograph. It was taken through a radially graded filter to compensate for the sharp decrease in electron density and coronal brightness with distance from the Sun. (Courtesy of Johannes Dürst and Antoine Zelenka, Swiss Federal Observatory.)

Focus 5.3 Magnetic and gas pressure

Throughout the solar atmosphere, a dynamic tension is set up between the charged particles and the magnetic field. In the photosphere and below, the gas pressure dominates the magnetic pressure, allowing the magnetic field to be carried around by the moving gas. Because the churning gases are ionized, and hence electrically conductive, they sweep the magnetic field along.

The gas density and pressure decrease very rapidly above the photosphere, whereas the magnetic pressure falls off less rapidly. So, the magnetic pressure greatly exceeds the gas pressure in the low corona, less than a million miles above the photosphere. Here strong magnetism wins - the hot particles are confined within coronal loops, and the magnetic field shapes the gas into the structures we observe. Nevertheless, the coronal loops are themselves tied into the underlying photosphere which is stirred up by mass motions. Moreover, the gas pressure dominates the situation once again further out in the corona, where the magnetic field decreases in strength.

A magnetic field tends to restrain a collection of electrons and protons, called a plasma, while the plasma exerts a pressure that opposes this. The pressure, P_m, produced by a magnetic field transverse to its direction is given by:

$$P_m = B^2 / (2\mu_0),$$

for a magnetic field of strength B in tesla, and the permeability of free space $\mu_0 = 4\pi \times 10^{-7}$ N A^{-2} (newtons per ampere squared). As expected, a stronger magnetic field applies a greater restraining pressure.

A hot plasma generates a gas pressure, P_g, owing to the motions of its particles. It is described by the expression

$$P_g = N kT,$$

where N is the particle number density, $k = 1.38 \times 10^{-23}$ J K^{-1} is Boltzmann's constant, and T is the temperature. Hotter particles move faster and create greater pressure to oppose the magnetic field, and a denser plasma also results in greater pressure.

The two kinds of pressure compete for control of the solar atmosphere. In the low solar corona, strong magnetic fields in active regions hold the hot, dense electrified gas within coronal loops. The magnetic and gas pressure become equal for a magnetic field, B, given by:

$$B = [(2\mu_0 k) NT]^{1/2} = [3.47 \times 10^{-29} NT]^{1/2} \text{ tesla (abbreviated T)}.$$

In order to contain the hot, dense plasma of a coronal loop, with $N = 10^{17}$ electrons per cubic meter and $T = 10^6$ K, the magnetic field must be stronger than $B = 0.002$ tesla $= 20$ gauss. The magnetic field strengths of coronal loops in active regions are therefore strong enough to hold this gas in, at least within the low corona near sunspots.

In contrast, coronal holes have relatively weak magnetic fields, and the normally constraining magnetic forces relax and open up, allowing the unencumbered outward flow of charged particles into interplanetary space. Far from the Sun, the magnetic fields of coronal loops also become too weak to constrain the outward pressure of the hot gas, and the loops expand or break open to allow electrons and protons to escape, contributing to the solar wind and carrying the magnetic fields away (see Section 7.2). Within the solar wind, the gas pressure of the electrons and protons is roughly equal to the magnetic pressure of the interplanetary magnetic field.

---

electrons, they supply most of the corona's mass. Because the corona is derived from the Sun's abundant hydrogen, there is one proton for every electron in the hot gas. The mass density in the low corona is about 10^{-12} kilograms per cubic meter.

The F corona is caused by sunlight scattered from solid dust particles in interplantary space. Both the K and F components of the corona decrease in intensity with increasing distance from the Sun, and beyond 2.5 solar radii from the Sun's center the F component is more intense than the K corona (see Fig. 5.19). Unlike the K corona, the spectrum of the F corona includes dark Fraunhofer absorption lines, so the F stands for Fraunhofer. The faint light of the F corona is not polarized, with any preferred direction, but the K corona is, so polarization is another way to distinguish between the two components.

When we look at the corona during a solar eclipse, we are seeing patterns of free electrons made visible because they scatter the light that strikes them, like motes in a sunbeam. And because these electrons are constrained and molded by magnetic fields, the shape of the corona reflects that of the magnetic fields. This is because the magnetic pressure of the magnetic loops exceeds the gas pressure in the low corona (Focus 5.2).

It is the other way around in the photosphere, where the gas pressure exceeds the magnetic pressure, so the gas can concentrate and transport the field. Motions in the photosphere therefore push the magnetic field around, producing changes in the corona above it.

White-light coronal photographs show that the electrons can be confined within helmet streamers (Fig. 5.20), which are peaked like old-fashioned, spiked helmets once fashionable in Europe. At the base of helmet streamers, electrified matter is densely concentrated within magnetized loops rooted in the photosphere. Further out in the corona, the streamers narrow into long stalks that stretch tens of billions of meters into space. These extensions confine material at temperatures of about two million degrees kelvin within their elongated magnetic boundaries.

The shape of the corona changes, depending on the amount of solar magnetic activity. At times of reduced activity, near a minimum in the 11-year activity cycle, the corona is relatively dim at the poles where faint plumes diverge out into interplanetary space, apparently outlining a global, dipolar magnetic field of about 0.001 tesla, or 10 gauss, in strength. This large-scale magnetism becomes pulled outward at the solar equator, confining hot material

Fig. 5.22 **Coronal streamers and comet** A coronagraph on *SOHO* shows bright helmet streamers near the Sun's equatorial regions (*right and left*) near the minimum in the Sun's magnetic-activity cycle. The edge of the visible solar disk, or photosphere, is indicated by the inner white circle, and the outer full circle marks the edge of the occulting disk of the coronagraph. A comet is also shown (*bottom left*) during its fiery plunge into the Sun; more than 200 Sun-grazing comets have been discovered with this instrument. (Courtesy of the *SOHO* LASCO consortium. *SOHO* is a project of international collaboration between ESA and NASA.)

in the equatorial regions. The bottoms of the coronal streamers can then straddle the Sun's equator, and the streamer stalks tend to point along the equatorial plane, forming a ring or belt of hot, dense gas that extends from the Sun. The streamers are sandwiched between regions of opposite magnetic direction or polarity, and are confined along an equatorial current sheet that is magnetically neutral.

Near the maximum in the activity cycle, the shape of the corona and the distribution of the Sun's extended magnetism can be much more complex. The corona then becomes crowded with streamers that can be found close to the Sun's poles (Fig. 5.21). At times of maximum magnetic activity, the width and radial extension of a streamer are smaller and shorter than at activity minimum. Near solar maximum, the global dipolar magnetic field of the Sun swaps its north and south magnetic poles, so a much more volatile corona can exist then.

Natural eclipses of the Sun occur every few years, and can then be seen from only a few, often remote, places on the globe. So, scientists decided to make their own artificial eclipses by putting occulting disks in their telescopes to mask the Sun's face and block out the photosphere's intense glare. Such instruments are called coronagraphs, since they let us see the corona. The first coronagraph was developed in 1930 by the French astronomer Bernard Lyot, and the corona is now routinely observed with coronagraphs at mountain sites.

As Lyot realized, coronagraph observations are limited by the bright sky to high-altitude sites where the thin, dust-free air scatters less sunlight. He therefore installed one at the Pic du Midi observatory in the Pyrenes. The higher and clearer the air, the darker the sky, and the better we can detect the faint corona around the miniature "moon" in the coronagraph. They work best in space, where almost no air is left. Modern solar satellites, such as the *SOlar and Heliospheric Observatory* or *SOHO*, use coronagraphs to get clear, edge-on views of the corona from outside our atmosphere (Fig. 5.22). Such satellites also use ultraviolet and X-ray telescopes to view the low corona across the face of the Sun, a development that followed the realization of the corona's million-degree temperature.

The corona's searing heat

The solar corona defies expectations, for it is hundreds of times hotter than the underlying photosphere, which is closer to the Sun's energy-generating core. The Sun's temperature rises to more than one million degrees kelvin just above the photosphere which is at a temperature of

Wavelength (nm)	Emitting ion	Wavelength (nm)	Emitting ion
332.8	Ca XII	530.3 (green line)	Fe XIV
338.8	Fe XIII	569.4 (yellow line)	Ca XV
360.1	Ni XVI	637.5 (red line)	Fe X
398.7	Fe XI	670.2	Ni XV
408.6	Ca XIII	706.0	Fe XV
423.1	Ni XII	789.1	Fe XI
511.6	Ni XIII	802.4	Ni XV

[a] Ca = calcium, Fe = iron, and Ni = nickel. Subtract one from the Roman numeral to get the number of missing electrons.

Table 5.3 **Intense coronal forbidden emission lines at optical wavelengths**[a]

5780 K. Heat simply should not flow outward from a cooler to a hotter region. It violates the laws of thermodynamics, the branch of physics that deals with the movement and transfer of heat. These laws indicate that it is physically impossible to transfer thermal energy by conduction from the underlying photosphere to the much hotter corona. The high temperature of the corona also defies common sense; after all, when you sit farther away from a fire it becomes colder, not hotter. It is as if a cup of coffee was put on a cold table and suddenly began to boil.

The corona's unexpected heat was suggested by the identification of emission lines, first observed during solar eclipses more than a century ago. During the solar eclipse of 7 August 1869, both Charles Young and William Harkness found that the spectrum of the solar corona is characterized by a conspicuous green emission line at a wavelength of 530.3 nm. Within half-a-century, eclipse observers had detected at least ten coronal emission lines and the number doubled in a few decades more, but not one of them had been convincingly explained.

Since none of the coronal features had been observed to come from terrestrial substances, astronomers concluded that the solar corona consisted of some mysterious ingredient, which they named "coronium". Belief in the new element lingered for many years, until it was realized that the coronal spectral lines are emitted by perfectly ordinary elements known on Earth, but in an unanticipated high-temperature state.

The solution to the coronium puzzle was provided by Walter Grotrian, of Potsdam, and the Swedish spectroscopist, Bengt Edlén, in 1939 and 1941 respectively. They attributed the coronal emission lines to terrestrial elements in an astonishingly high degree of ionization (Table 5.3). The atoms in the corona are stripped of a large number of electrons; the green line was, for example, attributed to iron atoms missing 13 electrons.

This meant that the corona had to be at an unexpectedly hot temperature. The coronal gas particles would have to be moving very fast, with searing temperatures of millions of degrees, to have enough energy to tear off so many electrons. The corona's high temperature was confirmed in 1946 by its radio emission, whose intensity corresponded to a temperature of a million degrees (Focus 5.4).

Spectral lines come from atoms and ions emitting or absorbing light when their electrons undergo an orbital transition. Since the atoms or ions have different numbers and arrangements of electrons, they undergo different transitions and produce different spectral lines. As it turned out, the coronal emission lines were due to relatively rare and highly improbable "forbidden" transitions. Collisions between the excited ions and electrons set free from atoms keep these transitions from happening in the terrestrial laboratory, so they are forbidden there.

The fact that the forbidden emission is observed in the corona implies not only an extremely high temperature but also an extraordinarily low electron density. Otherwise any ion capable of emitting the forbidden lines would be de-excited by free electron collisions before doing so. The electron density in the low corona, of up to a million billion (10^{15}) electrons per cubic meter, is in fact many millions of times less than a vacuum on Earth. The corona is also one hundred times more tenuous than the chromosphere, and an additional million times more rarefied than the photosphere (10^{23} particles per cubic meter).

We might expect that the rapidly thinning gas above the photosphere would cool with increasing distance from the Sun. For a while that is true, for the temperature drops from 5780 K to a minimum value of about 4400 K just below the chromosphere. But thereafter the drop in density with increasing distance from the Sun is matched by a dramatic increase in temperature.

The linkage between the chromosphere and the corona occurs in a very thin transition region, less than 100 thousand meters thick, where both the density and temperature change abruptly (Fig. 5.23). In the transition region, the temperature shoots up from ten thousand to more than a million degrees kelvin, but the density decreases as the temperature increases in such a way as to keep the gas pressure spatially constant. The corona then thins out and slowly cools with increasing distance from the Sun.

The hot corona extends to the Earth and beyond, and it remains very hot just outside our planet. The free electrons in the ionized gas are not attached to atoms, and they can therefore conduct the corona's intense heat far into space. A metal pan similarly becomes hot all over when it is heated from below on a stove. At the Earth's orbit, the coronal electrons have a mean temperature of about 0.14 million degrees kelvin.

If it is so hot just outside the Earth, why don't astronauts burn up when they venture outside their spacecraft? By the time that the corona has reached the Earth, its expansion into an ever greater volume has reduced the density to just

Focus 5.4 The non-flaring radio Sun and coronal magnetic fields

Soon after the end of World War II, Australian scientists, led by Joseph L. Pawsey, used war-surplus radar (radio detection and ranging) equipment to monitor the Sun's radio emission at a wavelength of 1.5 meters, showing that its intensity, though highly variable, almost never fell below a threshold. At least some of the variable component was due to explosive solar flares (Section 6.1), and the steady, non-varying threshold was attributed to electrons in the million-degree corona. That is, the brightness temperature derived from the observed flux density of the non-flaring radiation at a wavelength of 1.5 meters was approximately a million degrees.

The brightness temperature, T_B, of a source with radius, R, observed at a distance, D, is:

$$T_B = \frac{S\lambda^2}{2\pi k}\left(\frac{D}{R}\right)^2,$$

for a flux density, S, observed at a wavelength, λ and Boltzmann's constant $k = 1.380\,66 \times 10^{-23}$ J K^{-1}. This is known as the Rayleigh-Jeans expression for the brightness of a thermal (black-body) radiator. For the Sun, the mean distance is the astronomical unit, at 1.496×10^{11} m, and the solar radius $R_\odot = 6.955 \times 10^8$ m.

Very long radio waves are reflected by the corona, so the radius is a bit larger than that of the photosphere, but the calculation of the brightness temperature is correct by a factor of two or better. That was good enough to confirm the million-degree temperature of the corona in 1946.

The non-flaring, meter-wavelength radio emission of the outer corona is known as thermal bremsstrahlung. It arises when a free electron in the corona encounters a proton. The electrical attraction between the opposite charges deflects the electron from its path, causing it to emit the radio radiation. Since there are up to a million billion (10^{15}) electrons and protons in every cubic meter of the corona, and they are both very hot and moving rapidly, encounters between electrons and protons occur all the time. The corona is therefore continuously emitting radio radiation by the thermal bremsstrahlung process at meter wavelengths.

When the Sun is imaged at centimeter wavelengths, the brightest emission is from solar active regions with their exceptionally high electron density and strong magnetic fields. The free electrons in the low corona are then constrained by the powerful magnetism, producing a slowly-varying component of radio emission. It is attributed to gyroresonance radiation emitted by thermal electrons gyrating about the magnetic fields. The radiation is emitted at a frequency, ν, given by:

$$\nu = \frac{esB}{2\pi m_e} = s\nu_g \approx 2.8 \times 10^{10}\, sB \text{ Hz}$$

where the frequency $\nu = c/\lambda$ for wavelength, λ, with $c = 2.9979 \times 10^8$ m s^{-1}, the integer s is the harmonic of the gyrofrequency ν_g, and B is the magnetic field strength in tesla. For solar active regions, only the second or third harmonics with $s = 2$ or 3 are important. One can distinguish between harmonics using polarization measurements; the third harmonic is strongly circularly polarized and the second is not. The circularly polarized radio radiation at centimeter wavelengths is therefore attributed to gyroresonance radiation at the third harmonic $s = 3$.

The shortest wavelength, λ_c, generated by the third harmonic determines the strongest magnetic field, B, in the coronal region under consideration

$$B = \frac{0.000\,375}{\lambda_c} \text{ T}$$

where λ_c is in meters. The measured magnetic field strength at the million-degree coronal level directly above large sunspots is up to 0.2 T, and generally 75–80 percent of the magnetic field strength in the underlying photospheric sunspots measured by the Zeeman effect.

The intense circularly polarized radiation at centimeter wavelengths, generated by the thermal gyroresonance process in active regions, can pass through overlying regions of transverse magnetic fields, resulting in a change in the sense of circular polarization. The magnetic field strength in the region where the polarization inversion occurs is given by:

$$B \approx \frac{15\,700}{[\lambda_I(\lambda_I N_e L)^{1/3}]} \text{ T,}$$

where λ_I is the wavelength where the polarization inversion occurred, N_e is the electron density in the region of polarization inversion, and L is the characteristic scale length of the region. Observations indicate $\lambda_I = 0.06$ m, $N_e = 10^{14}$ electrons per cubic meter and $L = 10^{11}$ m, with a magnetic field strength of B = 0.003 T in this part of the low corona.

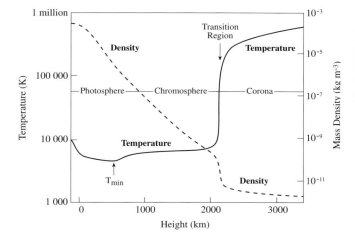

five million electrons per cubic meter. This gas is so thin, and the density so low, that there are not enough particles around to fry anything – not even an egg. So, properly speaking, the high coronal temperature applies only to the energies and velocities of the particles in the gas, and is thus called a kinetic temperature.

The unexpected temperature of the corona presents one of

Fig. 5.23 **Transition region** The temperature of the solar atmosphere decreases from values near 6000 K at the visible photosphere to a minimum value of roughly 4400 K about 500 km higher up. The temperature increases with height, slowly at first, then extremely rapidly in the narrow transition region, less than 100 km thick, between the chromosphere and corona, from about 10^4 to 10^6 K. The height is in kilometers, or km. where 1km = 1000 m. (Courtesy of Eugene Avrett, Smithsonian Astrophysical Observatory.)

Wavelength (nm)	Emitting ion	Formation temperature (K)	Wavelength (nm)	Emitting ion	Formation temperature (K)
0.178	Fe XXVI	> 100 000 000	28.42	Fe XV	2 100 000
0.186	Fe XXV	66 000 000	30.38	He II	83 000
0.917	Mg XI	6 300 000	33.5	Fe XVI	2 600 000
1.098	Fe XXIII	15 000 000	36.81	Mg IX	960 000
1.117	Fe XXIV	19 000 000	46.52	Ne VII	510 000
1.228	Fe XXI	11 000 000	49.94	Si XII	1 900 000
1.282	Fe XX	9 800 000	55.45	O IV	170 000
1.352	Fe XIX	8 500 000	62.49	Mg X	1 100 000
1.420	Fe XVIII	7 400 000	62.97	O V	240 000
1.502	Fe XVII	5 800 000	77.04	Ne VIII	630 000
2.160	O VII	2 100 000	97.70	C III	83 000
9.392	Fe XVIII	6 800 000	103.19	O VI	300 000
10.84	Fe XIX	7 900 000	117.6	C III	83 000
11.72	Fe XXII	12 000 000	120.65	Si III	60 000
12.87	Fe XXI	10 000 000	121.6	H I	~20 000
17.11	Fe IX	690 000	123.8	N V	190 000
17.45	Fe X	980 000	133.5	C II	38 000
18.04	Fe XI	1 200 000	139.4	Si IV	76 000
19.20	Fe XXIV	17 000 000	140.12	O IV	150 000
19.51	Fe XII	1 400 000	154.8	C IV	110 000
21.13	Fe XIV	1 900 000	164.0	He II	85 000
21.91	Fe XIV	1 800 000	165.7	C I	<10 000
24.92	Ni XVII	2 800 000	181.7	Si II	24 000
25.51	Fe XXIV	17 000 000	189.2	Si III	51 000
25.63	He II	85 000	190.9	C III	72 000
26.48	Fe XIV	1 900 000			

[a] C = carbon, Fe = iron, H = hydrogen, He = helium, Mg = magnesium, Ne = neon, O = oxygen, Si = silicon. Subtract one from the Roman numeral to get the number of missing electrons. [Courtesy of Kenneth P. Dere.]

Table 5.4 **Prominent solar emission lines in the transition region and low corona**[a]

the most puzzling enigmas of solar physics. Sunlight cannot heat the corona. It passes right through the corona without depositing substantial quantities of energy in it. But there must be some mechanism transporting energy from the photosphere, or below, out to the corona where it heats the gas.

Both the kinetic energy of motion and magnetic energy can flow from cold to hot regions. So writhing gases and shifting magnetic fields may be accountable. As we shall soon see, scientists now believe that at least some of the heat of the corona is related to the Sun's ever-changing magnetic fields. This recent understanding has resulted from extreme-ultraviolet and X-ray observations of the million-degree corona from space.

The ultraviolet and X-ray Sun

For studying the corona and identifying its elusive heating mechanism, scientists look at ultraviolet and X-ray radiation. This is because very hot material – such as that within the corona – emits most of its energy at these wavelengths. Also, the photosphere is too cool to emit intense radiation at these wavelengths, so it appears dark under the hot gas. As a result, the hot corona can be seen all across the Sun's face, with high spatial and temporal resolution, at ultraviolet and X-ray wavelengths.

The ultraviolet region of the electromagnetic spectrum has wavelengths of roughly 10^{-8} to 10^{-7} m, or 10 to 100 nm; the extreme ultraviolet is at the short wavelength part of this range. X-rays are even shorter, with wavelengths from 10^{-11} to 10^{-8} m, or 0.01 to 10 nm; soft X-rays are in the long-wavelength, low-energy part of this range with energies of 1 to 10 keV. Shorter wavelengths are generally

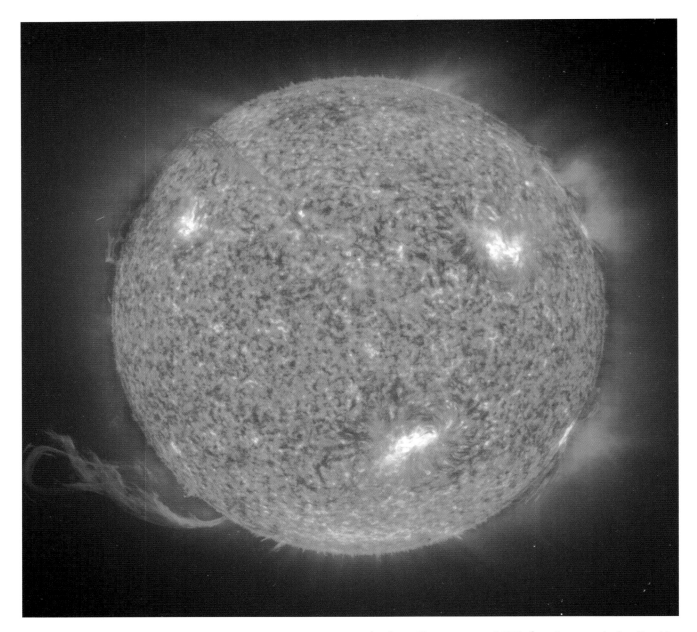

Fig. 5.24 **Chromosphere bright spots** Numerous ultraviolet bright spots are seen in this image from *SOHO*. It shows the chromosphere at temperatures of 60 000 to 80 000 K. Two intense active regions with numerous magnetic loops are also seen, as well as a huge eruptive prominence at the solar limb. This image was taken on 14 September 1997 in the extreme-ultraviolet resonance line of singly-ionized helium (He II) at 30.4 nm. (Courtesy of the *SOHO* EIT consortium. *SOHO* is a project of international cooperation between ESA and NASA.)

sensitive to hotter gases, so the corona is often observed at extreme-ultraviolet and soft X-ray wavelengths, and the chromo-sphere and transition region tend to be detected at ultraviolet wavelengths.

Since ultraviolet and X-rays are partially or totally absorbed by the Earth's atmosphere, they must be observed through telescopes in space. This has been done using a soft X-ray telescope on the *Yohkoh* spacecraft, and with ultraviolet and extreme-ultraviolet telescopes aboard the *SOlar and*

Heliospheric Observatory, or *SOHO* for short, and the *Transition Region And Coronal Explorer*, or *TRACE*.

To map out structures across the solar disk, ranging in temperature from six thousand to two million degrees kelvin, the spacecraft often use spectral lines. These lines appear when the Sun's radiation is displayed as a function of wavelength. Atoms in a hotter gas lose more electrons through collision, and so become more highly ionized. Because these different ions emit spectral lines at different wavelengths, they serve as a kind of thermometer, yielding the temperature where they are formed (Table 5.4).

Observations at different temperatures can also be used to focus on different layers of the solar atmosphere. As an example, the spectral line of singly ionized helium, He II, at 30.4 nm, is thought to be formed at 60 000 to 80 000 K, and it is therefore used to image structure in the lower part of the transition region, near the chromosphere (Fig. 5.24).

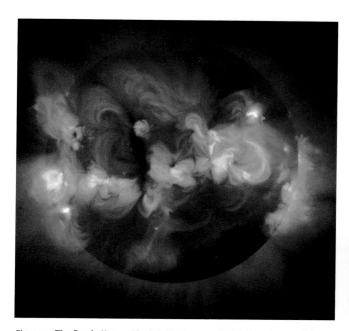

Fig. 5.25 **The Sun in X-rays** The bright glow seen in this X-ray image of the Sun is produced by ionized gases at a temperature of a few million degrees kelvin. It shows magnetic coronal loops which thread the corona and hold the hot gases in place. The brightest features are called active regions and correspond to the sites of the most intense magnetic field strength. This image of the Sun's corona was recorded by the Soft X-ray Telescope (SXT) aboard the Japanese *Yohkoh* satellite on 1 February 1992, near the maximum of the 11-year cycle of solar magnetic activity. Subsequent SXT images, taken about five years later near activity minimum, show a remarkable dimming of the corona when the active regions associated with sunspots have almost disappeared, and the Sun's magnetic field has changed from a complex structure to a simpler configuration – see Fig. 5.29. (Courtesy of Gregory L. Slater, Gary A, Linford, and Lawrence Shing, NASA, ISAS, the Lockheed-Martin Solar and Astrophysics Laboratory, the National Astronomical Observatory of Japan, and the University of Tokyo.)

The Sun is mottled all over in this ultraviolet perspective, like a cobbled road or a stone beach. Each stone is a continent-sized bubble of hot gas that flashes on and off in about ten minutes. The whole Sun seems to sparkle in the ultraviolet light emitted by these localized brightenings, known as blinkers. About 3000 of them are seen erupting all over the Sun, including the darkest and quietest places at the solar poles.

Images at extreme-ultraviolet and X-ray wavelengths have shown that the hottest and densest material in the low corona is concentrated in magnetic loops. Indeed, *Yohkoh's* soft X-ray images have demonstrated that the entire corona is "stitched" together by thin, bright, magnetized loops that shape, mold and constrain the million-degree gas (Fig. 5.25). Wherever the magnetism in these coronal loops is strongest, the coronal gas in them shines brightly at soft X-ray wavelengths (Fig. 5.26).

Since the hot gases are highly ionized, they cannot readily cross the intense, closed magnetic field lines. The electrons contained in these coronal loops have temperatures of 1 to 10 million (10^6 to 10^7) degrees kelvin and densities of 1 to

100 million billion (10^{15} to 10^{17}) electrons per cubic meter. The magnetized loops are often anchored in bipolar sunspots within solar active regions that occur more frequently at the maximum of the solar magnetic activity cycle.

High-resolution *TRACE* images at the Fe IX, Fe XII and Fe XV lines, respectively formed at 1.0, 1.5 and 2.0 million degrees kelvin, have demonstrated that there is a great deal of fine structure in the coronal loops (Fig. 5.27). They have pointed toward a corona comprised of thin loops that are naturally dynamic and continually evolving. These very thin loops are heated in their legs on a time span of minutes to tens of minutes, after which the heating stops or changes,

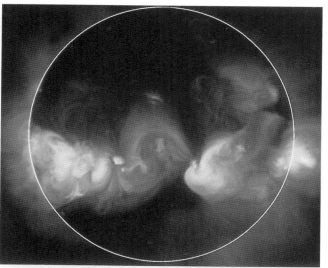

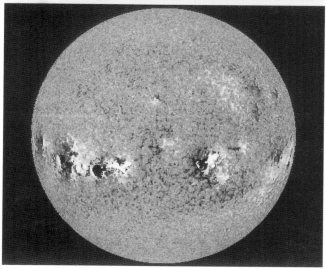

Fig. 5.26 **Magnetic shapes in X-rays** A coronal X-ray image (*top*) from the *Yohkoh* Soft X-ray Telescope on 26 December 1991, scaled to show the brightest parts, is compared with a magnetogram (*bottom*) from the National Solar Observatory, Kitt Peak, taken on the same day. In the magnetogram, black and white denote different magnetic directions or polarity, and the intensity is a measure of the magnetic field strength. Strong regions of opposite magnetic polarity are joined by magnetic loops that constrain the hot, dense gas that shines brightly in X-rays. This comparison shows that the structure and brightness of the X-ray corona are dictated by their magnetic roots in the photosphere. (Courtesy of David A. Falconer and Ronald L. Moore.)

Fig. 5.27 **Long thin loops heated from their base** The low solar corona and transition region are filled with bright, thin magnetized loops that extend for hundreds of millions of meters above the visible solar disk, or photosphere, spanning 30 or more times the diameter of the Earth. The coronal loops are filled with gas that is hundreds of times hotter than the photosphere. In this *TRACE* image, taken on 06 November 1999, the hot gas is detected in the ultraviolet light emitted by eight- and nine-times-ionized iron (Fe IX/X at 17.1 nm) formed at a temperature of about 1.0 million degrees kelvin. Such detailed *TRACE* images indicate that most of the heating occurs low in the corona, near the bases of the loops as they emerge from and return to the solar disk, and that the heating does not occur uniformly along the entire loop length. (Courtesy of Markus J. Aschwanden, the *TRACE* consortium, LMSAL, and NASA.)

suggesting the injection of hot material from somewhere near the loop footpoints in the photosphere or below. The high-resolution TRACE images indicate that the bright, thin loops are heated only in the lower 20 million meters of the corona, near the base of the looping magnetic fields (Fig. 5.27). The erratic changes in the rate of heating forces the loops to continuously change their internal structure.

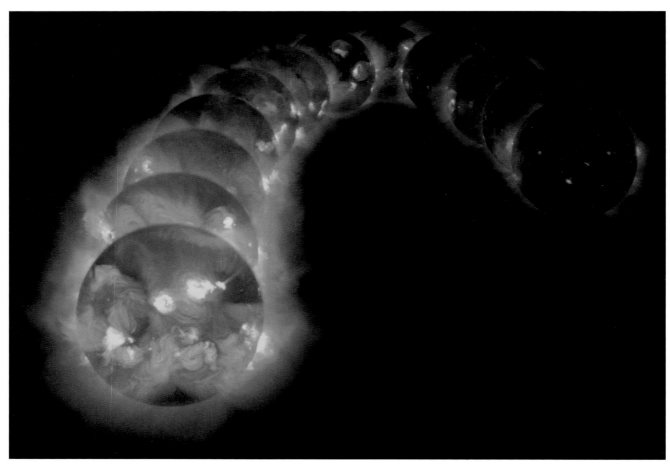

Fig. 5.28 **X-ray view of the solar cycle** Dramatic changes in the solar corona are revealed in this four-year montage of images from the Soft X-ray Telescope (SXT) aboard *Yokhoh*. The 12 images are spaced at 120-day intervals from the time of the satellite's launch in August 1991, at the maximum phase of the 11-year sunspot cycle (*left*), to late-1995 near the minimum phase (*right*). The bright glow of X-rays near activity maximum comes from very hot, million- degree coronal gases that are confined within powerful magnetic fields anchored in sunspots. Near the cycle minimum, the active regions associated with sunspots have almost disappeared, and there is an overall decrease in X-ray brightness by 100 times. (Courtesy of Gregory L. Slater and Gary A, Linford, NASA, ISAS, the Lockheed-Martin Solar and Astrophysics Laboratory, the National Astronomical Observatory of Japan, and the University of Tokyo.)

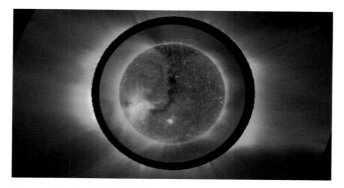

Fig. 5.29 **The ultraviolet Sun** This composite image, taken by two *SOHO* instruments and joined at the black circle, reveals the ultraviolet light of the Sun's atmosphere from the base of the corona to millions of kilometers above the visible solar disk. The region outside the black circle, obtained by UVCS (UltraViolet Coronagraph Spectrometer), shines in the ultraviolet light emitted by oxygen ions flowing away from the Sun to form the solar wind. The inner image, obtained by EIT (Extreme-ultraviolet Imaging Telescope), shows the ultraviolet light emitted by iron ions at a temperature near two million degrees kelvin. Dark areas, called coronal holes, are found at both poles of the Sun (*top* and *bottom*) and across the disk of the Sun; they are the places where the highest-speed solar wind originates. The structure of the corona is controlled by the Sun's magnetic field which forms the bright active regions on the solar disk and the ray-like structures extending from the coronal holes. (Courtesy of the *SOHO* UVCS consortium (*outer region*) and the *SOHO* EIT consortium (*inner region*). *SOHO* is a project of international cooperation between ESA and NASA.)

Feature	Largest extent (10^6 m)	Temperature (10^6 K)	Electron density (m^{-3})
Coronal holes	700 to 900	1.0 to 1.5	4×10^{14}
Active region loops	10	2 to 4	$(1 \text{ to } 7) \times 10^{15}$
X-ray bright point	5 to 20	2.5	1.4×10^{16}

Table 5.5 **Physical parameters of coronal features**

The ultraviolet and X-ray emission of the Sun vary significantly over the 11-year cycle of magnetic activity. The ultraviolet emission doubles from activity minimum to maximum, while the X-ray brightness of the corona increases by a factor of 100. At the cycle maximum, when the sunspots and their associated active regions are most numerous, bright coronal loops dominate the X-ray Sun; at activity minimum the bipolar sunspots and their connecting magnetic loops have largely disappeared, and the Sun is much dimmer in X-rays (Fig. 5.28). However, the corona still stays hot at a minimum in its 11-year activity cycle when active regions go away; the million-degree gas is just a lot more rarefied and less intense.

Not all magnetic fields on the Sun are closed loops. Some of the magnetic fields extend outward, within regions called coronal holes. These extended regions have so little hot material in them that they appear as large dark areas seemingly devoid of radiation at extreme-ultraviolet and X-ray wavelengths (Fig. 5.29). Coronal holes are nearly always present at the Sun's poles, and are sometimes found at lower solar latitudes. They are routinely detected by instruments aboard *SOHO*, *TRACE* and *Yohkoh*.

Coronal holes are not empty (Table 5.5). They are just more rarefied and cooler than other places in the corona, so their emission is faint. This is because most of the magnetic field lines in coronal holes do not form locally closed loops. Coronal holes are instead characterized by open magnetic fields that do not return directly to another place on the Sun,

allowing the charged particles to escape the Sun's magnetic grasp and flow outward into surrounding space.

Nevertheless, magnetic field lines are never broken, and always form a continuous thread. The magnetic fields in coronal holes are therefore closed in a technical sense, but they extend so far out into interplanetary space that they are effectively open. Solar material therefore spills out of the polar coronal holes, expelling a high-speed solar wind (Section 7.2).

The size and location of coronal holes vary with the solar magnetic-activity cycle. A few years before the cycle minimum, the poles of the Sun are covered by large coronal holes that occasionally extend from a pole across the solar equator. Near the maximum in the activity cycle, the polar holes shrink in size, and smaller holes appear nearer the solar equator.

Solving the heating crisis

As already noted, the temperature of the million-degree corona is not supposed to be hotter than the cooler photosphere immediately below it. Heat should not emanate from a cold object to a hot one any more than water should flow uphill. For more than half-a-century, scientists have been trying to identify the elusive heating mechanism that transports energy from the photosphere, or below, out to the corona. We know that sunlight will not do the trick, for the corona is transparent to most of it.

Sound waves generated within the turbulent convective zone can bring some energy up from inside the Sun, but most of the sound waves are reflected back inside the Sun when they encounter the steep density gradient at the photosphere's edge (Section 4.3). Those that manage to slip into the chromosphere accelerate and strengthen into shocks as they propagate outward through the increasingly rarefied gas, dissipating their energy there.

Spacecraft observations indicate that the lower layers of the chromosphere can be heated in this fashion, but that sound waves cannot play a role in heating the corona. The power in the sound, of about 10 W m^{-2}, is not nearly enough to energize the coronal gas (Table 5.6). Moreover, the sound waves do not travel that far. They are either reflected back inward or expend most of their energy before reaching the corona.

The discovery of a highly-structured corona, with intense X-ray emission from numerous magnetic loops, also argues against the presumably uniform acoustic heating of the corona.

There are all sorts of other ideas about why the outer parts of the Sun are so incredibly hot, but most of them go back to varying magnetism. The ubiquitous coronal loops can rise from inside the Sun, sink back down into it, or expand out into space. So, the corona is in a continuous state of change, constantly varying in brightness and structure on all detectable spatial and temporal scales. Its form is always adjusting to the shifting forces of magnetism, and the thin, long magnetized loops are in a state of continual agitation.

The X-ray emission from main-sequence stars similar to the Sun increases with their rotation speed, and presumably with enhanced magnetic dynamo action and stronger magnetic fields (Focus 5.5) So the coronas of these stars are most likely heated by their magnetism as well

Magnetic waves provide a method of carrying energy into the corona. The ever-changing coronal loops are always being jostled, twisted and stirred around by motions deep down inside the Sun where the magnetism originates. A tension acts to resist the motions and pull the disturbed magnetism back, generating waves that propagate along magnetic fields, somewhat like a vibrating string.

They are often called Alfvén waves after Hannes Alfvén who first described them mathematically. He pioneered the study of the interaction of hot gases and magnetic fields, receiving the Nobel Prize in Physics in 1970 for his discoveries in it.

Instruments on spacecraft have detected Alfvén waves out in the distant reaches of the solar corona (Section 7.3). Unexpectedly wide spectral lines could be the signature of such waves at lower levels in the solar atmosphere. The broadening of lines formed in the low corona is larger than that caused by the thermal speed of the radiating ions. This excess width has been attributed to non-thermal motions, such as Alfvén waves.

However, once you get energy into an Alfvén wave, it is difficult to get it out. So there may be a problem in depositing enough magnetic-wave energy into the coronal gas to heat it

up to the observed temperatures. Like radiation, the Alfvén waves seem to propagate right through the low corona without being noticeably absorbed or dissipated there.

Focus 5.5 Stellar coronas and magnetic fields

X-ray telescopes aboard modern spacecraft, such as the German RÖntgen SATellite, or ROSAT, have observed steady, quiescent, non-flaring, X-ray radiation from tens of thousands of stars, including virtually all nearby single, main-sequence stars of late spectral type F, G, K and M. These main-sequence stars have roughly the same mass, size and luminosity as the Sun. Like the Sun, their X-ray emission is attributed to the thermal radiation of electrons in stellar coronas with temperatures of 1 to 10 million degrees.

The visible light from these stars is emitted from photospheres with effective temperatures between 7200 K, for spectral type F0, and 3370 K, for type M4; the Sun has spectral type G2 and an effective temperature of 5780 K for its photosphere.

The Sun's coronal heating paradox therefore applies to other stars. They all have hot coronas overlying photospheres that are more than 100 times cooler than their coronas. None of these coronas can be heated by starlight. The heating problem is also aggravated by the greater absolute X-ray luminosity of these stars, up to $L_X = 10^{23}$ W compared with the Sun's value of $L_{X\odot} = 2 \times 10^{20}$ W.

The intensity of the coronal X-ray emission from stars of identical spectral type can differ by several orders of magnitude, which argues against heating by sound waves generated in their convective zones; stars of the same type should have similar convection.

Since the absolute X-ray luminosity increases with the stellar rotation speed, the coronal heating could be related to a magnetic dynamo in the rotating, stellar interiors. Enhanced dynamo action by rapid rotation will produce stronger magnetic fields and more intense X-ray radiation. So, magnetic fields most likely play a coronal heating role for nearby main-sequence stars of late spectral type, as they do when heating the Sun's corona (this section).

The X-ray emitting stars also emit quiescent, or non-flaring, radio radiation, and the brighter X-ray stars are the more luminous radio stars. Unlike the thermal radio emission from the Sun's corona (Focus 5.4), the radio emission from the other stars has to be non-thermal radiation. The thermal radio emission of the Sun's corona would be undetectable with even the world's largest radio telescope if the Sun was placed at the distance of the next-nearest star. In addition, the observed radio flux density of nearby stars requires implausible brightness temperatures that are ten thousand times hotter than the thermal X-ray emitting gas.

The long-lived (hours) quiescent radio emission from some of these stars is 100 percent circularly polarized, and this polarization is attributed to intense magnetic fields with strengths of about 0.03 T in the emitting regions. So, the non-thermal radio radiation is probably emitted by electrons accelerated to high speeds in strong, coronal magnetic fields. This magnetism seems to provide a link between the thermal heating of stellar coronas, with their X-ray radiation, and their non-thermal radio emission.

Loss mechanism	Active region ($W m^{-2}$)	Quiet region ($W m^{-2}$)	Coronal hole ($W m^{-2}$)
Conduction	100 - 10 000	200	60
Radiation	5 000	100	10
Solar Wind	< 100	< 50	500
Total	10 000	300	500

[a] One watt (W) is equivalent to one Joule per second, and the units of energy loss are watts per square meter, or $W m^{-2}$. By way of comparison, sunlight is emitted from the photosphere at the rate of 63 million (6.3×10^7) $W m^{-2}$.

Table 5.6 **Average coronal energy loss**[a]

Magnetic loops can heat the corona in another way – by coming together and releasing stored magnetic energy when they make contact in the corona. Motions twist and stretch the coronal magnetic fields, slowly building up their energy. When these magnetic fields are pressed together in the corona, they merge, join and self-destruct at the place where they touch, releasing their pent-up energy to heat the gas.

Magnetic fields can move into each other, but they never end. They just reform or reconnect in new magnetic orientations, so this method of coronal heating is termed magnetic reconnection.

Because the Sun's magnetism is always moving in, out and about, the ubiquitous coronal loops form and move all over the Sun. Magnetic reconnection can therefore occur when newly emerging magnetic fields rise through the photosphere to encounter pre-existing ones in the corona. It can also happen when internal motions force existing coronal loops together. In either case, the moving loops are charged with magnetic energy that can heat the gas.

The *SOHO* spacecraft has provided direct evidence for such a transfer of magnetic energy from the solar photosphere into the low corona. Images of the photosphere's magnetism, taken with *SOHO*, reveal tens of thousands of pairs of opposite magnetic polarity, each joined by a magnetic arch that rises above them. They form a complex, tangled web of magnetic fields low in the corona, dubbed the magnetic carpet (Fig. 5.30). The small magnetic loops rise up out of the photosphere and then disappear within hours or days. But they are continuously replenished by the emergence of new magnetic loops, rising up to form new magnetic connections all the time and all over the Sun.

Fig. 5.30 **Magnetic carpet** Magnetic loops of all sizes rise up into the solar corona from regions of opposite magnetic polarity (*black* and *white*) in the photosphere, forming a veritable carpet of magnetism in the low corona. Energy released when oppositely directed magnetic fields meet in the corona, to reconnect and form new magnetic configurations, is one likely cause for making the solar corona so hot. (Courtesy of the *SOHO* EIT and MDI consortia. *SOHO* is a project of international cooperation between ESA and NASA.)

Simultaneous comparisons with extreme-ultraviolet images from *SOHO* indicate that whenever the magnetism in the photosphere converges there is a brightening just above in the low corona, showing that energy flows from the magnetic loops when they interact and reconnect. The very strong electric currents in these magnetic "short circuits" could heat the corona to a temperature of several million degrees.

TRACE observations support the idea of a carpet of magnetic fields on a range of scales constantly emerging from the photosphere, causing a continuous rearrangement of magnetic fields in the corona and impressing on the solar atmosphere the fine structure in the varying photosphere magnetic fields. The *TRACE* images also suggest that the heating of coronal loops occurs near their base in the low corona, and that the loops are not heated uniformly along their entire length, like toaster wires, by electrical currents or waves traveling along the magnetic fields (see Fig. 5.28). The non-uniform heating of the loop footpoints is most likely related to magnetic interactions in these regions.

The idea of powerful energy release during magnetic reconnection is not a new one. It was proposed decades ago to account for sudden, brief, intense explosions on the Sun, called solar flares, that can release energies equivalent to billions of terrestrial nuclear bombs (Sections 6.1, 6.3). Converging flows in solar active regions apparently press oppositely-directed field lines together, releasing magnetic energy at the place that they join. The new, reconnected field lines can snap apart, accelerating and hurling energetic particles out into interplanetary space and down into the Sun.

Such bi-directional, collimated and explosive jets of material have been observed in ultraviolet images of the chromosphere outside active regions. The magnetic interaction of coronal magnetic loops, driven together by underlying convective motions, also energizes at least some of the bright "points" found in X-ray images of the Sun. Unlike sunspots and active regions, the X-ray bright "points" are uniformly distributed over the Sun, appearing at the poles and in coronal holes, some almost as large as the Earth. Hundreds and even thousands of them come and go, fluctuating in brightness like small flares, apparently energized by magnetic reconnection.

Scientists continue to search for numerous low-level flares that might contribute to the corona's heat. They have been called microflares, with "micro" designating small, but also dubbed nanoflares. If nanoflares were true to their name, a billion of them would be required to release the same amount of energy as a single powerful flare. All observers agree that the low-energy flares occur much more frequently than high-energy ones, but there is still disagreement over whether or not the low-energy flares occur often enough to significantly heat the corona. This brings us to a consideration of the truly powerful solar flares that can significantly impact interplanetary space and the terrestrial environment.

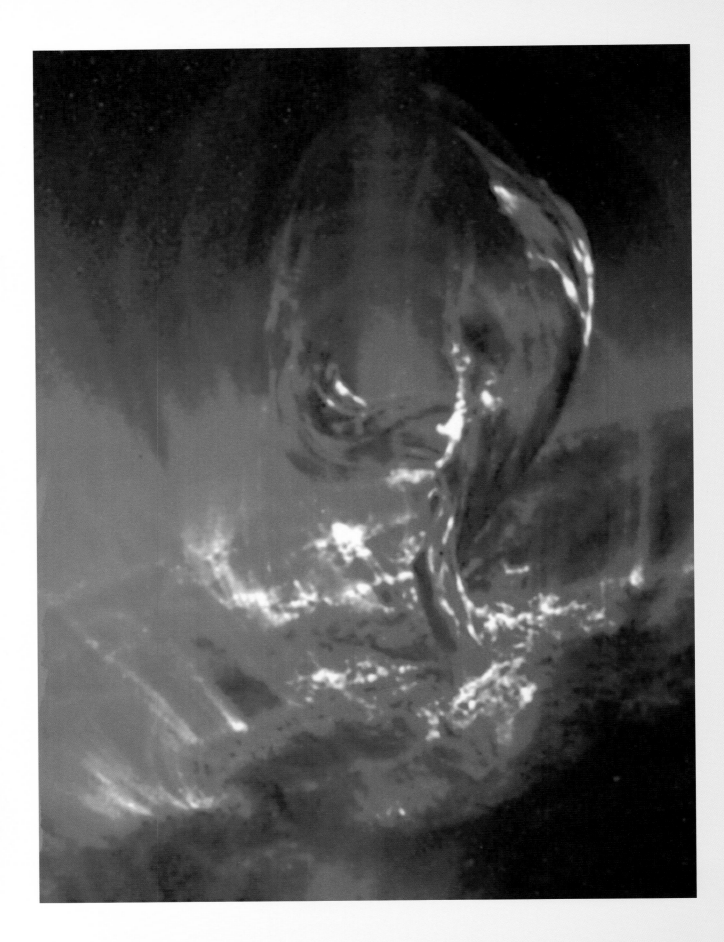

CHAPTER SIX

The explosive Sun

A filament erupts A filament is shown in the process of lifting off from the edge of the Sun – north is to the left. The dark matter is relatively cool, around 20 000 degrees kelvin, while the bright material is at a temperature of a million degrees kelvin or more. From top to bottom, the structure stands 120 million (1.20×10^8) meters tall. This image was taken from the *Transition Region And Coronal Explorer*, *TRACE*, on 19 July 2000 at a wavelength of 17.1 nm. (Courtesy of the *TRACE* consortium, the Stanford-Lockheed Institute of Space Research, and NASA.)

6.1 Solar flares

Active regions are the sites of sudden and brief explosions, called solar flares, that rip through the atmosphere above sunspots with unimagined intensity. In just 100 to 1000 seconds, the disturbance can release an energy of 10^{24} joule. A single flare then creates an explosion equivalent to 2.5 million terrestrial nuclear bombs, each with a destructive force of 100 megatons (10^{11} kg) of trinitrotoluene, or TNT. All of this power is created in a relatively compact explosion, comparable in total area to a sunspot, and occupying less than one-ten-thousandth (0.01 percent) of the Sun's visible disk.

For a short while, a flare can be the hottest place on the Sun, heating Earth-sized active regions to tens of millions of degrees kelvin. The explosion floods the solar system with intense radiation across the full electromagnetic spectrum, from the shortest X-rays to the longest radio waves.

Although flares appear rather inconspicuous in visible light, they can briefly produce more X-ray and radio radiation than the entire non-flaring Sun does at these wavelengths. We can use this radiation to watch the active-region atmosphere being torn asunder by the powerful explosions; and then view the lesion being stitched together again.

During the sudden and brief outbursts, electrons and protons can be accelerated to nearly the speed of light. Protons and helium nuclei are thrown down into the chromosphere, generating nuclear reactions there. The high-speed electrons and protons are also hurled out into interplanetary space where they can threaten astronauts and satellites. Shock waves can be produced during the sudden, violent release of flare energy, ejecting masses of hot coronal gas into interplanetary space. Some of the intense radiation and energetic particle emissions reach the Earth where they can adversely affect humans (Section 8.3).

Since flares occur in active regions, their frequency follows the 11-year magnetic-activity cycle. The rate of solar flares of a given energy increases by about an order of magnitude, or a factor of ten, from the cycle minimum to its maximum. At the cycle maximum, scores of small flares and several large ones can be observed each day. Yet, even at times of maximum solar activity, the most energetic flares remain infrequent, occurring only a few times a year; like rare vintages, they are denoted by their date. Flares of lesser energy are much more common. Those with half the energy of another group occur about four times as often.

Solar flares are always located near sunspots and occur more often when sunspots are most numerous. This does not mean that sunspots cause solar flares, but it does suggest that solar flares are energized by the powerful magnetism associated with sunspots. When the magnetic fields in a solar active region become contorted, magnetic energy is built up and stored in the low corona. When the pent-up energy is released it is often in the form of a solar flare. This energy is suddenly and explosively released at higher levels in the solar atmosphere just above sunspots.

Flares in the chromosphere

Our perceptions of solar flares have evolved with the development of new methods of looking at them. Despite the powerful cataclysm, most solar flares are not, for example, detected on the bland white-light face of the Sun. They are only minor perturbations in the total amount of emitted sunlight; every second the photosphere emits an energy of 3.85×10^{26} J, far surpassing the total energy emitted by any solar flare by at least a factor of one hundred.

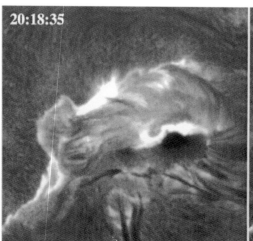
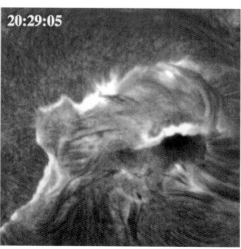

20:18:35   20:29:05

Fig. 6.1 **Hydrogen-alpha flare ribbons** A large solar flare observed in the red light of hydrogen alpha (Hα), showing two extended, parallel flare ribbons in the chromosphere. Each image is 200 million meters in width, subtending an angle of 300 seconds of arc or about one-sixth of the angular extent of the Sun. These photographs were taken at the Big Bear Solar Observatory on 29 April 1998. (Courtesy of Haimin Wang.)

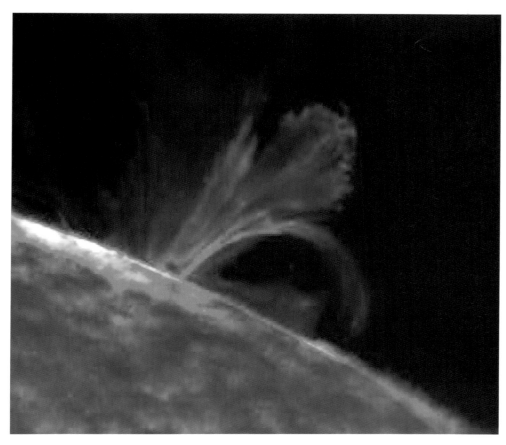

Fig. 6.2 **Lyman-alpha explosion** A flare moves to a height of 150 million meters at the edge of the Sun, shining in the light of the Lyman-alpha emission from hydrogen atoms at 121.6 nm. The emitting gas is at a temperature of 10 000 to 20 000 K, characteristic of the chromosphere. This image shows that flare radiation is emitted from numerous long, thin magnetic filaments. It was taken on 19 May 1998 from *TRACE*, a mission of the Stanford-Lockheed Institute for Space Research and part of the NASA Small Explorer Program. (Courtesy of the *TRACE* consortium and NASA.)

Thus, only exceptionally powerful solar flares can be detected in visible sunlight. These white-light flares, as they are now called, were first observed and carefully recorded by two Englishmen, Richard C. Carrington and Richard Hodgson, who noticed an intense flash of light near a complex group of sunspots on 1 September 1859, lasting just a few minutes.

Routine visual observations of solar explosions were made possible by tuning into the red emission of hydrogen alpha, designated Hα, at a wavelength of 656.3 nm, and rejecting all the other colors of sunlight. Light at this wavelength originates just above the photosphere, in the chromospheric layer of the solar atmosphere. For more than half a century,

Importance	Flare area (square degrees)	Flare area (10^{12} square meters)	Flare brilliance[a]
S (subflare)	Less than 2	Less than 300	f, n, or b
1	2.1 to 5.1	300 to 750	f, n, or b
2	5.2 to 12.4	750 to 1850	f, n, or b
3	12.5 to 24.7	1850 to 3650	f, n, or b
4	More than 24.7	More than 3650	f, n, or b

[a] f = faint; n = normal; b = bright.

Table 6.1 **Hydrogen-alpha classification of solar flares**

astronomers throughout the world have used filters to isolate the Hα emission, carrying out a vigilant flare patrol that continues today. Most solar observatories now have automated Hα telescopes, and some of them are used to monitor the Sun for solar flares by capturing images of the Sun every few seconds.

When viewed in this way, solar flares appear as a sudden brightening, lasting from a few minutes to an hour, usually in active regions with strong, complex magnetic fields. The Hα light is not emitted directly above sunspots, but is instead located between regions of opposite magnetic polarity in the underlying photosphere, near the line or place marking magnetic neutrality. Flares in the chromosphere often appear on each side of the magnetic neutral line as two extended parallel ribbons (Fig. 6.1). The two ribbons move apart as the flare progresses, and the space between them is filled with higher and higher shining loops while the lower ones fade away.

The hydrogen-alpha observations describe the shape, size, location and intensity of solar flares in the chromosphere. This leads to one flare classification scheme based on their extent and brightness in Hα (Table 6.1).

The powerful surge of flaring hydrogen light is also detected by spacecraft observing at ultraviolet wavelengths. Detailed magnetic filaments have been resolved in the Lyman-alpha, or Ly α, transition of hydrogen at 121.6 nm (Fig. 6.2).

Yet, a solar flare emits just a modest amount of its power in the chromosphere, and the chromospheric image

Importance	Peak flux at 0.1 to 0.8 nm [a] $(W\,m^{-2})$
A	10^{-8} to 10^{-7}
B	10^{-7} to 10^{-6}
C	10^{-6} to 10^{-5}
M	10^{-5} to 10^{-4}
X	10^{-4} and above

[a] A number following the letter gives the peak flux in the first unit. For example, X5.2 stands for a peak soft X-ray flux of 5.2×10^{-4}, or 0.000 52, W m^{-2}.

Table 6.2 **Soft X-ray classification of solar flares measured near the Earth**

provides only a two-dimensional, flatland picture without information about what is happening above or below it. Observations at X-ray and radio wavelengths provide new perspectives, leading to a more complete understanding of solar flares. We can tune into flares at these wavelengths and see them in different forms, providing a three-dimensional spatial view.

X-ray flares

Since solar flares are very hot, they emit the bulk of their energy at X-ray wavelengths. For a short while, a large flare can outshine the entire Sun in X-rays (Fig. 6.3). The hot X-ray flare then dominates the background radiation of even the

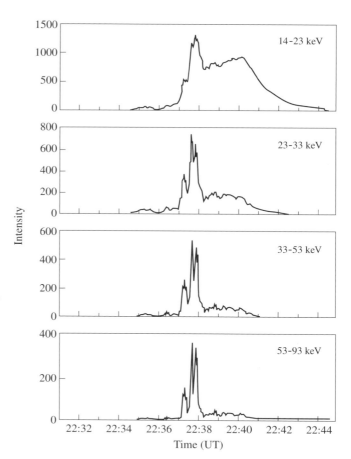

Fig. 6.4 **Impulsive and gradual phases of a flare** The time profile of a solar flare at hard X-ray energies, above 30 keV, is characterized by an impulsive feature that lasts for about one minute. The intensity scale provides the average counting rate, derived from 0.5-second integration, of the 64 imaging elements of the Hard X-ray Telescope (HXT) aboard *Yohkoh*. This impulsive phase coincides with the acceleration of high-speed electrons that emit non-thermal bremsstrahlung at hard X-ray wavelengths and non-thermal synchrotron radiation at centimeter radio wavelengths. The less-energetic emission shown here, below 30 keV, can be composed of two components, an impulsive component followed by a gradual one. The latter component builds up slowly and becomes most intense during the gradual decay phase of solar flares when thermal radiation dominates. At even lower soft X-ray energies (about 10 keV), the gradual phase dominates the flare emission. This data was taken on 15 November 1991. (Courtesy of NASA, ISAS, the Lockheed-Martin Solar and Astrophysics Laboratory, the National Astronomical Observatory of Japan, and the University of Tokyo.)

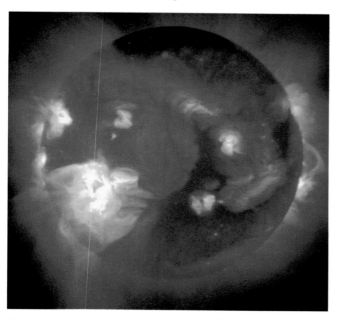

Fig. 6.3 **An X-ray flare** As shown in the lower left of this X-ray image, a solar flare can result in soft X-ray radiation that outshines the entire Sun at these wavelengths. Less luminous coronal loops are found in quiescent, or non-flaring, active regions on other parts of the Sun, and dark coronal holes are also present at both poles (*top* and *bottom*). This image was taken with the Soft X-ray Telescope (SXT) aboard *Yohkoh* on 25 October 1991. (Courtesy of Keith Strong, NASA, ISAS, the Lockheed-Martin Solar and Astrophysics Laboratory, the National Astronomical Observatory of Japan, and the University of Tokyo.)

brightest magnetic loops in the quiescent, or non-flaring, solar corona. Because any X-rays coming from the Sun are totally absorbed in the Earth's atmosphere, X-ray flares must be observed from satellites orbiting the Earth above our air.

Hotter gases radiate most intensely at shorter wavelengths, and the wavelength at which the brightness is a maximum, λ_{max}, varies inversely with the temperature, T. The exact relation, known as the Wien displacement law, is given by $\lambda_{max} = 0.002\ 897\ 75/T$ m (Section 1.2, Focus 1.1). A gas heated to ten million degrees kelvin will therefore be brightest at the X-ray wavelength of $\lambda_{max} = 3 \times 10^{-10}$ m = 0.3 nm.

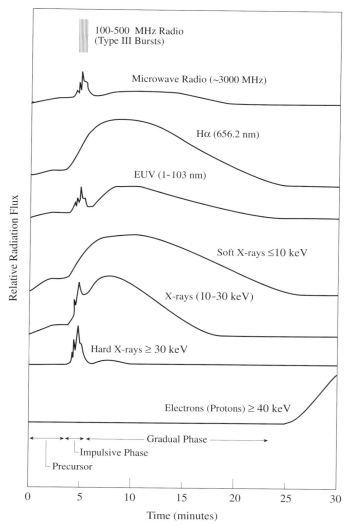

Fig. 6.5 Solar flares in varying perspectives During the early impulsive phase of a solar flare, electrons accelerated to high energies and very rapid speeds emit radio bursts, extreme-ultraviolet (EUV) radiation, and hard X-rays. The radio emission is at frequencies from 100 to 3000 MHz, or at wavelengths between 3 and 0.1 m; the hard X-rays have photon energies greater than 30 keV. The subsequent gradual phase is detected with soft X-rays, at energies of about 10 keV or less, as an after-effect of the impulsive radiation. The soft X-rays are the thermal radiation, or bremsstrahlung, of a gas heated to temperatures about ten million degrees kelvin.

Flares can be characterized by their brightness in X-rays, as observed by monitoring satellites near the Earth (Table 6.2). The biggest flares are X-class flares. M-class flares have one tenth the X-ray flux of an X-class one, and the C-class has one hundredth of the X-class flux.

The Space Environment Center of the National Oceanic and Atmospheric Administration, or NOAA, provides the peak soft X-ray flux for solar flares seen from their *Geostationary Operational Environmental Satellites*, or *GOES* for short. These satellites hover above points in the Earth's western hemisphere, orbiting at the same rate that the Earth spins. The flux of electrons and protons striking the satellites is also provided in NOAA's space-weather information, available on the web at **www.sec.noaa.gov**. Space-weather forecasts are discussed further in Section 8.3.

Researchers describe X-rays by the energy they carry. There are soft X-rays with relatively low energy and modest penetrating power. The hard X-rays have higher energy and greater penetrating power. The wavelength of radiation is inversely proportional to its energy, so hard X-rays are shorter than soft X-rays.

The energy of the X-ray radiation is a measure of the energy of the electrons that produce it. In fact, X-rays with a given energy are produced by electrons with roughly the same amount of energy. The energy of both the high-speed electrons and the flaring X-rays is often specified in kilo-electron volts, denoted keV. One keV is equivalent to an energy of 1.6×10^{-16} J, and corresponds to a wavelength of 12.4×10^{-10} m, or 1.24 nm (Focus 1.1). Soft X-rays have energies between 1 and 10 keV, and hard X-rays lie between 10 and 100 keV.

The time profiles of a solar flare depend upon the choice of observing wavelength; when combined they provide detailed information about the physical processes occurring during the lifetime of a solar flare. Hard X-rays are emitted during the impulsive onset of a solar flare, while the soft X-rays gradually build up in strength and peak a few minutes after the impulsive emission (Fig. 6.4). The soft X-rays are therefore a delayed effect of the main flare explosion.

The energetic electrons that produce the impulsive, flaring hard X-ray emission also emit radiation at microwave (centimeter) and radio (meter) wavelengths (Fig. 6.5). The similarity in the time profiles of the microwave and hard X-ray bursts, on time-scales as short as a second, suggests that the electrons that produce both the hard X-rays and the microwaves are accelerated and originate in the same place. Impulsive flare radiation at both the long, hard X-ray wavelengths and the short microwave wavelengths is apparently produced by the same population of high-speed electrons, with energies of 10 to 100 keV.

The *Solar Maximum Mission*, or *SMM* obtained pioneering images in the 1980s of the hard X-rays emitted during the impulsive phase of solar flares. The two-component, or double, hard X-ray sources were concentrated at the footpoints of the coronal loops detected at soft X-ray wavelengths. This is explained if the hard X-rays are generated by energetic, non-thermal electrons hurled down the two legs of a coronal loop into the low corona and dense chromosphere. The discovery of two-centimeter microwave emission at the two footpoints of coronal loops at about the same time supported the hard X-ray double-footpoint evidence.

The slow, smooth rise of the soft X-ray afterglow resembles the time integral of the rapid, impulsive hard X-ray and microwave bursts. This relationship is known as the Neupert effect, named after Werner M. Neupert who first noticed it in 1968 when comparing satellite observations of soft X-ray flares with impulsive microwave bursts observed from the ground.

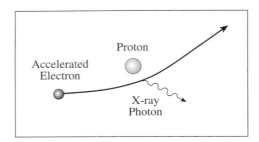

Fig. 6.6 **Bremsstrahlung** When a hot electron moves rapidly and freely outside an atom, it inevitably moves near a proton in the ambient gas. There is an electrical attraction between the electron and proton because they have equal and opposite charge, and this pulls the electron toward the proton, bending the electron's trajectory and altering its speed. The electron emits electromagnetic radiation in the process.

Precise measurements of soft X-ray spectral lines, from the SMM, and P78-1 satellites in the early-1980s, revealed a Doppler shift to shorter wavelengths, suggesting an upward motion of the hot thermal gas at velocities as high as 4×10^5 m s^{-1}. This rise in heated material was explained by a theory called chromospheric evaporation, although it has nothing to do with the evaporation of any liquid. Initially cool material in the chromosphere is heated by downflowing, or precipitating, flare electrons that emit the hard X-rays and microwaves. The heated gas expands upward into the low-density corona along magnetic loops that, after filling, shine brightly in soft X-rays. When replenished and full, the post-impulsive flare loops contain plasma heated to temperatures of up to 40 million (4.0×10^7) degrees kelvin.

The soft X-rays emitted during solar flares are thermal radiation, released by virtue of the intense heat and dependent upon the random thermal motions of very hot electrons. At such high temperatures, the electrons are set free from atoms and move off at high speed, leaving the ions (primarily protons) behind. When a free electron moves

Focus 6.1 Thermal electron density from soft X-ray bremsstrahlung

The greater the number of electrons, the stronger the thermal bremsstrahlung. To be precise, the thermal bremsstrahlung power, P, emitted from a plasma increases with the square of the electrons density, N_e, and the volume of the radiating source, V. It also depends upon the temperature of the electrons, T_e. A formula for the bremsstrahlung power is:

$$P = const \times N_e^2 \, V \, T_e^{1/2} \times \text{Gaunt factor},$$

where the Gaunt factor is also a function of the electron temperature. Scientists can use this expression with measurements of the X-ray power during a flare to determine the density of electrons participating in the radiation, assuming that they completely fill the observed volume. Electron density values of $N_e \approx 10^{17}$ electrons per cubic meter are often obtained. Densities of 10^{17} to 10^{18} electrons per cubic meter have also been derived from the ratios of density-sensitive soft X-ray emission lines detected during solar flares.

through the surrounding material, it is attracted to the oppositely charged protons. The electron is therefore deflected from a straight path and changes its speed during its encounter with the proton, emitting electromagnetic radiation in the process (Fig. 6.6). This radiation is called *bremsstrahlung* from the German word for "braking radiation". Scientists use measurements of the flaring X-ray power to infer the density of the electrons emitting the bremsstrahlung (Focus 6.1).

Bremsstrahlung can be emitted at all wavelengths, from long radio waves to short X-rays, but during solar flares it becomes very intense at X-ray wavelengths. As we shall soon seen, the high-speed electrons accelerated during solar flares emit powerful radio-wavelength radiation by a different synchrotron process that involves electrons moving at nearly the velocity of light in the presence of magnetic fields. This radio synchrotron radiation is much more intense than the thermal bremsstrahlung at radio wavelengths during solar flares. In contrast, the free electrons in the outer quiescent, or non-flaring, solar corona do not move fast enough to emit synchrotron radiation, and they instead emit observable thermal bremsstrahlung at long radio wavelengths (Focus 5.4).

The hard X-rays marking the flare onset are also due to bremsstrahlung, but they are produced by electrons that have much higher energies than those emitting the thermal soft X-ray bremsstrahlung. These energetic, high-speed, non-thermal electrons are believed to be accelerated above the tops of coronal loops, and to radiate energy by non-thermal bremsstrahlung as they are beamed down along the looping magnetic channels into the low corona and chromosphere. Only about 0.000 01 or 10^{-5}, of the electron's energy is lost by this non-thermal bremsstrahlung; most of the energy is lost during collisions with ambient, non-flaring electrons.

Observations from the *Yohkoh* spacecraft, launched on 30 August 1991, have confirmed and extended this understanding of X-ray flares. They have shown exactly where both the soft X-rays and hard X-rays are coming from, and confirmed the overall Neupert effect/chromospheric evaporation scenario. According to this picture, solar flare energy release occurs mainly during the rapid, impulsive phase, when charged particles are accelerated and hard X-rays are emitted. The subsequent, gradual phase, detected by the slow build up of soft X-rays, is viewed as an atmospheric response to the energetic particles generated during the impulsive hard X-ray phase.

The *Yohkoh* spectrometer showed that the upflow of heated material coincides in time with the impulsive phase of flares detected as hard X-ray bursts and microwave bursts. *Yohkoh*'s Soft X-ray Telescope, or SXT, additionally demonstrated that the initial site of chromospheric evaporation coincides with the hard X-ray bursts at the footpoints of magnetic loops. Non-thermal electrons are apparently sent down along the closed magnetic loops into the initially low-temperature chromosphere, and the heated material subsequently expands upward to create the gradual soft X-ray phase seen in the corona.

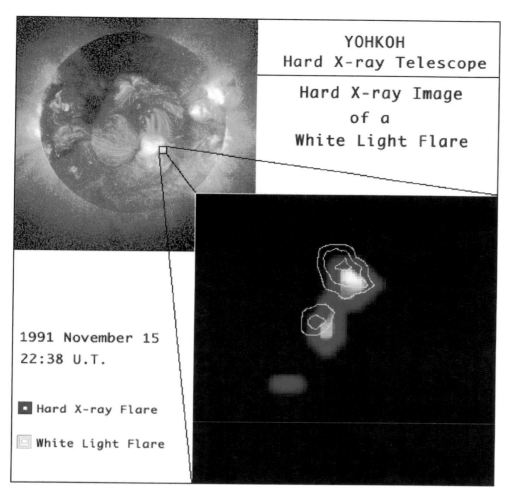

Fig. 6.7 **Double hard X-ray sources** Hard X-ray flares often appear as double sources that are aligned with the photosphere footpoints of a flaring magnetic loop detected at soft X-ray wavelengths. These footpoints can also be the sites of white-light flare emission. The time profiles of this flare, detected on 15 November 1991, show that the increase of white-light emission matches almost exactly that of the hard X-ray flux. This and the simultaneity of hard X-ray emission from the two footpoints establish that non-thermal electrons transport the impulsive-phase energy along the flare loops. These soft X-ray, hard X-ray and white-light images of the solar flare were taken with telescopes aboard the *Yohkoh* mission. (Courtesy of NASA, ISAS, the Lockheed-Martin Solar and Astrophysics Laboratory, the National Astronomical Observatory of Japan, and the University of Tokyo.)

With *Yohkoh*, the double-source, loop footpoint structure of impulsive hard X-ray flares was confirmed with unprecedented clarity. It established a double-source structure for the hard X-ray emission of roughly half the flares observed in the purely non-thermal energy range above 30 keV. The other half of the flares detected with *Yohkoh* were either single sources, that could be double ones that are too small to be resolved, or multiple sources that could be an ensemble of double sources. As subsequently discussed (Section 6.3), a third hard X-ray source is sometimes detected near the apex of the magnetic loop joining the other two; this loop-top region marks the primary energy release site and the location of electron acceleration due to magnetic interaction.

Two white-light emission patches were also detected by *Yohkoh* during at least one flare, at the same time and place as the hard X-ray sources (Fig. 6.7). This shows that the rarely-seen, white-light flares can also be produced by the downward impact of non-thermal electrons, and demonstrates their penetration deep into the chromosphere.

Flare-accelerated ions are also hurled down into the lower solar atmosphere, generating gamma rays there. Gamma rays are even more energetic than X-rays, exceeding 100 keV in energy. The hard X-ray and gamma-ray time profiles of solar flares are time-coincident within the accuracy of measurement, demonstrating the simultaneous acceleration of relativistic electrons and energetic ions in solar flares to within a few seconds. Measurements of the high-energy electrons are obtained from the hard X-rays, produced by non-thermal bremsstrahlung during solar flares, and the flare-accelerated ions are observed using gamma rays. In large solar flares, the onset of the impulsive phase emission is simultaneous for radiation with energies from about 40 keV (hard X-rays) to about 40 MeV (gamma rays). This provides important constraints to theories for how particles are accelerated in solar flares.

Gamma rays from solar flares

Protons and heavier ions are accelerated to high speed during solar flares, and beamed down into the chromosphere where they produce nuclear reactions and generate gamma rays, the most energetic kind of radiation detected from solar flares. Thus, for a few minutes, during an impulsive phase of a solar flare, nuclear reactions occur in the low, dense solar atmosphere; they also occur all the time deep down in the Sun's energy-generating core that is even denser and hotter. Like X-rays, the gamma rays are totally absorbed in our air and must be observed from space.

The protons slam into the dense, lower atmosphere, like a

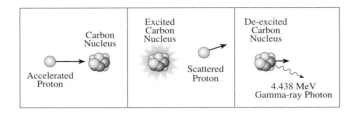

Figure 6.8 **Gamma rays during solar flares** Energetic protons, accelerated during solar flares, can encounter the nuclei of carbon and other elements found in the solar atmosphere. The nucleus is excited to a higher energy level during each collision. It then emits gamma radiation (called a gamma-ray photon) with a specific energy characteristic of the nucleus involved.

bullet hitting a concrete wall, shattering nuclei in a process called spallation. The nuclear fragments are initially excited, but then relax to their former state by emitting gamma rays (Focus 6.2). Other abundant nuclei are energized by collision with the flare-accelerated protons, and emit gamma rays to get rid of the excess energy (Fig. 6.8). The excited nuclei emit gamma rays during solar flares at specific, well-defined energies between 0.4 and 7.1 MeV. One MeV is equivalent to a thousand keV and a million electron volts, so the gamma rays are ten to one hundred times more energetic than the hard X-rays and soft X-rays detected during solar flares.

During bombardment by flare-accelerated ions, energetic neutrons can be torn out of the nuclei of atoms. Whenever one of these neutrons is eventually captured by a hydrogen nucleus (proton) in the photosphere, the neutron and proton form a deuteron nucleus, D, with the delayed emission of a gamma ray line at 2.223 MeV. It is typically the strongest gamma-ray line in solar flare spectra.

Particles of anti-matter are also produced during solar flares, in the form of positrons, e^+, but they cannot stay around very long in our material world. The positrons almost instantaneously annihilate with their material counterparts, the electrons denoted by e^-, producing radiation at 0.511 MeV. It is the second-strongest-gamma ray line from solar flares.

Narrow gamma-ray lines (≤ 100 keV in width) have been observed during solar flares from deuteron formation, or neutron capture, electron-positron annihilation, and excited carbon, nitrogen, oxygen and heavier nuclei (Table 6.3), mainly in the 1980s using the Gamma Ray Spectrometer (GRS) aboard the *Solar Maximum Mission (SMM)* satellite.

When protons with energies above 300 MeV interact with the abundant hydrogen in the solar atmosphere, they can produce short-lived fundamental particles, called mesons, whose decay also leads to gamma-ray emission. The decay of neutral mesons produces a broad gamma-ray peak at 70 MeV, but the decay of charged mesons leads to bremsstrahlung giving a continuum of gamma rays with energies extending to several MeV.

Neutrons with energies above 1000 MeV can also be produced during solar flares. Such energetic neutrons have been directly measured in space near Earth in the 1980s from the SMM and in the 1990s from the *Compton Gamma Ray Observatory*. In the most energetic flares showing meson decay gamma-ray emission, the associated relativistic neutrons can reach the Earth and produce a signal in ground-level neutron monitors.

Non-thermal electrons that have been accelerated during solar flares are additionally hurled out into interplanetary space, emitting intense radio emission in the process. The

Element	Energy (MeV)	Element	Energy (MeV)
Delayed lines		*Proton excitation*	
$e^+ + e^-$ (positron and electron) (pair annihilation)	0.511*	^{14}N (nitrogen)	5.105* 2.313
^{2}H (= D) (deuterium) (neutron capture)	2.223*	^{20}Ne (neon)	1.634* 2.613 3.34
Spallation		^{24}Mg (magnesium)	1.369* 2.754
^{12}C (carbon) ^{16}O (oxygen)	4.438* 6.129* 6.917* 7.117* 2.741	^{28}Si (silicon)	1.779* 6.878
		^{56}Fe (iron)	0.847* 1.238* 1.811
Alpha Excitation			
^{7}Be (beryllium) ^{7}Li (lithium)	0.431* 0.478*		

[a] The most-prominent lines are marked with an asterisk * and have been detected in the flare of 27 April 1981. The energy is given in MeV, where 1 MeV = 1000 keV = 1.6022×10^{-13} J or 1.6022×10^{-6} erg.

Table 6.3 **Some important gamma-ray lines from solar flares**[a]

Focus 6.2 Nuclear reactions on the Sun

Nuclear reactions during solar flares produce gamma-ray lines, emitted at energies between 0.4 and 7.1 MeV. They result when either flare-accelerated protons or flare-accelerated helium nuclei, known as alpha particles, collide with nuclei in the dense atmosphere below the acceleration site.

These reactions are often written using letters to denote the nuclei, a Greek letter γ to denote gamma-ray radiation, and an arrow → to specify the reaction; nuclei on the left side of the arrow react to form products given on the right side of the arrow. The letter p is used to denote a proton, the nucleus of a hydrogen atom, and the Greek letter α signifies an alpha particle, which is the nucleus of the helium atom.

The collision of a flare-associated proton, p, or alpha particle, α, with a heavy nucleus in the dense solar atmosphere may result in a spallation reaction that causes the nucleus to break up into lighter fragments that are left in excited states denoted by an asterisk *. They subsequently de-excite to emit gamma-ray lines. For example, the collision of a flaring proton, p, with an oxygen, ^{16}O, or a neon, ^{20}Ne, nucleus produces excited carbon, $^{12}C^*$, and oxygen, $^{16}O^*$, by the reactions:

$$p + {}^{16}O \rightarrow {}^{12}C^* + \alpha$$

and

$$p + {}^{20}Ne \rightarrow {}^{16}O^* + \alpha,$$

with de-excitation and emission of a gamma-ray line, γ, of energy, $h\nu$, by:

$$^{12}C^* \rightarrow {}^{12}C + \gamma \qquad (h\nu = 4.438 \text{ MeV})$$

and

$$^{16}O^* \rightarrow {}^{16}O + g \qquad (h\nu = 6.129 \text{ MeV}).$$

A flare-accelerated proton, p, can bounce off nucleus, N, one of the abundant nuclei in the solar atmosphere, exciting it by the reaction:

$$N + p \rightarrow N^* + p'$$

where the unprimed and primed sides respectively denote the incident and scattered proton, p. The excited nucleus N* reverts to its former unexcited state by emitting a gamma-ray line, γ, in a reaction denoted by:

$$N^* \rightarrow N + \gamma.$$

Elements and energies of prominent gamma-ray lines resulting from proton excitation of abundant solar elements are given in Table 6.3.

Heavy excited nuclei can also be generated by the fusion of flare-associated particles with lighter nuclei in the atmosphere. Examples include beryllium, $^7Be^*$, and lithium, $^7Li^*$, produced by flaring alpha particles, α, and ambient helium, ^4He.

$$\alpha + {}^4He \rightarrow {}^7Be^* + n$$

and

$$\alpha + {}^4He \rightarrow 7Li^* + p,$$

with the excited nuclei emitting a gamma-ray line, γ, of energy, $h\nu$, by de-excitation:

$$^7Be^* \rightarrow {}^7Be + \gamma \qquad (h\nu = 0.431 \text{ MeV})$$

and

$$^7Li^* \rightarrow {}^7Li + \gamma \qquad (h\nu = 0.478 \text{ MeV}).$$

Gamma-ray lines are produced at 0.511 MeV when matter and anti-matter annihilate each other during solar flares, disappearing in a puff of radiation by:

$$e^- + e^+ \rightarrow \gamma + \gamma \qquad (h\nu = 0.511 \text{ MeV}),$$

where e^- and e^+ respectively denote an electron and a positron.

Another common gamma-ray line is the feature at 2.223 MeV associated with neutron, n, capture by hydrogen nuclei, ^1H, to produce excited deuterons, or excited deuterium nuclei denoted by $^2H^*$, that relax with the emission of gamma rays, γ.

$$^1H + n \rightarrow {}^2H^*$$

with

$$^2H^* \rightarrow {}^2H + \gamma \qquad (h\nu = 2.223 \text{ MeV}).$$

The neutrons are produced when flare-associated ions hit atmospheric nuclei and break them apart. The neutron-capture line emission is delayed, typically by one hundred seconds, from the time of flare acceleration since the neutrons must adjust, or thermalize, before they are captured by the abundant hydrogen nuclei in the photosphere.

---

expulsion of these energetic electrons has been confirmed by direct *in-situ* measurements in interplanetary space (Section 7.1, Fig. 7.1).

Solar radio bursts

The radio emission of a solar flare is often called a radio burst to emphasize its brief, energetic and explosive characteristics. During such outbursts, the Sun's radio emission can increase up to a million times normal intensity in just a few seconds, so a solar flare can outshine the entire Sun at radio wavelengths. Although the radio emission of a solar flare is much less energetic than the flaring X-ray

emission, the solar radio bursts provide an important diagnostic tool for specifying the magnetic and temperature structures at the time. They additionally provide evidence for electrons accelerated to very high speeds, approaching that of light, as well as powerful shock waves.

Radio bursts do not occur simultaneously at different radio frequencies, but instead drift to later arrival times at lower frequencies. This is explained by a disturbance that travels out through the progressively more rarefied layers of the solar atmosphere, making the local electrons in the corona vibrate at their natural frequency of oscillation, called the plasma frequency (Focus 6.3).

Australian radio astronomers pioneered this type of

Focus 6.3 Exciting plasma oscillations in the corona

At the high million-degree temperature of the solar corona, electrons are stripped from abundant hydrogen atoms by innumerable collisions, leaving electrons and protons that are free to move about. The electrons have a negative charge, and the protons are positively charged by an equal amount. The mixture of electrons and protons in the solar corona is called a plasma, and it has no net charge.

When a flare-associated disturbance, such as an electron beam or a shock wave, moves though the coronal plasma, the local electrons are displaced with respect to the protons, which are more massive than the electrons. The electrical attraction between the electrons and protons pulls the electrons back in the opposite direction, but they overshoot the equilibrium position. The light, free electrons therefore oscillate back and forth when a moving disturbance passes through the corona.

The natural frequency of oscillation, called the plasma frequency, depends on the local electron concentration, with a lower plasma frequency at smaller coronal electron densities. The plasma frequency therefore decreases with diminishing electron density at greater distances from the Sun. The exact expression for the plasma frequency, ν_p, is:

$$\nu_p = \left[\frac{e^2 N_e}{4\pi^2 \, \varepsilon_0 \, m_e}\right]^{1/2} = [81 N_e]^{1/2} \text{ Hz}$$

where the electron density, denoted by N_e, is in units of electrons per cubic meter, the electron charge $e = 1.6022 \times 10^{-19}$ coulomb, the electron mass $m_e = 9.1094 \times 10^{-31}$ kg, the constant $\pi = 3.141\ 59$, and the permittivity of free space $\varepsilon_0 = 10^{-9}/(36\pi)$ farad m^{-1}.

Low in the solar corona, where $N_e \approx 10^{14}$ m^{-3}, the plasma frequency is $\nu_p \approx 9 \times 10^7$ Hz = 90 MHz, where one MHz is a million Hz or a million cycles per second. The product of frequency and wavelength is equal to the velocity of light $c = 2.9979 \times 10^8$ m per second, so a frequency of 90 MHz corresponds to a wavelength of 3.33 m. Thus, a solar flare can set the corona oscillating at radio wavelengths of a few meters and less.

As the explosive disturbance moves out through the progressively more rarefied layers of the corona, it excites radiation at lower and lower radio frequencies. Ground-based radio telescopes can identify these radio signals by changing the frequency to which their receiver is tuned. As an example, the radiation at a frequency of 200 MHz, or at 1.5-m wavelength, might arrive about a minute before the 20 MHz (15 m) outburst.

With an electron density model of the solar atmosphere (Fig. 6.9), the emission frequency can be related to height, and combined with the time delays between frequencies to obtain the outward velocity of the moving disturbance. Electron beams that produce type III radio bursts (Table 6.4) are moving at velocities of up to half the velocity of light, or 150 million meters per second. Outward moving shock waves that generate type II radio bursts move at a slower speed, at about a million meters per second.

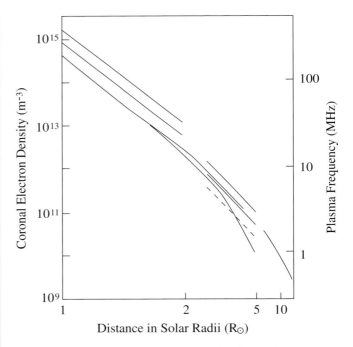

Fig. 6.9 **Plasma frequency and coronal electron density** The plasma frequency (*right axis*) that corresponds with the electron density (*left axis*) in the corona, as inferred from measurements during total eclipses of the Sun. The dashed line is for a coronal hole. As the corona expands out to larger distances from the Sun, and into an ever-larger volume of space, it becomes thinner and less dense, and the plasma frequency is smaller.

investigation in the 1950s, using swept-frequency receivers to distinguish at least two kinds of meter-wavelength radio bursts (Table 6.4). Designated as type II and type III bursts, they both show a drift from higher to lower frequencies, but at different rates. The most common bursts detected at meter wavelengths are the fast-drift type III bursts that provide evidence for the ejection of very energetic electrons from the Sun, with energies of about 100 keV. (An electron moving at half the velocity of light, c, has a kinetic energy of $0.5 \, m_e V^2 = 1.6022 \times 10^{-14}$ J or 100 keV, where m_e is the mass of the electon and the velocity $V = 0.5 \, c$.) These radio bursts last for only a few minutes at the very onset of solar flares and extend over a wide range of radio frequencies (Fig. 6.10).

The slow-drift type II radio bursts drift to lower frequencies at a leisurely pace, corresponding to an outward velocity of about a million meters per second. They are excited by shock waves set up at the time of a solar explosion and moving out into space. Spatially-resolved radio interferometric observations indicate that the type II bursts are very large, with angular extents that can become comparable to that of the Sun.

Solar radio astronomers usually measure frequency in units of MHz, where 1 MHz is equal to a million, or 10^6, Hz. A frequency of 300 MHz corresponds to a wavelength of 1 m. Ground-based instruments observe the dynamic spectra, or frequency drift, for solar radio bursts between 10 MHz and 8000 MHz. Solar radiation is reflected by the Earth's ionosphere at frequencies below 10 MHz, or wavelengths longer than 30 m, and cannot reach the ground.

Type I bursts (Noise storms)	Long-lived (hours to days) sources of radio emission with brightness temperatures from ten million to a billion (10^7 to 10^9) degrees kelvin. Although noise storms are the most common type of activity observed on the Sun at meter wavelengths, they are not associated with solar flares. Noise storms are attributed to electrons accelerated to modest energies of a few keV within large-scale magnetic loops that connect active regions to more distant areas of the Sun.
Type II bursts	Meter-wavelength type II bursts have been observed at frequencies between 0.1 and 100 MHz. A slow drift to lower frequencies at a rate of about 1 MHz per second suggests an outward motion at about a million meters per second and has been attributed to shock waves.
Type III bursts	The most common flare-associated radio bursts at meter wavelengths, observed from 0.1 to 1000 MHz. Type III bursts are characterized by a fast drift from high to low frequency, at a rate of up to 100 MHz per second. They are attributed to beams of electrons thrown out from the Sun with kinetic energies of 10 to 100 keV, and velocities of up to half the velocity of light, or 150 million meters per second. The U-type bursts are a variant of type III bursts that first decrease in frequency and then increase again, indicating upward and downward electron motion along closed magnetic field lines.
Type IV bursts	Broad-band continuum radiation lasting for up to one hour after impulsive flare onset. The radiation from a moving type IV burst is partly circularly polarized, and has been attributed to synchrotron emission from energetic electrons trapped within magnetic clouds that travel out into space with velocities from several hundred thousand to a million meters per second.
Centimeter bursts	Impulsive continuum radiation at centimeter wavelengths that lasts just a few minutes. These microwave bursts are attributed to the gyrosynchrotron radiation of high-speed electrons accelerated to energies of 100 to 1000 keV. The site of acceleration is located above the tops of coronal loops.
Millisecond spikes	Radio flares can include literally thousands of spikes, each lasting a few milliseconds, suggesting sizes less than a million meters across and brightness temperatures of up to a million billion (10^{15}) degrees kelvin, requiring a coherent radiation mechanism.

[a]A frequency of one MHz corresponds to a million Hz, or 10^6 Hz. An energy of 1 keV corresponds to 1.6022×10^{-16} J or 1.6022×10^{-9} erg.

Table 6.4 **Types of solar radio bursts**[a]

Terrestrial long-wavelength radio communication utilizes this reflective capability of the ionosphere to get around the curvature of the Earth.

Spacecraft lofted above the ionosphere are used to track high-speed electrons or shock waves at remote distances from the Sun, where the density and plasma frequency are low. They have monitored the corona's plasma radiation at frequencies from 0.01 to 10 MHz for more than three decades (see Section 7.1, Fig. 7.1).

A type III radio burst emits non-thermal radiation, and cannot be due to the thermal radiation of a hot gas. Thermal radiation is emitted by a collection of particles that collide with each other and exchange energy frequently, giving a distribution of particle energy that can be characterized by a single temperature. Impulsive solar flares do not have enough time to achieve this equilibrium, and the flaring electrons are far too energetic for a thermal process to be at work. To make electrons travel at half the speed of light, with a kinetic energy of 100 keV, a thermal gas would have to be heated to implausible temperatures of one billion degrees, or about 70 times hotter than the center of the Sun. The non-thermal radio bursts are instead accelerated by different

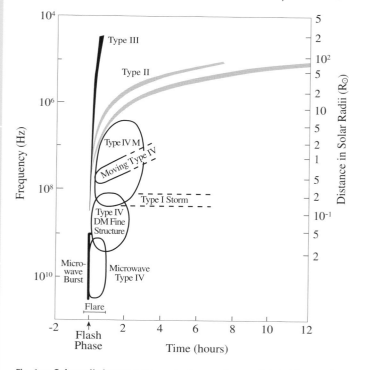

Fig. 6.10 **Solar radio bursts** A large solar flare can be associated with several different kinds of intense radio emission, depending on the frequency (*left vertical axis*) and time after the explosion (*bottom axis*). The impulsive, or flash, phase of the solar flare, starting at 0 hours, normally lasts about 10 minutes and is associated with a powerful microwave burst. Dynamic spectra at frequencies of about 10^8 Hz, show type II and type III bursts that drift from high to low frequencies as time goes on, but at different rates depending on the type of burst. The distance scale (*right vertical axis*) is in units of the Sun's radius of 696 million meters; it is the distance at which the coronal electron density yields a plasma frequency corresponding to the frequency on the left-hand side.

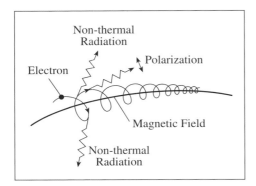

Fig. 6.11 **Synchrotron radiation** High-speed electrons moving at velocities near that of light emit a narrow beam of synchrotron radiation as they spiral around a magnetic field. This emission is sometimes called non-thermal radiation because the electron speeds are much greater than those of thermal motion at any plausible temperature. The name "synchrotron" refers to the man-made, ring-shaped synchrotron particle accelerator where this type of radiation was first observed; a synchronous mechanism keeps the particles in step with the acceleration as they circulate in the ring.

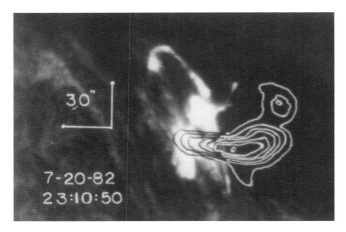

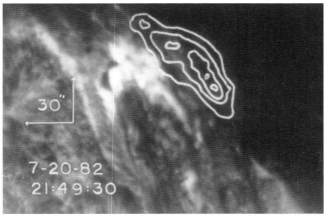

Fig. 6.12 **Site of radio burst** Electrons are rapidly accelerated just above the tops of coronal loops during the early stages of the flare, emitting the powerful loop-like radio signals mapped here with white contours. The underlying hydrogen-alpha (Hα) emission is shown in the accompanying photographs. These 10-second snapshot radio maps were obtained with the Very Large Array (VLA) at a wavelength of 20 cm, or a frequency of 1420 MHz. The angular extent of each 20-cm flaring loop is about one minute of arc, or one-thirtieth of the angular width of the visible solar disk.

processes.

Other non-thermal radio outbursts associated with solar flares, and designated type IV bursts, were discovered at the Nancay Observatory in France. The best-studied component is the so-called moving type IV bursts observed at frequencies from roughly 10 MHz to 100 MHz. They are explained by energetic electrons trapped in a magnetic cloud that is propelled outward into the solar atmosphere at a speed of 10^5 to 10^6 m s^{-1}. Like the non-thermal electrons that produce type III bursts, the electrons giving rise to type IV bursts are traveling at speeds far exceeding those of thermal motions at any plausible temperature.

Focus 6.4 Coherent radio bursts from the Sun and other active stars

Radio astronomers use brightness temperature, T_B, to quantify the emission from radio sources. For a source of radius, R, distance, D, and radio flux density, S, at wavelength, λ, we have $T_B = S\lambda^2 D^2/(2\pi k R^2)$, where k is Boltzmann's constant (Focus 5.4). The brightness temperature is that temperature the source would have if it was emitting thermal radiation from a hot gas. For a given flux density, wavelength and distance, a smaller source has a higher brightness temperature.

Exceptionally small sizes, and correspondingly high brightness temperatures, are inferred for very rapid radio bursts on the Sun and other active stars. In some extreme instances, solar flares at decimeter wavelengths include radio spikes with a rapid rise as short as a few milliseconds and similarly brief duration. Since nothing can move faster than the velocity of light, it sets an upper limit on any travel-time. Rapid changes over time-scales of a few milliseconds therefore restrict the size of the radiating volume to relatively small spatial scales of less than a few million meters across. The radio spikes must have brightness temperatures as high as a million billion (10^{15}) degrees kelvin if their intense radiation is emitted from such a small source.

Yet, such high temperatures on the Sun or other active stars are impossible to achieve by heating any gas, and have only been realized in the earliest, big-bang stages of the expanding Universe where material particles, as we know them, did not exit. The high brightness temperatures of millisecond radio spikes are explained by coherent radiation of ordinary particles working together in an organized manner, rather than by the random, incoherent motions of heated gas particles.

Rapid radio flares from dwarf M stars require sources much smaller than the visible stars in size, implying brightness temperatures exceeding 10^{15} K, so they also require a coherent radiation mechanism. The high circular polarization (100 percent) of these stellar radio bursts indicates an intimate connection with stellar magnetic fields. The available radio and X-ray evidence for flares on other late type stars, including RS CVn binaries, suggests that they also have highly structured magnetic coronas with active regions that are much smaller than the visible stars, and coherent non-thermal radio bursts.

The high-speed electrons that emit type III or type IV bursts spiral around the magnetic field lines in the low corona, moving rapidly at velocities near that of light, and send out radio waves called synchrotron radiation after the man-made synchrotron particle accelerator where it was first observed (Fig. 6.11). Unlike the thermal radiation of a very hot gas, the non-thermal synchrotron radiation is most intense at long, invisible radio wavelengths rather than short X-rays.

Radio bursts at meter wavelengths occur at altitudes far above the site where the flare energy is released and particles are accelerated. The high-speed electrons originate much lower in the corona, and are observed at decimeter-wavelengths (1 decimeter = 0.1 meter = 10 centimeter). Decimeter-wavelength bursts have been observed by Swiss radio astronomers at frequencies up to 8000 MHz, and at wavelengths as short as 3.75 centimeters, moving downward from the acceleration regions into the low corona.

A giant array of radio telescopes, located near Socorro, New Mexico and called the Very Large Array (Section 9.2), can zoom in at the very moment of a solar flare, taking snapshot images with just a few seconds exposure (Fig. 6.12). It has pinpointed the location of the impulsive decimeter radiation and the electrons that produce it. These radio bursts are triggered low in the Sun's atmosphere, unleashing their vast power just above the apex of magnetic arches, called coronal loops, that link underlying sunspots of opposite magnetic polarity. Some of the energetic electrons are confined within the closed magnetic structures, and are forced to follow the magnetic fields down into the chromosphere. Other high-speed electrons break free of their magnetic cage, moving outward into interplanetary space along open magnetic field lines and exciting the meter-wavelength type III bursts.

To complete our inventory of solar radio bursts, there are millisecond spikes whose rapid rise times and short duration require an entirely different, coherent radiation mechanism that is also needed to explain some radio bursts on other active stars (Focus 6.4).

6.2 Coronal mass ejections and eruptive prominences

Another dramatic, magnetically energized type of solar explosion is called a Coronal Mass Ejection, or CME for short. This is a giant magnetic bubble that rapidly expands to rival the Sun in size, and hurls billions of tons of million-degree gas into interplanetary space at speeds of about 4×10^5 m s^{-1}, reaching the Earth in about four days. Their associated shocks accelerate and propel vast quantities of high-speed particles ahead of them.

CMEs are detected during routine visible-light observations of the corona from spacecraft such the *SOlar and Heliospheric Observatory*, or *SOHO*. With a disk in the center to block out the Sun's glare, the coronagraph sees huge pieces of the corona that are blasted out from the edge of the occulted photosphere (Figs. 6.13, 6.14). The mass ejections are seen as bright moving loop-like features in the corona, whose white light is scattered by free electrons.

Bright regions in coronagraph images have the greatest concentration of electrons and therefore contain excess coronal mass, while dark regions have fewer electrons and less mass. Each time a mass ejection rises out of the corona, it carries away 5 to 50 billion tons (5×10^{12} to 5×10^{13} kg) of coronal material. A CME therefore rips a giant piece out of the corona, and this causes a scar-like reduction in its soft X-ray emission. The expelled mass inferred from this large-scale, soft X-ray dimming of the soft X-ray corona is comparable to that estimated for coronal mass ejections (see Focus 6.5).

Coronal mass ejections are huge. Their average angular span of 45 degrees along the disk edge implies a size near the Sun that is comparable to that of the visible solar disk, and they can expand to even larger sizes further out.

Nearly everything we know about CMEs has been learned

Fig. 6.13 **Coronal mass ejection** A huge coronal mass ejection is seen in this coronagraph image, taken on 27 February 2000 with the Large Angle Spectrometric COronagraph (LASCO) on *SOHO*. The white circle denotes the edge of the photosphere, so this mass ejection is about twice as large as the visible Sun. The dark area corresponds to the occulting disk of the coronagraph that blocks intense sunlight and permits the corona to be seen. About one hour before this image was taken, another *SOHO* instrument, the Extreme-ultraviolet Imaging Telescope or EIT, detected a filament eruption lower down near the solar chromosphere. (Courtesy of the *SOHO* LASCO consortium. *SOHO* is a project of international cooperation between ESA and NASA.)

in just a few decades. They were discovered on 14 December 1971 and 8 Februrary 1972 using the white-light coronagraph aboard NASA's *Seventh Orbiting Solar Observatory* (*OSO 7*). Since then, thousands of CMEs have been identified

Satellite	OSO 7	Skylab	P78-1	SMM	SOHO
Coronagraph		ATM[a]	Solwind	C/P[b]	LASCO/C3[c]
Observation time	Oct. 71 to June 74	May 73 to Feb. 74	March 79 to Sept. 85	Feb.–Sept. 80 Apr. 84 to Nov. 89	Feb. 96 on
Field of view[d]	3 to 10 R$_\odot$	1.6 to 6 R$_\odot$	2.6 to 10 R$_\odot$	1.6 to 8 R$_\odot$	3.7 to 30 R$_\odot$
Resolution[e] (seconds of arc)	75	8	75	6.4	56

[a] Apollo Telescope Mount; [b] designates Coronagraph/Polarimeter; [c] LASCO consists of three coronagraphs, designated C1, C2 and C3, with respective overlapping fields of view of 1.1 to 3, 1.7 to 6, and 3.7 to 32 solar radii and respective pixel angular resolutions of 5.6, 11.2 and 56.0 seconds of arc; [d] here R$_\odot$ denotes the solar radius, or 696 million (6.96×10^8) meters; [e] an angular resolution of 1 second of arc corresponds to 725 thousand (7.25×10^5) meters at the Sun.

Table 6.5 **Space-borne orbiting coronagraphs**[a]

using data from the NASA's *Skylab* in 1973-74, the U. S. Air Force's *P78-1* satellite (1979-1985), NASA's *Solar Maximum Mission* (SMM, 1980 and 1984-1989), and the Large Angle Spectrometric COronagraph (LASCO) of the *SOlar and Heliospheric Observatory* (SOHO, 1996 on) – Table 6.5. LASCO observes the corona from 1.1 to 30 solar radii from Sun center, looking closer to, and further from, the Sun than all previous space-borne coronagraphs.

These spacecraft observations have established the essential physical properties of coronal mass ejections, showing that they are big, massive, fast and energetic (Table 6.6).

The Sun is not wasting away because of coronal mass ejections, in spite of their awesome mass. Coronal mass ejections typically occur only once a day, on average, and therefore account for only about five percent of the mass loss due to the steady, perpetual solar wind (Focus 6.5).

By the time that they are a few solar radii above the photosphere, CMEs have reached a cruising velocity of about 4×10^5 m s^{-1}. At that speed, the expelled mass can reach the Earth in about 100 hours, carrying with it an average kinetic energy of 10^{23} to 10^{24} J (Focus 6.5). The amount of energy that a CME liberates in producing the motions of the expelled mass and in lifting it against the Sun's powerful gravity is roughly comparable to the energy of a typical solar flare. However, most of the energy of a mass ejection goes into the expelled material, while a flare's energy is mainly transferred into accelerated particles that subsequently emit intense X-ray and radio radiation.

Since the white-light brightness varies in proportion to the electron density and does not depend on the temperature, the coronagraph images reveal the density structure of the corona. Sequential coronagraph images record changes in the density of the corona, and thereby reveal transient expulsions of matter at the apparent edge of the Sun. These CMEs balloon out of the solar atmosphere at high speed, moving through the coronagraph's field of view in just a few hours. Such events work only in one direction, always moving away from the Sun into interplanetary space and never falling back in the reverse direction.

Coronal mass ejections usually have expanding

Focus 6.5 Mass, mass flux, energy and time delay of coronal mass ejections

Coronal mass ejections are detected as localized increases in the brightness of white-light coronagraph images. Integration of the brightness increase, that depends only on the electron density, N_e, permits evaluation of the total mass, M, of the ejection. For a sphere of radius, R, we have:

$$M = 4\pi R^3 N_e m_p /3,$$

where the proton mass $m_p = 1.67 \times 10^{-27}$ kg. The corona is a fully ionized, predominantly (90 percent) hydrogen plasma, so the number densities of protons and electrons are equal, but since the protons are 1,836 times more massive than the electrons, the protons dominate the mass.

For a mass ejection with an electron, or proton, density of $N_e = 10^{13}$ electrons per cubic meter, that has grown as large as the Sun, with $R = 6.96 \times 10^8$ m, this expression gives

$$M = 10^{13} \text{ kg} = 10 \text{ billion tons},$$

where one ton is equivalent to 1000 kilograms. Because it looks at regions further away from the Sun than previous coronagraphs were able to do, the LASCO coronagraph observes CMEs that sometimes evolve into bigger structures than those seen before, leading to mass estimates up to 50 billion tons.

At the rate of one ejection per day, and 10^{13} kg per ejection, this amounts to a mass flow rate of about 10^8 kg s^{-1}, since there are 86 400 seconds per day. By way of comparison, the flux of the solar wind, discussed in the next chapter, is about 5×10^{12} protons m^{-2} s^{-1}, or 8.3×10^{-15} kg m^{-2} s^{-1} near the Earth. If this flux is typical of that over the entire Sun-centered sphere, with an average Sun-Earth distance of $D = 1.5 \times 10^{11}$ m, we can multiply by the sphere's surface area, $4\pi D^2$, to obtain a solar-wind mass flux of about 2×10^9 kg s^{-1}. Thus, coronal mass ejections contribute roughly five percent of the solar-wind mass flux.

The kinetic energy of a coronal mass ejection with a speed of $V = 4 \times 10^5$ m s^{-1} and a mass $M = 10^{13}$ kg is:

$$\text{kinetic energy} = MV^2/2 \approx 10^{24} \text{ J} = 10^{31} \text{ erg}.$$

This is comparable to the energies of large solar flares that lie between 10^{21} and 10^{25} J.

At a speed of $V = 4 \times 10^5$ m s^{-1}, the time, T, to travel the average distance, D, from the Sun to the Earth is:

$$T = D/V = 3.75 \times 10^5 \text{ s} = 104 \text{ hours} = 4.34 \text{ days}.$$

Characteristic	Value
Average angular width (heliocentric)	45 degrees
Largest mass ejected	5×10^{12} to 5×10^{13} kg (5 to 50 billion tons)
Frequency of occurrence [b]	3.5 events per day (activity maximum) 0.2 events per day (activity minimum)
Mass flow rate	About 2×10^8 kilograms per second
Speed range of leading edge	5×10^4 to 1.2×10^6 m s^{-1}
Average speed of leading edge	4×10^5 m s^{-1}
Average time to reach Earth	About 100 hours
Average kinetic energy	Approximately 10^{23} to 10^{24} J

[a] Also see Focus 6.5.
[b] LASCO's improved sensitivity enabled it to detect 0.8 events per day during activity minimum.

Table 6.6 **Physical properties of coronal mass ejections near the Sun**[a]

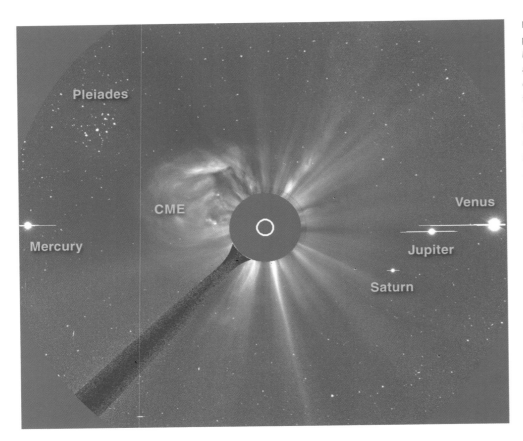

Fig. 6.14 **Coronal mass ejection, planets and Pleiades** A Coronal Mass Ejection, or CME, four planets and the bright stars of the Pleiades cluster are all captured in this image from *SOHO*, taken with the LASCO instrument on 15 May 2000. The CME is seen in the foreground of the 15-degree-wide field of view, leaving the Sun whose position and relative size are indicated by the central white circle. Four of the five naked-eye planets are seen in a rare alignment, as bright disks bisected by a line which is an instrumental effect caused by the intense sunlight reflected from the planet. *SOHO* is about 150 billion meters away from the Sun, while Mercury, Venus, Jupiter and Saturn are, respectively, about 18, 110, 780 and 1400 billion meters beyond the Sun. The Pleiades star cluster is 3.86 million trillion (3.86×10^{18}) meters (408 light-years) away. (Courtesy of the *SOHO* LASCO consortium. *SOHO* is a project of international collaboration between ESA and NASA.)

curvilinear shapes that resemble the cross-sections of loops, shells or filled bubbles, suggesting magnetically closed regions that are sporadically blown out by the eruption. The upper portions of the magnetic loops are sometimes carried out by the highly-ionized material, while remaining attached and rooted to the Sun at both ends. In other situations, the expelled material stretches the magnetic field until it snaps, taking the coiled magnetism with it and lifting off into space like a hot-air balloon that breaks its tether. Whenever a big, closed loop of magnetism is unable to hold itself down, a coronal mass ejection takes off.

Mass ejections erupt from the Sun as self-contained structures of hot material and magnetic fields. They apparently result from a rapid, large-scale restructuring of magnetic fields in the low corona. The spatial distribution of CMEs over the 11-year cycle of magnetic activity is similar to that of the other large structures on the Sun, the coronal streamers and their underlying filaments. Near activity minimum the coronal mass ejections are largely confined to equatorial regions, but near maximum they can be observed at all solar latitudes.

Like sunspots, solar flares and other forms of solar activity, CMEs occur with a frequency that varies in step with the 11-year cycle. A few coronal mass ejections balloon out of the corona per day, on average, during activity maximum, and the rate decreases by about an order of magnitude by sunspot minimum. The rate also depends on the sensitivity of the coronagraph; the *SOHO* LASCO instrument observed

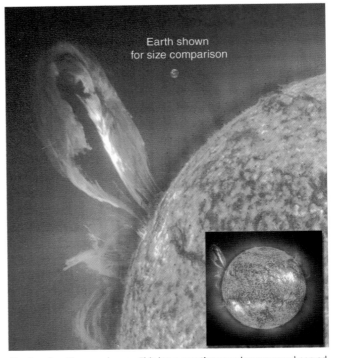

Fig. 6.15 **Eruptive prominence** This large erupting prominence was observed in the extreme-ultraviolet light of ionized helium (He II at 30.4 nm) on 24 July 1999. The comparison image of the Earth shows the enormous extent of the prominence; the inset full-disk solar image indicates that the eruption looped out for a distance almost equal to the Sun's radius. (Courtesy of the *SOHO* EIT consortium. *SOHO* is a project of international collaboration between ESA and NASA.)

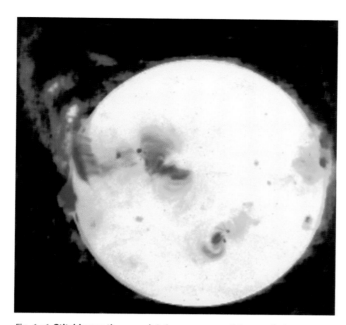

Fig. 6.16 **Stitching up the wound** A time sequence of three radio images show an eruptive prominence above the solar disk at the north-east (*top left*) during a 1.5-hour period. A negative soft X-ray image shows an arcade of loops formed by magnetic reconnection at a slightly later time, as if the Sun was stitching itself back together after the eruption. The radio images are from the Nobeyama Radioheliograph, while the X-ray image was obtained with the Soft X-ray Telescope (SXT) on the *Yohkoh* satellite. (Courtesy of Shinzo Enome, Nobeyama Radio Observatory.)

about 0.8 CMEs per day at activity minimum.

A powerful mass ejection, with its associated shock, energetic particles and magnetic fields, can pummel the Earth's magnetic environment in space, sometimes with devastating consequences for our planet. When it hits the Earth, a mass ejection can produce powerful geomagnetic storms, intense auroras, electrical power blackouts, and other threatening consequences (Section 8.3).

Although CMEs are often observed as magnetic bubbles ejected from one side of the Sun, these events are not likely to collide with the Earth. A coronagraph can detect an Earth-directed CME as a gradually expanding, Sun-centered ring around the Sun. Such a halo coronal mass ejection may signal a future terrestrial hit. Still, you cannot tell from the halo itself whether the mass ejection is targeted for the Earth or moving away from it in the opposite direction.

Coronal mass ejections often exhibit a three-part structure - a smooth, bright outer loop or bubble of enhanced density, followed by a dark cavity of low density, within which sits an erupted prominence. The leading bright loop or shell is the coronal mass ejection that opens up and lifts off like a huge umbrella in the solar wind, piling the corona up and shoving it out like a snowplow. About 70 percent of the coronal mass ejections are associated with, and followed by, eruptive prominences (Fig. 6.15).

The close proximity of coronal mass ejections and erupting prominences in space and time indicates that they are consequences of a similar instability and restructuring of the large-scale (global) magnetic fields in the corona. Restraining coronal magnetic fields may be blown open or carried away by the coronal mass ejection, which precedes the prominence eruption

Quiescent prominences, or filaments, can hang above the photosphere for weeks or months (Section 5.2). Then the supporting magnetism becomes unhinged. The quiescent prominence becomes completely unstable, and it erupts. Instead of falling down under gravity, the stately structures rise up and break away from the Sun, as though propelled by a loaded, twisted spring. The eruptive prominence ascends, and disappears in tens of minutes to several hours. The eruption of an old quiescent prominence is thus sometimes called a *disparition brusque*, French for "sudden disappearance". Some of the prominence material escapes from the Sun, releasing a mass equivalent to that of a small mountain in just a few hours, while some of the material descends to the chromosphere along helical arches.

In two-thirds of the cases, the prominence reforms after the explosive convulsion in the same place and with much the same shape over the course of 1 to 7 days. It is as if some irritation builds up beyond the limit of tolerance, and the magnetic structure tosses off the pent-up frustration, like a dog shaking off in the rain.

Closed magnetic loops apparently support the long, thin prominence, or filament, like parallel hammocks, at heights of up to 100 million meters, or ten times the diameter of the Earth. This arcade of closed loops is anchored in the Sun, but is opened up at the top by the rising filament, like taking the cork out of a bottle of champagne. The magnetism subsequently reconnects and closes up again beneath the erupting prominence, forming a new arcade of closed loops.

We can see the magnetic backbone of an erupting prominence regroup and close up again in soft X-ray images, retaining a memory of its former stability. When the prominence erupts, it is replaced by a row of bright X-ray emitting loops (Fig. 6.16), aligned like the bones in your rib cage or the arched trestle in a rose garden. First observed from *Skylab*, the X-ray loops bridge the magnetic neutral line between opposite polarity regions in the photosphere, stitching together and healing the wound inflicted by emptying that part of the corona.

Yohkoh's Soft X-ray Telescope has investigated these beautiful, arched, post-ejection structures in detail. The loops form progressively from one end of the arcade to the other, and they also rise vertically with time. The enhanced X-ray emission is first seen after the onset of the mass ejection, and remains long after the ejection has departed from the low corona. These long-duration (hours) soft X-ray events, are due to global magnetic restructuring that proceeds along the ejection path.

Roughly 40 percent of coronal mass ejections are accompanied by solar flares that occur at about the same time and place. In fact, the flare-associated, meter-wavelength type II and IV radio bursts suggested the expulsion of mass from the corona long before coronal mass ejections were discovered using white-light coronagraphs. Yet, the physical size of coronal mass ejections is huge compared with flares or even the active regions in which flares occur. Moreover, the accompanying flares can occur at any time before, during or after the departure or "lift-off" of the mass ejection, and the relative locations of the two phenomena also show no systematic ordering.

So, most mass ejections are not initiated by solar flares and most flares are not caused by the ejections. After all, the flares are much more common. Nevertheless, it seems that the two types of solar explosions do involve similar processes, and they can result from the same magnetic activity in the corona.

What causes the Sun's magnetism to suddenly erupt with enough force to drive a large section of the corona out against the restraining force of solar gravity? The triggering mechanism seems to be related to large-scale interactions of magnetic fields in the low solar corona. This magnetism is always slowly evolving. It is continuously emerging from inside the Sun, and disappearing back into it, driven by the Sun's 11-year cycle of magnetic activity. The release of a coronal mass ejection appears to be one way that the solar atmosphere reconfigures itself in response to these slow magnetic changes. The physical processes involved in storing the magnetic energy in the solar corona and releasing it to launch coronal mass ejections or ignite solar flares are next discussed.

6.3 Theories for explosive solar activity

Powerful solar flares involve the explosive release of incredible amounts of energy, sometimes amounting to as much a million billion billion (10^{24}) joule in just a few minutes. A substantial fraction of this energy goes into accelerating electrons and protons to very high speeds. Comparable amounts of energy are released in expelling matter during a CME, and they are most likely powered by similar processes to those that drive solar flares.

To explain how solar explosions happen, we must first know where their colossal energy comes from. The only plausible sources of energy for these powerful outbursts are the strong magnetic fields in the low solar corona (Focus 6.6). After all, solar flares occur in active regions where the strongest magnetic fields are found. Both solar flares and CMEs are also synchronized with the Sun's 11-year cycle of magnetic activity, becoming more frequent and violent when sunspots and intense magnetic fields are most commonly observed.

Once the source of explosive energy has been established, we must explain where and why that energy is suddenly and rapidly let go. The energy release for solar flares has to occur in the low corona where the energetic particles are

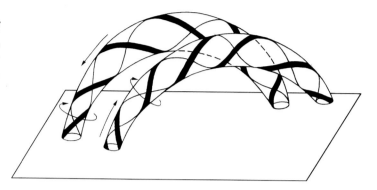

Fig. 6.17 **Magnetic interaction** A pair of oppositely-directed, twisted coronal loops come in contact, releasing magnetic energy to power a solar flare. Arrows indicate the direction of the magnetic field lines and the sense of twist. Such magnetic encounters can occur when newly emerging magnetic fields rise through the photosphere to merge with pre-existing ones in the corona, or when the twisted coronal loops are forced together by underlying motions.

accelerated and the magnetic fields are strong enough to provide the necessary energy. Coronal mass ejections similarly require intense magnetic fields to be sufficiently energized, and their enormous size suggests an origin above the photosphere and chromosphere.

The free magnetic energy needed to power the solar explosions is stored in the low corona in the form of non-potential magnetic field components, or, equivalently, as electric current systems. We can see how this excess "free" magnetic energy might be produced in the corona by considering a magnetic loop that links regions of opposite magnetic polarity or direction in the underlying photosphere. The line dividing the two polarities is called the magnetic neutral line. The magnetic fields have a potential configuration if they connect the two polarities in the shortest, most direct path, and therefore run perpendicular to the magnetic neutral line. They can be distorted into a non-potential shape when motions at or below the photosphere shear and twist the looping fields, creating more magnetic energy than the potential form. This extra energy, called free magnetic energy, can be released during explosions on the Sun.

The free magnetic energy accumulates in the corona, but it comes from the dynamo below. Differential rotation and turbulent convective churning shuffle the photospheric footpoints of coronal loops, and these loops become sheared, twisted and braided. All of this distortion creates large electric current densities and non-potential magnetic fields within the coronal gas.

But what triggers the instability and suddenly ignites the explosions from magnetic loops that remain unperturbed for

---

Focus 6.6 Energizing solar flares

How much energy is released during a typical solar flare? The total flare energy, E_f, expended in producing electrons with energies, E_e, of about 30 keV, or 4.8×10^{-15} J, within a sphere of R with an electron density of $N_e \approx 10^{17}$ electrons m^{-3} (Focus 6.1) is:

$$E_f = 4\pi R^3 E_e N_e / 3 \approx 2 \times 10^{24} \text{ J},$$

where the radius of a compact flare is $R = 10^7$ m, or about twice the Earth's radius. Such a flare subtends an angular radius of 14 seconds of arc when viewed from the Earth.

Magnetic energy stored in the low corona powers these flares. The magnetic energy, E_m, for a magnetic field strength, B, in a radius, R, is:

$$E_m = [4\pi / (6\mu_0)] B^2 R^3 = 1.66 \times 10^6 \, B^2 R^3 \text{ J},$$

where the permeability of free space is $\mu_0 = 4\pi \times 10^{-7}$ henry m^{-1}, the radius is in meters and the magnetic field strength in tesla. To provide the total flare energy, $E_f = 2 \times 10^{24}$ J in a sphere of radius $R = 10^7$ m, a magnetic field of about 0.03 tesla is required. Solar astronomers often use the c.g.s. unit of gauss, where 1 gauss = 10 000 tesla, so the required magnetic field change in the corona is roughly 300 gauss. Lower magnetic field strengths might suffice if the volume is larger and the electron density smaller.

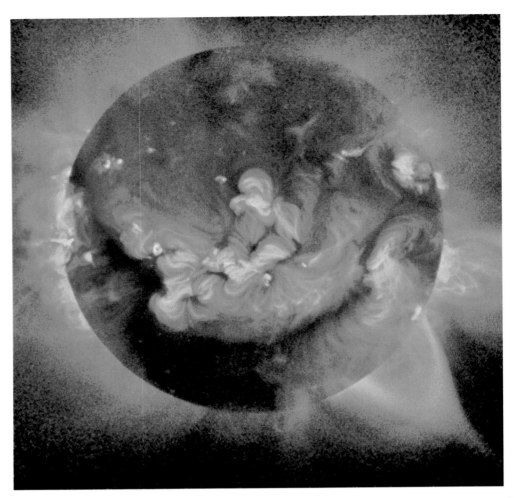

Fig. 6.18 **Cusp geometry** A large, soft X-ray cusp structure (*lower right*) is detected after a coronal mass ejection on 25 January 1992. The cusp, seen edge-on at the top of the arch, is the place where the oppositely-directed magnetic fields, threading the two legs of the arch, are stretched out and brought together. Several similar images have been taken with the Soft X-ray Telescope (SXT) aboard *Yohkoh*, showing that magnetic reconnection is a common method of energizing solar explosions. (Courtesy of Loren W. Acton, NASA, ISAS, the Lockheed-Martin Solar and Astrophysics Laboratory, the National Astronomical Observatory of Japan, and the University of Tokyo.)

long intervals of time? They might be triggered when magnetized coronal loops, driven by motions beneath them, meet to touch each other and connect (Fig. 6.17). If magnetic fields of opposite polarity are pressed together, an instability takes place and the fields partially annihilate each other. Nevertheless there is still some magnetism left. The magnetic field lines are never permanently broken, and they simply reconnect into new magnetic configurations. The non-potential components of the magnetic fields are destroyed in this reconnection process, and their free magnetic energy is used to energize solar explosions.

So, we now think of these powerful outbursts as stemming from the interaction of coronal loops. They are always moving about, like swaying seaweed or wind-blown grass, and existing coronal loops may often be brought into contact by these movements. Magnetic fields coiled up in the solar interior, where the Sun's magnetism is produced, could also bob into the corona to interact with pre-existing coronal loops. In either case, the coalescence leads to the rapid release of free magnetic energy through magnetic reconnection.

The explosive instability has been compared to an earthquake. According to this analogy, the moving roots or footpoints of a sheared magnetic loop resemble two tectonic plates. As the plates move in opposite directions along a fault line, they grind against each other and build up stress and energy. When the stress is pushed to the limit, the two plates cannot slide further, and the accumulated energy is released as an earthquake. That part of the fault line then lurches back to its original, equilibrium position, waiting for the next earthquake. After an explosive convulsion on the Sun, the magnetic fields similarly regain their composure, fusing together and becoming primed for the next outburst.

The rapid, explosive loss of equilibrium has additionally been compared to avalanches in a sand pile or on a ski slope, to the quick snap of a rubber band that has been twisted too tightly, and to the sudden flash and crack of a lightening bolt.

The Soft X-ray Telescope, or SXT, aboard the *Yohkoh* spacecraft has demonstrated that magnetic interactions can indeed strike the match that ignites solar explosions. Sequential SXT images indicate that magnetized loops can become sheared and twisted, while also converging within active regions or on larger scales. SXT also witnessed the coalescence of emerging, twisted loops with overlying ones. The magnetic distortions suggest the build up of energy that is dissipated at the place where the loops meet.

Yohkoh's SXT even revealed the probable location of the magnetic reconnection site, showing that the rounded magnetism of a coronal loop can be pulled into a peaked shape at the top (Fig. 6.18). The sharp, cusp-like feature marks the place where oppositely directed field lines stretch

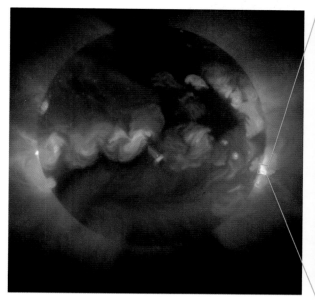

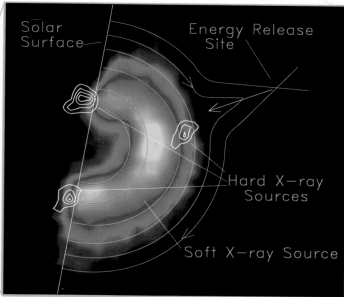

Fig. 6.19 Hard X-ray flare in the corona White contour maps show three impulsive hard X-ray sources from high-energy electrons accelerated during a solar flare, superposed on the loop-like configuration of soft X-rays emitted during the flare gradual or decay phase. In addition to the double, footpoint sources, a hard X-ray source exists in the low corona above the corresponding soft X-ray magnetic loop structure, with an intensity variation similar to those of the other two hard X-ray sources. This indicates that the flare is energized from a site near the loop top. These simultaneous images were taken on 13 January 1992 with the Hard X-ray Telescope (HXT) and the Soft X-ray Telescope (SXT) aboard *Yohkoh*. (Courtesy of Satoshi Masuda, NASA, ISAS, the Lockheed-Martin Solar and Astrophysics Laboratory, the National Astronomical Observatory of Japan, and the University of Tokyo.)

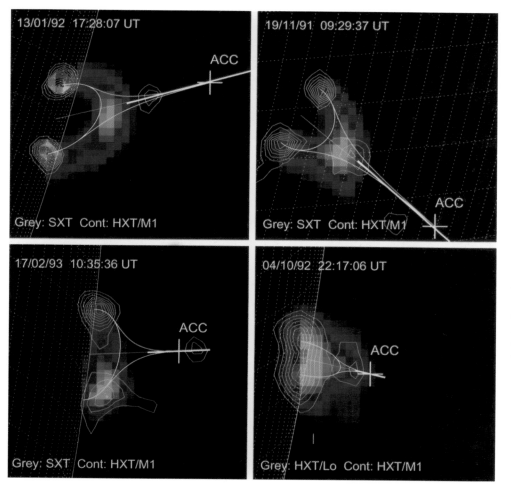

Fig. 6.20 Particle acceleration site Time-of-flight localization of the acceleration site (*labeled with a cross*) indicates that the energetic flaring electrons originate above the coronal loops detected at soft X-ray wavelengths and the hard X-ray bursts (*contours*) located at the loop footpoints. These images were obtained with the Hard X-ray (HXT) and Soft X-ray (SXT) Telescopes aboard *Yohkoh* at the dates and times indicated on each image. The precise timing of the hard X-rays was obtained with the large-area hard X-ray detectors aboard the *Compton Gamma Ray Observatory* (*CGRO*), (Courtesy of Markus J. Aschwanden, NASA and ISAS, the Lockheed-Martin Solar and Astrophysics Laboratory, the National Astronomical Observatory of Japan, and the University of Tokyo.)

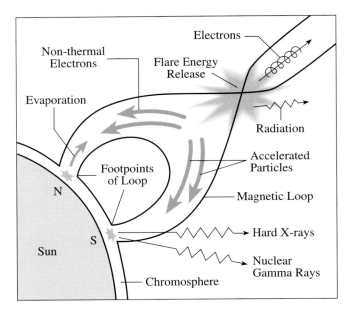

Fig. 6.21 **Solar flare model** A solar flare is powered by magnetic energy released from a magnetic interaction site above the top of the loop shown schematically here. Electrons are accelerated to high speed, generating a burst of radio energy as well as impulsive loop-top hard X-ray emission. Some of these non-thermal electrons are channeled down the loop and strike the chromosphere at nearly the speed of light, emitting hard X-rays by electron-ion bremsstrahlung at the loop footpoints. When beams of accelerated protons enter the dense, lower atmosphere, they cause nuclear reactions that result in gamma-ray spectral lines and energetic neutrons. Material in the chromosphere is heated very quickly and rises into the loop, accompanied by a slow, gradual increase in soft X-ray radiation. This upwelling of heated material is called chromospheric evaporation.

out nearly parallel to each other and are brought into close proximity. Here the magnetism comes together, merges and reconnects, releasing the energy needed to power a solar explosion. Many long-lived (hours), gradual soft X-ray flares show cusp-shaped loop structures suggesting magnetic reconnection, and they are often associated with coronal mass ejections.

Evidence for magnetic reconnection during the compact, short-lived (minutes) impulsive solar flares has been obtained by observing them at the limb of the Sun. Co-aligned *Yohkoh* images show a compact, impulsive hard X-ray source well above and outside the corresponding soft X-ray loop structure, in addition to the double-source footpoint hard X-ray emission (Fig. 6.19). The hard X-rays at the loop footpoints are produced when high-speed, non-thermal electrons collide with dense material in the chromosphere. However, there is no similar dense material out in the

tenuous corona, so the hard X-ray emission out there has to be emitted during the electron acceleration process. The loop-top source most likely represents the site where oppositely-directed magnetic fields meet and electrons are accelerated to high energy. These electrons rapidly move down toward the footpoints, explaining the similarity of the time variations of all three hard X-ray sources.

As previously noted (Section 6.1), solar radio astronomers have, in the meantime, shown that flare energy release and the acceleration of high-energy electrons occurs near the tops of coronal loops. The demarcation region between downward-directed and upward-directed electron beams, observed during some type III radio bursts, pinpoints the acceleration site. The electron density at this place, inferred from the plasma frequency, is about 10^{16} electrons m^{-3}. This is one to two orders of magnitude, or 10 to 100 times, less than the density of the soft X-ray flare loops, indicating that the acceleration site is above these loops. Moreover, electron time-of-flight measurements with the *Compton Gamma Ray Observatory* (CGRO) satellite, confirm the existence of a coronal acceleration site for flares observed with both CGRO and *Yohkoh* (Fig. 6.20).

In summary, a well-developed magnetic theory for solar explosions has received substantial observational verification. These violent outbursts originate in the low solar corona, where free magnetic energy, associated with non-potential magnetic fields, is stored. Thanks to the detailed views afforded by radio telescopes on the ground and X-ray observatories in space, we now know that the explosions are frequently triggered in compact structures just above the tops of coronal loops. Magnetic fields of opposite magnetic polarity are probably driven together there.

During an impulsive solar flare, the free magnetic energy released during magnetic reconnection is converted to charged particle kinetic energy. In less than a second, electrons are accelerated to nearly the speed of light, producing intense radio signals. Protons are likewise accelerated, and both the electrons and protons can be hurled down into the Sun and out into space. The downward moving beams strike the denser chromosphere below, producing nuclear reactions and creating X-rays and gamma rays (Fig. 6.21).

Coronal mass ejections are also probably energized and expelled by a similar process, leaving cusp-shaped soft X-ray loops and arcades of loops behind as signatures of the magnetic reconnection. As we shall next see, the magnetic contortions that act as the prelude to this energy release may provide warning signals of impending solar flares or coronal mass ejections that can threaten satellites and humans in space.

6.4 Predicting explosions on the Sun

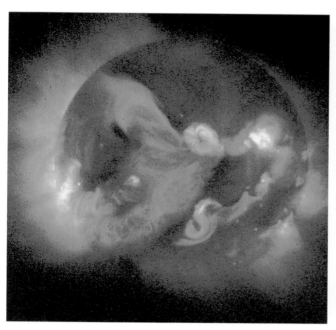

Fig. 6.22 **Magnetic shapes** This X-ray image shows loop-like emission, with contorted, twisted geometry, as well as bright, compact, active-region coronal loops and relatively faint and long magnetic loops. It was taken on 11 June 1992 with the Soft X-ray Telescope (SXT) aboard *Yohkoh*. (Courtesy of NASA, ISAS, the Lockheed-Martin Solar and Astrophysics Laboratory, the National Astronomical Observatory of Japan, and the University of Tokyo.)

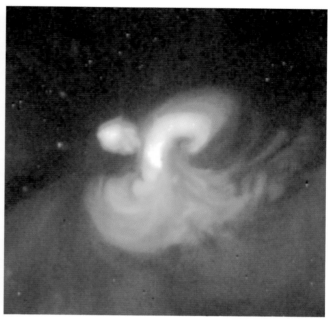

Fig. 6.23 **The Sun getting ready to explode** When the coronal magnetic fields get twisted into an S, or sigmoid, shape, they become dangerous, like a coiled rattlesnake waiting to strike. Statistical studies indicate that the appearance of such a large S or inverted S shape in soft X-rays is likely to be followed by an explosion in just a few days. This image was taken on 8 June 1998 with the Soft X-ray Telescope (SXT) aboard *Yohkoh*. (Courtesy of Richard C. Canfield, NASA, ISAS, the Lockheed-Martin Solar and Astrophysics Laboratory, the National Astronomical Observatory of Japan, and the University of Tokyo.)

The low solar corona is in a constant state of agitation and metamorphosis. Coronal loops are magnetically reconfigured as they twist and writhe in response to internal differential rotation and convective motions (Fig. 6.22). Yet, the coiled magnetic fields hold their energy in place, and remain without substantial change for days, weeks and even months at a time, like a rattlesnake waiting to strike. Then they suddenly and unpredictably go out of control, igniting an explosion that tears the magnetic cage apart and breaks its grip.

The solar explosions emit energetic particles, intense radiation, powerful magnetic fields and strong shocks that can have enormous practical implications when directed toward the Earth. They can disrupt radio navigation and communication systems, pose significant hazards to astronauts, satellites and space stations near Earth, and interfere with power transmission lines on the ground (Section 8.3). National space environment centers and defense agencies therefore continuously monitor the Sun from ground and space to forecast threatening activity.

The ultimate goal is to learn enough about solar activity to predict when the Sun is about to unleash its pent-up energy. Such space-weather forecasts will probably involve magnetic changes that precede solar flares and coronal mass ejections, especially in the low corona where the energy is released or at lower levels in the photosphere where we can watch internal motions pulling and twisting the magnetism about.

Scientists may have discovered how to predict the sudden and unexpected outbursts. When the bright, X-ray emitting coronal loops are distorted into a large, twisted sigmoid (S or inverted S) configuration, a coronal mass ejection from that region becomes more likely (Fig. 6.23). In some instances, a coronal mass ejection occurs just a few hours after the magnetic fields have snaked past each other in a sinuous S-shaped feature. The mass ejection arrives at the Earth three or four days later. In the meantime, just after the mass has been expelled from the Sun, the X-ray-emitting region dramatically changes shape, exhibiting the tell-tale, cusp-like signature of magnetic reconnection (Fig. 6.24) and X-ray fading or dimming due to the mass removal. In other words, the magnetism gets stirred up into a complex, stressed and twisted situation before it explodes, and relaxes to a more stable configuration after the outburst.

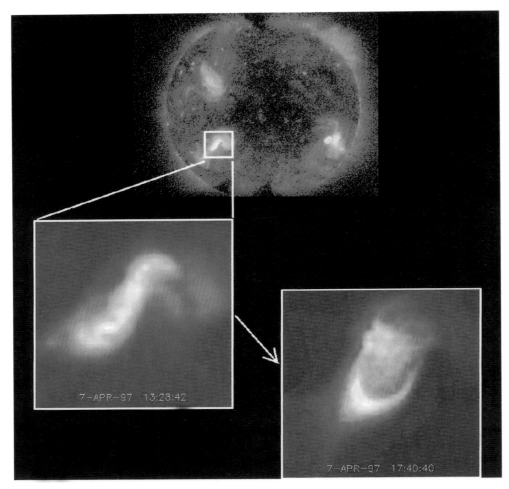

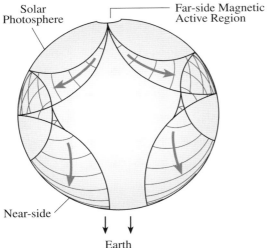

Fig. 6.24 **Sigmoid and Cusp**
The full-disk X-ray image shows the Sun with a twisted, sigmoid present on 7 April 1997. It produced a halo Coronal Mass Ejection (CME) on the following day. The inset (*left*) shows the soft X-ray sigmoid before eruption of the CME. The other inset (*right*) shows the soft X-ray cusp and arcade formed just after the CME took place. These images were taken with the Soft X-ray Telescope (SXT) on *Yohkoh*. (Courtesy of Richard C. Canfield, Alphonse C. Sterling, NASA, ISAS, the Lockheed-Martin Solar and Astrophysics Laboratory, the National Astronomical Observatory of Japan and the University of Tokyo.)

Fig. 6.25 **Looking through the Sun** The arcing trajectories of sound waves from the far side of the Sun are reflected internally once before reaching the front side, where they are observed with the Michelson Doppler Imager (MDI) aboard *SOHO*. Sound waves returning from a solar active region on the hidden back side of the Sun have a round-trip travel time about twelve seconds shorter than the average of six hours. This timing difference permits scientists to detect potentially threatening active regions on the far side of the Sun before the Sun's rotation brings them around to the near side that faces the Earth. (Courtesy of the *SOHO* MDI/SOI consortium. *SOHO* is a project of international cooperation between ESA and NASA.)

It is also critically important to know if the material sent out from the solar outbursts is headed toward Earth. Coronal mass ejections that are expelled from near the visible edge of the Sun are not likely to impact Earth, but threaten other parts of space. Mass ejections are most likely to hit the Earth if they originate near the center of the solar disk, as viewed from the Earth, and are sent directly toward the planet. As previously noted, the outward rush of such a mass ejection appears in coronagraph images as a ring or halo around the occulting disk. The Earth-directed ones may be preceded by coronal activity at ultraviolet and X-ray wavelengths near the center of the solar disk as viewed from the Earth. The mass ejection is itself sometimes associated with waves running across the Sun, like tidal waves or tsunami going across the ocean.

The high-energy electrons that accompany solar flares follow the spiral pattern of the interplanetary magnetic field (Section 7.1, Fig. 7.1), so they must be emitted from active regions near the west limb and the solar equator to be magnetically connected with the Earth. Solar flares emitted from other places on the Sun are not likely to hit Earth, but they could be headed toward interplanetary spacecraft, the Moon, Mars or other planets.

Nowadays scientists can use sound waves to see right through the Sun to its hidden, normally-invisible, back side, enabling them to monitor active regions before they rotate to face the Earth. The new technique, dubbed helioseismic holography, examines a wide ring of sound waves that emanate from a region on the side of the Sun facing away from the Earth (the far side) and reach the near side that faces the Earth (Fig. 6.25). When a large active region is present on the back side of the Sun, its intense magnetic fields compress the gases there, making them slightly lower and more dense than the surrounding material. A sound wave that would ordinarily take 6 or 7 hours to travel from the near side to the far side of the Sun and back again takes approximately 12 seconds less when it bounces off the compressed active region on the far side. When near-side, photosphere oscillations are examined by SOHO's Michelson Doppler Imager, or MDI, they can detect the quicker return of these sound waves.

Solar astronomers might be able to use this technique to monitor the structure and evolution of large regions of magnetic activity as they cross the far side of the Sun, thereby revealing the regions that are growing in magnetic complexity or strength and seem primed for explosion. Since the solar equator rotates with a period of 27 days, when viewed from the Earth, this can give at least seven day's extra warning of possible bad weather in space before the active region swings into view and pummels the Earth.

These solar explosions produce strong, violent gusts in a seemingly eternal wind of charged particles and magnetic fields that is always flowing from the Sun. We now turn to a description of this solar wind.

CHAPTER SEVEN

The Sun's winds

Stellar winds Hot stars with powerful winds shape interstellar gas (*red*) and dust (*green*) into a comet-like apparition. It is called a cometary globule, or CG for short, and designated CG 4. (Courtesy of David Malin, Anglo-Australian Telescope Board.)

7.1 The fullness of space

The space between the planets, once thought to be a tranquil, empty void, is swarming with hot, charged invisible pieces of the Sun. They expand and flow away from the Sun, forming a perpetual solar wind. The relentless wind was inferred from comet tails, suggested by theoretical considerations, and fully confirmed by direct *in-situ* measurements from spacecraft in the early-1960s (Section 1.6).

The reason that space looks so empty is that the Sun's wind is exceedingly tenuous, even at its origin near the visible Sun. By the time that it reaches the Earth's orbit, the solar wind has been further diluted by expanding into the increasing volume of space. There are about five million electrons and five million protons per cubic meter in the solar wind near the Earth. By way of comparison, there are 25 million billion

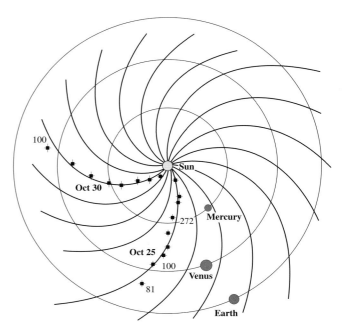

Fig. 7.1 **Spiral path of interplanetary electrons** The trajectory of flare electrons in interplanetary space as viewed from above the polar regions using *Ulysses*. As the high-speed electrons move out from the Sun, they excite radiation at successively lower plasma frequencies; the numbers denote the observed frequency in kilohertz, or kHz. Since the flaring electrons are forced to follow the interplanetary magnetic field, they do not move in a straight line from the Sun to the Earth, but instead move along the spiral pattern of the interplanetary magnetic field, shown by the solid curved lines. The squares and crosses show *Ulysses* radio measurements of type III radio bursts on 25 October 1994 and 30 October 1994. The approximate locations of the orbits of Mercury, Venus and the Earth are shown as circles. (Courtesy of Michael J. Reiner. *Ulysses* is a project of international collaboration between ESA and NASA.)

Parameter	Fast wind	Slow wind
Composition, temperature and density	Uniform	Highly variable
Proton density, N_p (m^{-3})	3×10^6	10.7×10^6
Proton speed, V_p (m s^{-1})	$(6.67 \times 10^5)^b$	3.48×10^5
Proton flux, $F_p = N_p V_p$ (m^{-2} s^{-1})	1.99×10^{12}	3.66×10^{12}
Proton temperature, T_p (K)	2.8×10^5	5.5×10^4
Electron temperature, T_e (K)	1.3×10^5	1.9×10^5
Helium temperature, T_α (K)	7.3×10^5	1.7×10^5
Helium to proton abundance, A^a	0.036 (constant)	0.025 (very variable)

[a] Measurements are referred to a distance of 1 AU = 1.496 × 10^{11} m. The helium ion to proton abundance $A = N_\alpha/N_p$, where N is the number density and the subscripts α and p respectively denote the helium ions and the protons.
[b] The *Helios 1* and 2 spacecraft traveled near the ecliptic where the slow solar wind dominates the flow, and this led to an underestimate of the velocity of the high-speed component. It has a speed of about 750 thousand meters per second, or 7.5 × 10^5 m s^{-1}.

Table 7.1 **Average solar-wind parameters measured from *Helios 1* and *2* between December 1974 and December 1976, normalized to the distance of the Earth's orbit at 1 AU**^{a}

billion (2.5×10^{25}) molecules in every cubic meter of our transparent air at sea level. The density of the solar wind is so low that if we could go out into space and put our hands in it, we would not be able to feel it.

The Sun's continuous wind moves at supersonic speeds near the Earth. It travels with two main velocities, like an automobile with one high gear and one low gear. There is a fast component moving at about 750 thousand meters per second, and a slow one with about half that speed.

The twin spacecraft, *Helios 1* and *Helios 2*, provided *in-situ* analysis of the solar wind in the ecliptic, the plane of the Earth's orbit around the Sun, during the 1970s and 1980s, showing that the two kinds of solar-wind flow, the fast and slow ones, have different physical properties (Table 7.1). The fast wind is always there, lasting for years without substantial changes in composition, speed or temperature, but the slow

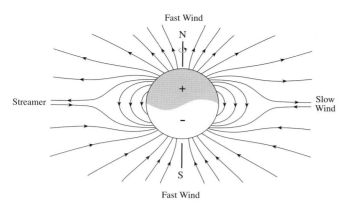

Fast Wind

N

Streamer

Slow
Wind

S

Fast Wind

Fig. 7.2 **Coronal magnetism at activity minimum** Near the minimum in the
11-year cycle of solar magnetic activity, the coronal magnetic field lines have
a dipole-type geometry and an equatorial current sheet. The high-speed
wind escapes along the open magnetic field lines in the northern and
southern polar regions, denoted by N and S respectively. At the equator,
where the slow wind originates, the magnetic field lines have been pulled
outward by the solar wind into oppositely-directed, parallel magnetic fields
separated by a neutral current sheet.

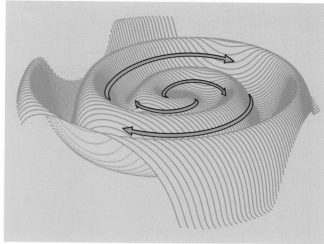

Fig. 7.3 **Spiral, warped current sheet** Near the minimum in the Sun's 11-year
activity cycle, the Sun's magnetic field is primarily directed outward from one
pole and inward in the other. The oppositely-directed magnetic field lines
meet near the solar equator, forming a thin, wave-like neutral current sheet
that divides magnetic fields directed away from the Sun and those directed
toward it. The neutral current sheet winds out in the Sun's equatorial plane,
with a spiral shape and a warped structure when viewed from near the Earth,
that resembles the skirt of a ballerina or a whirling dervish. As it rotates with
the Sun, the current sheet sweeps regions of opposite magnetic polarity past
the Earth. (Courtesy of J. Randy Jokipii, and the *Advanced Composition
Explorer* mission.)

wind is more variable. In the high-speed wind, the protons
are a few times hotter than the electrons, and the helium
nuclei, or alpha particles, are even hotter than the protons. It
is the other way around in the slow wind, where the lighter
electrons are hotter than the heavier protons.

The energy transported by the solar wind is dominated by
the massive protons. The *Helios* spacecraft showed that the
proton density is high whenever the wind is slow, and that
the proton density is low when the wind speed is high. The
product of the proton density and velocity, or the proton
flux, is about the same in the fast and slow winds, with a
value of between 1.5 and 4.0 million million $[(1.5 \text{ to } 4.0) \times 10^{12}]$ protons $m^{-2} s^{-1}$ at the Earth's distance from the Sun.

The charged particles in the solar wind drag the Sun's
magnetic fields with them. While one end of the
interplanetary magnetic field remains firmly rooted in the
photosphere and below, the other end is extended and
stretched out by the radial expansion of the solar wind. The
Sun's rotation bends this radial pattern into an interplanetary
spiral shape within the plane of the Sun's equator, coiling the
magnetism up like a tightly-wound spring. This spiral pattern
has been confirmed by tracking the radio emission of high-
energy electrons emitted during solar flares (Fig. 7.1), as
well as by spacecraft that have directly measured the
interplanetary magnetism in the ecliptic.

The shape of the interplanetary magnetic field depends on
the Sun's 11-year cycle of magnetic activity (Section 5.1). Near
activity minimum, the large-scale, global magnetism of the
Sun can be described as a simple magnet with north and south
poles where large, unipolar coronal holes are located. The
northern hole is of one magnetic polarity, or direction, and the
southern one of opposite polarity. The negative and positive
field lines meet near the solar equator, where a magnetically
neutral layer, called a current sheet, is dragged out into space
by the outflowing wind (Fig. 7.2). Near the Sun, the current

sheet coincides with a belt of coronal streamers that seem to
meander across the star like the seam of a baseball.

Magnetometers aboard the *Ulysses* spacecraft showed that
the wind's magnetic field is uniform at all latitudes, at least
out near the distance of the Earth's orbit and near the
minimum of the 11-year solar activity cycle. This is what one
would expect from a dipolar magnetic field with a neutral
current sheet wrapped around the solar equator, where the
streamers appear at activity minimum. The dipole is
stretched way out at its middle, resulting in two polar
monopoles whose magnetic field lines do not cross the

Property	Solar wind	Sun's radiation
Mass loss rate, (kg s⁻¹)	10^9	5×10^9
Total mass loss, (M☉) (in 5 billion years)	5×10^{-5}	2.5×10^{-4}
Energy flux (at 1 AU), (J m⁻² s⁻¹)	1.6×10^{-4}	1.4×10^3
Momentum flux (at 1 AU), (kg m⁻¹ s⁻¹)	8×10^{-10}	5×10^{-6}
Despin time	10^{16} years	10^{12} years

[a] The M☉ denotes the Sun's mass of ~ 2×10^{30} kg. The mean
distance of the Earth from the Sun is one astronomical unit, or
1 AU, where 1 AU = 1.496×10^{11} m.

Table 7.2 **Relative importance of solar wind and radiation**[a]

equatorial regions. By way of comparison, the dipole magnetic field near the surface of the Earth is more concentrated over the poles, with magnetic fields that come together at the poles and spread out between them.

Measurements from other spacecraft have repeatedly demonstrated that the interplanetary magnetic field is oriented into a simple spiral pattern in the ecliptic, if averaged over a sufficiently long time. The magnetic fields on the two sides of the equatorial current sheet point in opposite directions, either toward or away from the Sun. This magnetic orientation is preserved throughout most of an 11-year activity cycle. The magnetic field that is pointed toward the Sun is directed away during the next cycle, and *vice versa*, returning back to the original direction every 22 years.

Because the Sun's magnetic dipole axis is tilted with respect to its rotation axis, spacecraft near Earth detect a warped current sheet (Fig. 7.3). That is, the plane of the Earth's orbit, the ecliptic, is nearly coincident with the Sun's equatorial plane, so they are both inclined with respect to the plane of the Sun's magnetic equator and the current sheet. As the Sun rotates, the current sheet wobbles up and down, like the folds in the skirt of a whirling dervish, sweeping regions of opposite magnetic polarity past the Earth.

The global pattern of the interplanetary magnetic field changes dramatically near the maximum in the 11-year activity cycle. There is a switch in overall magnetic polarity; the north pole becomes the south and *vice versa*. So, the dipolar, warped current-sheet model does not apply at activity maximum. The main current sheet then moves about and secondary sheets are often present.

Space is also filled with the Sun's radiation, and it is more important than the solar wind in removing mass from the Sun (Table 7.2). Every second the Sun blows away a billion (10^9) kilograms in its solar wind. That is a tremendous mass loss, but it is five times less than the amount consumed every second during the nuclear-fusion reactions that make the Sun shine. To supply the Sun's present luminosity, hydrogen must be converted into helium, within the Sun's energy-generating core, with a mass loss of about five million tons (5×10^9 kg) every second (Section 3.3). This mass is carried off by radiation. In supplying the radiation, the Sun will

eventually consume all the hydrogen in its core, in about seven billion years; it will then expand into a giant star (Section 3.5, Fig. 3.9). By that time, the Sun will have lost only about 0.01 percent (10^{-4}) of its mass by the solar wind at the present rate.

Cosmic rays are another ingredient of space. They are energetic, charged, sub-atomic remnants of other stars that enter the solar system from interstellar space, raining down on the Earth in all directions and traveling at nearly the velocity of light. Since they are moving so fast, cosmic rays contain phenomenal amounts of energy. That energy is usually measured in units of electron volts, abbreviated eV; for conversion use $1 \text{ eV} = 1.602 \times 10^{-19}$ J.

The greatest flux of cosmic rays arriving at Earth occurs at about 1 GeV, or at a billion (10^9) electron volts of energy. In contrast, a solar-wind proton moving at a velocity of $V = 5 \times 10^5$ m s^{-1} has a kinetic energy of $0.5 m_p V^2 = 1.6 \times 10^{-16}$ J = 1000 eV, where the mass of the proton is $m_p = 1.67 \times 10^{-27}$ kg. At this energy, a cosmic-ray proton must be traveling at 88 percent of the speed of light while a proton in the Sun's wind moves about 500 times slower.

Focus 7.1 Energizing charged particles within space

Some ions found in the solar wind move far too swiftly to have been accelerated at the wind's source. When the fast winds overtake the more sluggish flows in their path, they produce shock waves that locally accelerate ions to high speeds, like ocean waves taking a surfer for a ride. Such shock waves, originating at places where the slow and fast winds interact, have been observed many times with the *Ulysses* spacecraft

Other heavy nuclear particles, such as helium, nitrogen and oxygen, come from interstellar space, but they are ionized and accelerated within the solar system. Because the compositions of these particles are peculiar, when compared to other types of low-energy cosmic rays, scientists dubbed them anomalous cosmic rays.

The anomalous material begins its life as electrically neutral, uncharged interstellar atoms, and then drifts into the solar system at low velocities, where ultraviolet sunlight tears off just one electron from each atom and thereby ionizes the atom. The singly-ionized atoms have increased mobility in the interplanetary magnetic field, compared with other cosmic rays that have lost most of their electrons by more violent processes, often associated with the death throes of stars.

The newborn ions are then picked up by the magnetic fields entrained in the outflowing solar wind, and taken for a ride out to the edge of the solar system. There they can rebound at higher velocities due to shock waves, something like a tennis ball bouncing off a wall. *Ulysses* discovered a vast population of the pick-up ions, many for the first time, showing that they are also accelerated where the fast and slow winds interact. These particles have additionally been used to determine the abundance of elements in the interstellar space that they come from.

Type	Flux (particles m^{-2} s^{-1})
Hydrogen (protons)	640
Helium (alpha particles)	9
Electrons	<13
Carbon, nitrogen, oxygen	6

[a] The flux is in units of nuclei m^{-2} s^{-1} for particles with energies greater than 1.5 billion (1.5×10^9) eV per nucleon, arriving at the top of the atmosphere from directions within 30 degrees of the vertical.

Table 7.3 **Average fluxes of primary cosmic rays at the top of the atmosphere**[a]

Focus 7.2 Gyration radius

If a charged particle approaches a magnetic field in a direction perpendicular to the field, a magnetic force pulls it into a circular motion about the magnetic field line. Since the particle can move freely in the direction of the magnetic field, it spirals around it with a helical trajectory (see Fig. 6.11). The radius of this circular gyration, designated by R_g, is given by:

$$R_g = [m/(Ze)] [V_\perp / B],$$

where $V_\perp$, is the velocity of the particle in a direction perpendicular to the field, B is the magnetic field strength, and m and Ze respectively denote the mass and charge of the particle. Thus, a stronger magnetic field tightens the gyration into smaller coils, and faster particles will gyrate in larger circles.

For an electron with mass $m_e = 9.1094 \times 10^{-31}$ kg and charge $e = 1.602 \times 10^{-19}$ C, with Z = 1.0, the corresponding gyration radius is:

$$R_g \text{ (electron)} = 5.7 \times 10^{-12} [V_\perp / B] \text{ m},$$

where $V_\perp$ is in meters per second and B is in tesla, and for a proton it is:

$$R_g \text{ (proton)} = 1.05 \times 10^{-8} [V_\perp / B] \text{ m}.$$

where the mass of the protons is $m_p = 1836 m_e$.

These expressions only apply if the velocity is not close to the velocity of light, c. At high particle velocities approaching that of light, the radius equation is multiplied by the a Lorentz factor $\gamma = [1 - (V/c)^2]^{-1/2}$, which becomes unimportant at low velocities when $\gamma = 1$. For cosmic-ray protons of high velocity and large energies, $E = \gamma m_p c^2$, our equation becomes:

$$R_g \text{(energetic proton)} = 3.3 E/B \text{ m},$$

when the energy E is in units of GeV, where 1 GeV = 1.602 $\times 10^{-10}$ J, and the magnetic field strength is in tesla. When a 1-GeV proton encounters the terrestrial magnetic field, whose strength is roughly 10^{-4}T, its radius of gyration is $R_g \approx 10^4$ m, which is hundreds of times smaller than the Earth whose mean radius is 6.37×10^6 m. That means that even the very energetic protons will gyrate about the magnetic field, and move toward Earth's magnetic poles.

The cosmic rays consist of hydrogen nuclei (protons), helium nuclei (alpha particles) and heavier nuclei (Table 7.3). The number of cosmic-ray electrons is less than 2 percent of the number of cosmic-ray protons, and the electrons are far less abundant than the protons at a given energy. Also, despite their incredible energy, there are far fewer cosmic rays than solar-wind particles in the vicinity of the Earth. The cosmic-ray protons at the top of our atmosphere have a flux of 640 protons $m^{-2} s^{-1}$.

At the Earth's orbit, protons in the solar wind have a flux that is about ten billion (10^{10}) times larger than the cosmic-ray flux. The most-abundant cosmic-ray protons near Earth have an energy of 10^{-10} J but a local energy density of about 10^{-13} J m^{-3}, or about one-ten-thousandth (10^{-4}) the kinetic energy density of protons in the solar wind, at about 10^{-9} J m^{-3}.

Cosmic rays do not all have the same energy, and there are fewer of them with higher energy. The number of particles with kinetic energy E is proportional to $E^{-\alpha}$, where the index $\alpha = 2.5$ to 2.7 for energies from a billion to a million million (10^9 to 10^{12}) electron volts. At higher energies there are relatively few particles. At lower energies the magnetized solar wind acts as a valve for the more-abundant lower-energy cosmic rays, controlling the amount entering the solar system.

When solar activity is at the peak of its 11-year cycle, the flux of low-energy cosmic rays detected at the top of the Earth's atmosphere is least, and the cosmic-ray flux is greatest near the minimum of the solar cycle of magnetic activity. Whenever the number of cosmic rays near the Earth increases, the Sun's activity decreases, and vice versa. This intriguing relation was discovered by Scott Forbush in the early-1950s, and it is therefore sometimes called the Forbush effect. It is caused by interplanetary magnetic fields that act as a barrier to the electrically-charged cosmic rays that are inbound from the depths of space. During the maximum in the solar cycle, stronger solar magnetic fields are carried out into interplanetary space by the Sun's wind, deflecting more cosmic rays from Earth-bound paths. Less-extensive interplanetary magnetism, during a minimum in the 11-year cycle of magnetic activity, lowers the barrier to the cosmic particles and allows more of them to arrive at Earth.

Thus, nearby space contains particles with different energies, originating at the Sun, in other dying stars, or in space itself. Most of them flow out of the Sun with the solar wind; these have relatively low energies associated with the production of the wind, unless accelerated during transient explosions in the corona. The cosmic rays are distinguished by both their low density and very high energies. They are most likely hurled into space when an entire star explodes. Other types of rapidly-moving particles are energized in interplanetary space (Focus 7.1).

The Earth's magnetic field deflects all but the most-energetic charged particles away from its equatorial regions and toward the Earth's magnetic poles. As a result, their intensity is least near the Earth's equator, and increases at higher latitudes. This is because a charged particle gyrates around the terrestrial magnetic field and moves towards Earth's magnetic poles (Focus 7.2).

7.2 Where do the Sun's winds come from?

The solar wind has never stopped blowing during the more than three decades that it has been observed with spacecraft. Two winds are always detected – a fast, uniform wind blowing at about 750 thousand meters per second, and a variable, gusty slow wind, moving at about half that speed.

One method of studying the velocities and origin of the solar wind is called radio scintillation. This technique uses two or more radio telescopes to observe very distant, cosmic radio sources that fluctuate, or scintillate, when their radio waves pass through the solar wind. The stars we see in the dark night sky similarly twinkle when seen through the Earth's wind-blown atmosphere. The velocity of the solar wind can be inferred from the time it takes the fluctuating radio signal to move between two telescopes. You could similarly determine the speed of a wind-blown cloud by seeing how long it takes for its shadow to move along the ground.

The radio-scintillation data indicate that the high-speed and slow-speed winds do not originate from the same place on the Sun. They show that the slow component of the solar wind is confined to low latitudes within the Sun's equatorial regions near the minimum in the 11-year activity cycle. At this time, the average wind velocity increases from the solar equator to higher latitudes. At activity maximum, the slow winds seem to emanate from all over the Sun, and the high-speed winds are less prevalent.

The blinking radio signals also indicate that the slow-speed wind blows hard and soft, with gusts and squalls like terrestrial gales. In contrast, a relatively steady and uniform, fast wind seems to spill out at higher latitudes, at least near the minimum of the 11-year activity cycle.

Comparisons of X-ray images of coronal holes (Section 5.3, Fig. 5.30) with in-situ satellite wind-velocity measurements near Earth, indicated, in the 1970s, that some of the high-speed wind is gushing out of polar coronal holes. Whenever a high-speed stream in the solar wind swept past Earth, a coronal hole rotated into alignment with our planet.

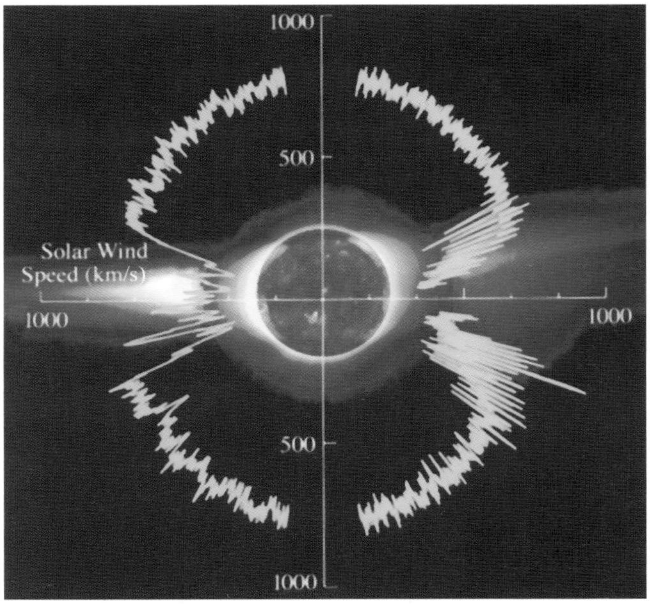

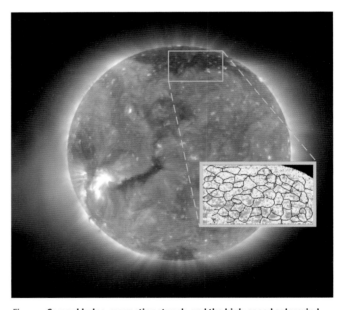

Fig. 7.5 **Coronal holes, magnetic network, and the high-speed solar wind** Dark coronal holes at the Sun's polar regions (*top* and *bottom*) are the source of much of the high-speed solar wind. The inset provides a close-up, Doppler velocity map of the million-degree gas at the base of the corona where the fast solar wind originates, taken in the extreme-ultraviolet light of ionized neon (the Ne VIII line at 77.0 nm). Dark blue represents an outflow, or blueshift, at a velocity of 10^4 m s^{-1}; it marks the beginning of the high-speed solar wind. Dark red indicates a downflow at the same speed. Superposed are the edges of the "honey-comb"-shaped pattern of the magnetic network, where the strongest outward flows (*dark blue*) are found. The relationship between the outflow velocities and the network suggests that the high-speed wind emanates from the boundaries and boundary intersections of the magnetic network. These observations were taken on 22 September 1996. [Courtesy of the *SOHO* EIT consortium (*full disk*) and the *SOHO* SUMER consortium (*velocity inset*). *SOHO* is a project of international collaboration between ESA and NASA.]

Fig. 7.4 **Fast and slow winds** A composite picture of an ultraviolet image of the solar disk and a white-light image of the solar corona form a backdrop for a radial plot of solar-wind speed versus latitude. These velocity data, obtained by *Ulysses* between 1991 and 1996, show that a fast wind escapes from the polar regions where coronal holes are found, while a slow wind is associated the Sun's equatorial regions that contain coronal streamers (Courtesy of the *Ulysses* mission, a project of international collaboration between ESA and NASA.)

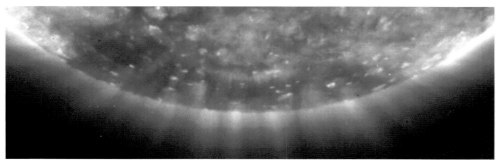

Fig. 7.6 **Polar plumes** Close-up, extreme-ultraviolet views of the south polar coronal hole (*large dark region*) obtained with the *SOHO* EIT instrument on 8 May 1996. The fast component of the solar wind emanates from these coronal holes. Plumes can be seen emerging from tiny bright spots, but they are not the main source of the high-speed winds. These images were taken in the extreme-ultraviolet light of highly ionized iron - the Fe XII emission line at 19.5 nm (*top*), formed at a temperature of about 1.5 million degrees kelvin, and the Fe IX/X emission lines near 17.1 nm (*bottom*), formed at a temperature of about 1.0 million degrees kelvin. (Courtesy of the *SOHO* EIT consortium. *SOHO* is a project of international collaboration between ESA and NASA.)

Yet, until recently, spacecraft measurements were always made near the ecliptic, and the radio-scintillation data only hinted at what the flows looked like above the Sun's poles. Then, in 1994-95, the *Ulysses* spacecraft made measurements all around the Sun, at a distance comparable to that of the Earth and near a minimum in the Sun's 11-year activity cycle. *Ulysses'* velocity data conclusively proved that a relatively uniform, fast wind pours out at high latitudes near the solar poles, and that a capricious, gusty, slow wind emanates from the Sun's equatorial regions and the plane of the Earth's orbit around the Sun, the ecliptic (Fig. 7.4).

For the winds to be maintained, they must be replenished by hot gases welling up from somewhere on the Sun. However, since *Ulysses* never passed closer to the Sun than the Earth does, simultaneous observations with other satellites were required to tell exactly where the winds come from. Fortunately, the *Ulysses* data were obtained near activity minimum with a particularly simple corona characterized by marked symmetry and stability. There were pronounced coronal holes at the Sun's north and south poles, and its equator was encircled by coronal streamers.

Comparisons of *Ulysses'* high-latitude passes with *Yohkoh* soft X-ray images showed that coronal holes were then present at the poles of the Sun, as they usually are during activity minimum. Much, if not all, of the high-speed solar wind therefore seems to come from the open magnetic fields in coronal holes, at least during the minimum in the 11-year cycle of magnetic activity. The coronal holes have comparatively weak and open magnetic fields that stretch radially outward with little divergence, providing a fast lane for the electrified wind. The high-speed solar wind squirts out of the nozzle-like coronal holes, like water out of a fire hose.

The detailed structure of a polar coronal hole has only recently been investigated by the SUMER instrument on *SOHO*. It showed that the high-speed outflow is concentrated at the boundaries of the magnetic network formed by underlying supergranular convection cells (Fig. 7.5). These edges are places where the magnetic fields are concentrated into inverted magnetic funnels that open up into the overlying corona (Section 5.2, Fig. 5.14). The strongest high-speed flows apparently gush out of the crack-like edges of the network, like grass or weeds growing in the dirt where paving stones meet. Thus, *SOHO* has for the first time discovered one of the exact sources of the fastest winds.

It was once thought that polar plumes might be the main source of the high-speed wind. Since these long, narrow features are the brightest things around in the dark coronal holes, something had to be energizing them (Fig. 7.6). However, careful *SOHO* measurements indicate that the fast winds are pouring out of the entire coronal hole, with no substantial difference between the narrow plumes and the inter-plume regions. The honeycomb-shaped boundaries of the magnetic network are apparently the main localized source of the high-speed wind in coronal holes.

The fast wind is not only found above polar coronal holes during solar activity minimum. At roughly the Earth's distance from the Sun, *Ulysses* found that the slow wind is localized near the ecliptic, and that the fast winds are everywhere else, even at lower latitudes nearer the solar equator and outside the radial projection of the coronal hole edges (Fig. 7.4). The charged particles could be guided to low latitudes by magnetic fields that originate in coronal holes and bend outward toward the equator (Fig. 7.2). This would be consistent with the curved helmet shapes of streamers found near the solar equator. The high-speed wind can additionally shoot straight out of coronal holes that extend down to low latitudes, even crossing the equator, and directly feed low-latitude regions.

Although much, if not all, of the fast solar wind gushes out of coronal holes, they might not necessarily be the only source of the high-speed wind. Some of the fast wind could be squirted

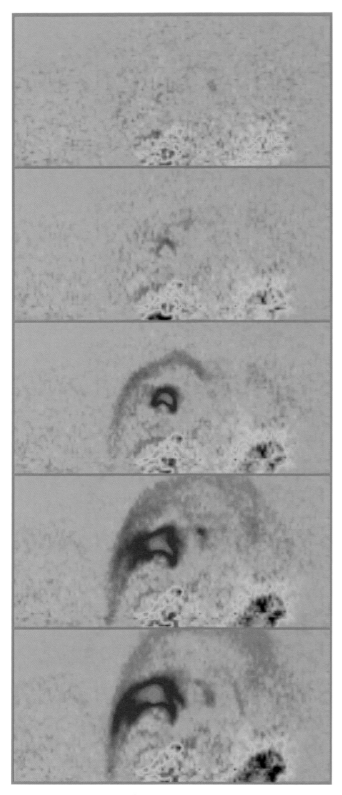

Fig. 7.7 Expanding magnetic loops in an active region Difference images reveal the expansion of active-region loops in a time sequence, running from top to bottom, lasting about half-an-hour. The fainter loops near the top are expanding with an apparent velocity of about 4×10^4 m s^{-1}. These images were taken on 22 April 1992 with the Soft X-ray Telescope (SXT) on *Yohkoh*. (Courtesy Alan McAllister, NASA, ISAS, the Lockheed-Martin Solar and Astrophysics Laboratory, the National Astronomical Observatory of Japan, and the University of Tokyo.)

out along narrow, straight, open magnetic channels all over the Sun and not just from coronal holes. Scintillation of spacecraft radio-communication signals indicates that the corona might be filled with such straight, nozzle-like filamentary structures. However, such features are controversial, and additional observations are required to fully confirm them.

Comparisons of *Ulysses* data with coronagraph images pinpointed the equatorial streamers as the birth place of the slow and sporadic wind during the minimum in the 11-year activity cycle. Hot gas is bottled up in the closed coronal loops at the bottom of the helmet streamers (Section 5.3, Fig. 5.21). The capricious slow wind can therefore only leak out along elongated, stretched-out streamer stalks. The part that manages to escape seems to vary in strength as the result of the effort.

Oppositely-directed magnetic fields run in and out of the Sun on each side of the long, narrow stalks, providing an open channel for the slow flow once it gets out. SOHO's LASCO instrument has observed spurt-like blobs of material moving out along the stalks, like water working its way down a clogged pipe in your bathtub or sink. Observations of scintillating radio signals from spacecraft confirm the localized nature of the slow wind at activity minimum, suggesting that it is associated with the narrow stalks of coronal streamers near the solar equator.

A streamer could get so stretched out and constricted that it pinches itself off and snaps at just a few solar radii from Sun center. The lower parts of the streamer would then close down and collapse, and the outer disconnected segment would be propelled out to form a gust in the slow solar wind.

Coronal loops are found down at the very bottom of streamers, and the expansion of these magnetized loops may provide the energy and mass of the slow component of the solar wind. Sequential soft X-ray images, taken from the *Yohkoh* spacecraft, have shown that magnetic loops expand out into space, perhaps contributing to the slow wind. When viewed at the Sun's apparent edge, near the photosphere, coronal loops are seen rising upwards at speeds of some tens of thousands of meters per second (Fig. 7.7).

SOHO's LASCO has shown that small magnetic loops can be nearly continuously emerging into the Sun's equatorial regions at activity minimum. They expand into the global magnetic field at the current sheet, and establish new magnetic connections with it. The reconnection could happen again and again, becoming the engine that drives the slow solar wind. The small expanding loops also carry material with them, perhaps accounting for the moving material detected further out in the low-latitude, slow-speed solar wind.

Thus, expanding coronal loops and equatorial streamers seem to be the well-springs of the slow component of the solar wind, at least during the minimum in the 11-year magnetic-activity cycle. The magnetic network within coronal holes apparently provides the source for the fast wind during this part of the activity cycle. Now that we know the probable sources of the two winds, we turn our attention to how they are accelerated away from the Sun to such high velocities.

7.3 Getting up to speed

By the time that the solar wind has reached the Earth, it is moving along at supersonic velocities of hundreds of thousands of meters per second. What forces propel it to such high velocities? The expansion of the hot corona is responsible for some of it.

The corona's expansion will begin slowly near the Sun, where the solar gravity is the strongest, and then continuously accelerate out into space as it breaks away from the Sun, gaining speed with distance and reaching supersonic velocities. Since there is a limit to the amount of energy being pumped into it, the solar wind will eventually reach a limiting asymptotic or terminal velocity, and then cruise along at a roughly constant speed.

The slow wind naturally reaches terminal velocities of a few hundred thousand meters per second as the million-degree corona expands away from the Sun. Additional energy must be deposited in the low corona to give the fast wind an extra boost and double its speed. In technical terms, the fast wind has a velocity and mass flux density that are too high to be explained by heat transport and classical thermal conduction alone.

You have to look down into the bottom of the corona to investigate the regions where the corona is heated and the solar wind is accelerated. *Yohkoh*'s X-ray observations show that the coronal electrons become fully heated at a height of between 0.2 and 0.5 solar radii, or between 140 and 348 million meters, above the photosphere (Fig. 7.8). In addition, the electron temperatures in coronal holes are several hundred thousand degrees kelvin cooler than the temperatures of electrons in coronal streamers at the same height. *Ulysses* measurements of ion temperatures also indicate that the fast polar wind originates in a relatively low-temperature region in the corona.

Images from *SOHO*'s Large Angle Spectrometric COronagraph, or LASCO, suggest that the slow wind takes a long time to get up to speed (Fig. 7.9). Blobs moving along the stalks of helmet streamers have to move out to 20 or 30 solar radii from Sun center to accelerate to speeds of 300 or 400 thousand meters per second. In contrast, the high-speed wind is accelerated relatively close to the Sun. Radio-scintillation measurements indicate that the polar wind

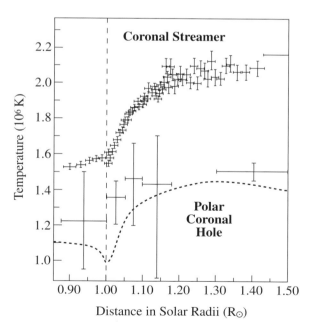

Fig. 7.8 **Rising temperatures near the photosphere** The increase in temperature with height in a coronal streamer (*top*) and a coronal hole (*bottom*) have been inferred from data taken with the Soft X-ray Telescope on *Yohkoh*. Here the distance, *r*, from Sun center is specified in units of the Sun's radius. At a given height in the low corona, the temperature of coronal streamers is hotter than the temperature of coronal holes. Both regions seem to be fully heated by between 1.3 and 1.5 solar radii from Sun center. (*Yohkoh* is a project of international cooperation between ISAS and NASA.)

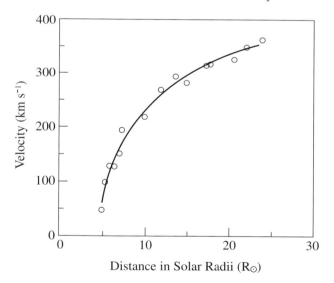

Fig. 7.9 **Blobs accelerate slowly** Time-lapse coronagraph images show prominent features, sometimes called blobs, that move radially out in equatorial regions from helmet streamers. Their speed typically doubles between 5 and 25 solar radii from the Sun, with a slow acceleration of about 4 m s^{-2} through most of this distance. This data was taken on 30 April 1996 with the *SOHO* LASCO instrument, at a time near a minimum in the 11-year solar activity cycle. The solid line denotes the best fit to the data using an exponential function starting at a distance of 4.5 solar radii from Sun center and with an asymptotic terminal speed of 4.187×10^5 m s^{-1} at a distance of about 15 solar radii from Sun center. Courtesy of the *SOHO* LASCO consortium. (*SOHO* is a project of international collaboration between ESA and NASA.)

reaches terminal speeds of 750 thousand meters per second within just 10 solar radii or less (Fig. 7.10). Thus, the fast wind accelerates quickly, like a racing horse breaking away from a starting gate.

Another *SOHO* coronagraph, known as UVCS for UltraViolet Coronagraph Spectrometer, has measured temperatures and velocities within the source regions of the solar wind from 1.2 to 10 solar radii from Sun center. It has used the Doppler shifts of ultraviolet spectral lines to show that the high-speed solar wind, emerging from coronal holes, accelerates to supersonic velocity within just 2.5 solar radii from Sun center.

SOHO's UVCS has additionally demonstrated that heavier particles in polar coronal holes move faster than light particles in coronal holes. Above two solar radii from the Sun's center, oxygen ions have the higher outflow velocity, approaching 500 thousand meters per second in the holes, while hydrogen moves at about half this speed (Fig. 7.11). In contrast, within equatorial regions where the slow-speed wind begins, the lighter hydrogen moves faster than the oxygen, as one would expect for a gas with thermal equilibrium among different types of particles.

If the outflow velocities in coronal holes measure temperature, then the temperatures of the protons are a few million degrees kelvin, and the oxygen ions have searing temperatures of hundreds of millions of degrees kelvin. So, the oxygen ions would be more than ten times hotter than the center of the Sun (Focus 7.3). However, the oxygen ions and the protons are not in thermal equilibrium, so the

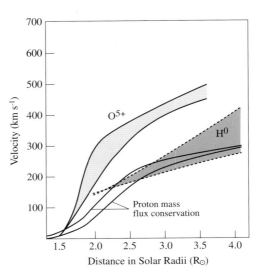

Fig. 7.11 **Heavier ions move faster in coronal holes** Outflow speeds at different distances over the solar poles for hydrogen atoms (*dark gray*, H^0) and ionized oxygen (*light gray*, O^{5+}). These data, taken in late-1996 and early-1997 with the *SOHO* UVCS instrument, show that the heavier oxygen ions move out of coronal holes at faster speeds than the lighter protons, and that the oxygen ions attain supersonic velocities within 2.5 solar radii from Sun center. The solid lines denote the proton outflow speed derived from mass flux conservation; for a time-steady flow, the product of the density, speed and flow-tube area should be constant. (Courtesy of the *SOHO* UVCS consortium. *SOHO* is a project of international collaboration between ESA and NASA.)

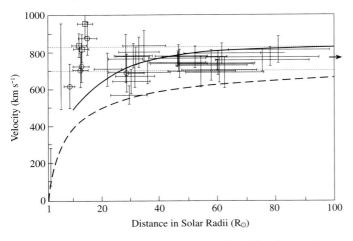

Fig 7.10 **Fast winds accelerate rapidly** The fast solar wind accelerates to its high velocity very close to the Sun, within at least 10 solar radii. Interferometric observations of interplanetary radio scintillations are marked with circles and squares. The vertical bar on each data point is the 90 percent confidence limit, and the horizontal bar indicates the distance range over which the scintillation estimate is averaged. The upper and lower bounds of the *Ulysses* measurements are plotted as horizontal dotted lines, and the mean *Ulysses* fast-wind speed is marked with an arrow at 100 solar radii. The flow speed from a wave-assisted acceleration model is plotted as a dashed line, and the apparent scintillation velocity calculated from this model is plotted as a heavy solid line. The point nearest the Sun is estimated from *Spartan-201* coronagraph measurements. (Adapted from R. Grall et al, Nature 379, 429 (1996).)

Focus 7.3 Temperature, mass and motion

The kinetic temperature, T, of a particle is obtained by equating the thermal energy of the particle to its kinetic energy of motion, or by:

$$\text{thermal energy} = \tfrac{3}{2}kT = \tfrac{1}{2}mV^2_{\text{thermal}} = \text{kinetic energy,}$$

or equivalently:

$$T = \frac{mV^2_{\text{thermal}}}{3k},$$

where m is the mass of the particle and Boltzmann's constant $k = 1.380\,66 \times 10^{-23}\,\text{J K}^{-1}$.

For hydrogen in coronal holes, $m = 1.67 \times 10^{-27}$ kg, and the measured velocities of $V = 2.5 \times 10^5$ m s^{-1} imply a temperature of $T = 2.5 \times 10^6$ K, or 2.5 million degrees kelvin if the velocities are thermal. Oxygen ions are 16 times heavier than the protons and moving twice as fast, so they would be 64 times hotter. However, the particles are not in equilibrium, so our basic assumptions are incorrect. In thermal equilibrium, the particles interact with each other often enough to have the same uniform temperature, which is not the observed situation in coronal holes. Moreover, at a given temperature the thermal velocity will decrease with increasing particle mass, and that is also not observed. So, there has to be some non-thermal process that preferentially accelerates the heavier particles in coronal holes (Focus 7.4).

speeds do not measure temperature in the usual sense. The particles are not near enough to each other, and do not have enough time to jostle together and smooth out their velocity or temperature differences. In contrast, frequent collisions within coronal streamers, where the density is greater, would adjust particle temperatures to similar values, while also wiping out any memory of the initial acceleration mechanism and erasing signatures of it.

The amazing thing is that the heavier oxygen ions move faster than the lighter hydrogen in coronal holes. That violates common sense. It would be something like watching people jogging around a race track, with heavier adults running much more rapidly than lighter, slimmer youngsters. Something is unexpectedly and preferentially energizing the heavier particles in coronal holes.

Magnetic waves might preferentially accelerate the heavier ions by pumping up their gyrations around the open magnetic fields. More-massive ions gyrate with lower frequencies where the magnetic waves are most intense (Focus 7.4), thereby absorbing more magnetic-wave energy and becoming accelerated to higher speeds.

The ponderous magnetic waves remind us of the waves in a stormy ocean that push heavy logs to shore. The lighter shells twist and spiral about in the pounding surf, rarely reaching the beach. That is why the heaviest debris is sometimes found left on the beach after high tide.

The magnetic waves will produce rapid gyrations in the direction perpendicular to the fields and little extra motion along them, something like a hula-hoop that moves in and out from your hips but not up and down them. UVCS measurements confirm that the oxygen ions are moving at high speed perpendicular to the field lines, and at much lower speeds along them, and a similar velocity anisotropy is found for the protons. So, the polar corona and fast solar wind might be heated by magnetic waves.

Ulysses has detected magnetic fluctuations, attributed to Alfvén waves (see Section 5.3,), blowing further out in the winds far above the Sun's poles. They are changes in the direction of the magnetic field, and not variations in its strength, with periods of 10 to 20 hours. When these Alfvén waves are added to the heat-driven wind, they might provide an extra boost that pushes the polar winds to higher speeds.

The magnetic waves probably block high-energy cosmic rays coming into the Sun's polar regions, repelling them back into outer space (Fig. 7.12). The incoming cosmic rays meet an opposing force, like a swimmer entering the surf on a distant shore or one trying to swim upstream against the current of a powerful river. To put it in more scientific language, the Alfvén waves are very long, so they can resonate with the energetic cosmic rays and oppose their entry into the polar regions.

Focus 7.4 Gyration frequency

If a magnetic field of strength, **B**, acts on a particle with charge eZ, for electron charge e, and velocity, **V**, the particle experiences a force, **F**, called the Lorentz force:

$$\mathbf{F} = eZ\,(\,\mathbf{V} \times \mathbf{B}\,),$$

and from Newton's law for a particle of mass, m, and momentum, m**V**, we have:

$$m\frac{dV}{dt} = \mathbf{F} = eZ\,(\mathbf{V} \times \mathbf{B}),$$

where we have assumed that gravitational forces are negligible.

The motion of the charged particle is a circle, and it does not change the particle's kinetic energy. If $V_{\perp}$ denotes the component of velocity perpendicular to the magnetic field, we can rewrite our equation as:

$$\frac{d^2 V_{\perp}}{dt^2} = \frac{eZB}{m}\frac{dV_{\perp}}{dt} = \left(\frac{eZB}{m}\right)^2 V_{\perp},$$

which describes circular motion with a gyration radius (Section 7.1, Focus 7.2 and Section 5.1, Focus 5.2) of $R_g = mV_{\perp}/(eZB)$. The period, P, of the circular orbit is $P = 2\pi R_g/V_{\perp} = 2\pi\,m/(eZB)$, and the frequency, ν_g, given by:

$$\nu_g = 1/P = eZB/\,(2\pi m) = 2.8 \times 10^{10}\,(Zm_e/m)\,B\text{ Hz},$$

where m_e denotes the electron mass, the magnetic field strength B is in units of tesla, and one hertz (Hz) is equivalent to one cycle per second.

In the context of the acceleration of particles by waves in coronal holes, there is more power in the lower frequencies, and heavier particles gyrate at these lower frequencies.

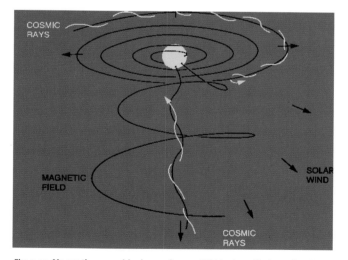

Fig. 7.12 **Magnetic waves block cosmic rays** Within the ecliptic, or the plane of the solar equator, the magnetic fields of the solar wind are wound up in a spiral pattern due to the rotation of the Sun. Since the solar rotation velocity is lower at higher latitudes, the magnetic fields are oriented more or less radially above the solar poles. Scientists therefore predicted that the abundance of cosmic-ray ions would increase over the Sun's polar regions. However, *Ulysses* did not find the expected increase of cosmic rays, apparently because strong magnetic waves above the poles act to repel the high-energy ions back into space. (Courtesy of the *Ulysses* mission, a project of international cooperation between ESA and NASA.)

7.4 Termination of the solar wind

All of the planets are immersed in the solar wind that becomes increasingly rarefied as it spreads out into space. It moves past the planets and beyond the most distant comets. Thus, the entire solar system is bathed in the hot gale that blows from the Sun, creating a large cavity in interstellar space called the heliosphere (Section 1.6, Fig. 1.16). Within the heliosphere, physical conditions are dominated – established, maintained, modified and governed – by the magnetized and electrified solar wind.

Contemporary spacecraft have measured the shape and content of the heliosphere. One instrument on board *SOHO*, the Solar Wind ANisotropies (SWAN), examines interstellar hydrogen atoms sweeping through our solar system from elsewhere. The Sun's ultraviolet radiation illuminates this hydrogen, much the way that a street lamp lights a foggy mist at night. Since solar-wind particles tear the hydrogen atoms apart, their ultraviolet glow outlines the asymmetric shape of the Sun's winds and also establishes its flux.

Focus 7.5 Edge of the solar system

The solar wind carves out a cavity in the interstellar medium known as the heliosphere. The radius of the heliosphere can be estimated by determining the standoff distance, or stagnation point, in which the ram pressure, P_W, of the solar wind falls to a value comparable to the interstellar pressure, P_I. As the wind flows outward, its velocity remains nearly constant, while its density decreases as the inverse square of the distance. The dynamic pressure of the solar wind therefore also falls off as the square of the distance, and we can use the solar wind properties at the Earth's distance of 1 AU to infer the pressure, P_{WS}, at the stagnation point distance, R_S. Equating this to the interstellar pressure we have:

$$P_{ws} = P_{1AU} \times \left(\frac{1AU}{R_S}\right)^2 = (m_p N_{1AU} V_{1AU}^2) \times \left(\frac{1AU}{R_S}\right)^2 = P_I,$$

where the proton's mass $m_p = 1.67 \times 10^{-27}$ kg, the number density of the solar wind near the Earth is about $N_{1AU} = 5 \times 10^6$ protons m^{-3} and the velocity there is about $V_{1AU} = 4 \times 10^5$ m s^{-1}.

To determine the distance to the edge of the solar system, R_S, we also need to know the interstellar pressure, and that is the sum of the thermal pressure, the dynamic pressure, and the magnetic pressure in the local interstellar medium. It is about $P_I = (1.3 \pm 0.2) \times 10^{-13}$ N m^{-2}. The estimate obtained from the equation is $R_S = 100$ AU, far beyond the orbits of the known outer planets. However, the estimates by different authors give a broad range for the distance to the edge of the solar system, depending on the uncertain values of various components of the interstellar pressure.

SWAN's measurements indicate that the solar wind is more intense in the equatorial plane of the Sun than over the north or south poles, which is consistent with *Ulysses'* measurements of the latitudinal variations of wind speed and density.

The Sun's wind thins out as it expands. By the time it has reached the Earth's orbit, there are about five million protons and five million electrons per cubic meter in the solar wind, which is nearly a perfect vacuum by terrestrial standards. As it spreads into a greater volume, the density of the solar wind decreases even further, as the inverse square of the distance from the Sun, and eventually blends with the gas between the stars.

How far does the Sun's influence extend, and where does it all end? Somewhere out there the solar wind ebbs and the cold of interstellar space begins. Eventually the solar wind is no longer dense or powerful enough to repel the ionized matter and magnetic fields coursing between the stars (Focus 7.5). This turbulent boundary, called the heliopause, marks the outer boundary of the heliosphere and the edge of our solar system. There is a celestial standoff out there between the solar wind and interstellar forces, like two gunfighters facing off at sundown.

The size of the heliosphere has been inferred from the twin *Voyager* spacecraft, cruising far beyond the outermost planets (Fig. 7.13). At the time of writing they are more than 21 years old and approaching 80 AU distance from the Sun. Strong shock waves, associated with intense explosions on the Sun, have plowed into the cold interstellar gas at the heliopause, generating a hiss of radio noise detected by the remote *Voyagers*. Thirteen months before the spacecraft detected the radio hiss, unusually intense eruptions on the Sun generated one of the largest interplanetary disturbances ever observed. From the measured speed of the disturbance, and the time it took to travel to the heliopause and generate the radio signals, the heliopause has been located somewhere between 110 and 160 AU, or roughly a hundred times further from the Sun than the Earth. That is where the solar system ends, in a gigantic distant wall of compressed gas that fences off our Sun from the rest of the cosmos.

The distance to the turbulent edge of our solar system can also be inferred by observing certain anomalous cosmic rays, dubbed ACRs, that are distinguished by their unusual composition and origin. Unlike the other types of low-energy cosmic rays, which come fully energized and ionized from interstellar space (Section 7.1), the anomalous material begins its life as electrically neutral, uncharged interstellar atoms. They drift into the solar system at low velocities, where ultraviolet sunlight ionizes them. The newborn ions are then picked up by the magnetic fields entrained in the outflowing solar wind, and carried out to the edge of the solar system. There they rebound at higher velocities due to shock waves at the termination of the solar wind. *Voyager 1* and 2 have used observations of the amounts of the ACRs at different distances in the outer heliosphere to determine the distance to the termination shock of 85± 5AU, which lies inside of the heliopause. In late-2000, *Voyager 1* was already at 79 AU from the Sun, and it is expected to cross the shock before 2004; *Voyager 2* will cross it by 2010.

Closer to home, space physicists are concerned about the impact of the Sun's winds, powerful solar explosions, and varying solar radiation on the Earth and its environment in space.

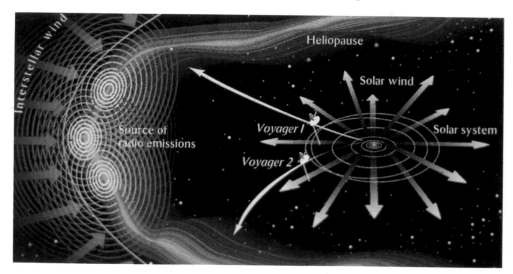

Fig. 7.13 **Edge of the solar system**
Distant *Voyager 1* and *2* spacecraft have picked up signals thought to come from the heliopause, where the solar system ends and interstellar space begins. At the heliopause, the pressure of the solar wind balances that of the interstellar medium. When a dense, high-speed gust of solar particles, and its associated shock wave, hits the heliopause, they generate an intense radio hiss detected by the spacecraft. This enabled scientists to estimate that the heliopause is between 110 and 160 times farther away from the Sun than is Earth, or at 110 to 160 AU. The solar wind is expected to change speed and magnetic field strength before it hits the heliopause, in a termination shock located at about 85 AU. *Voyager 1* will cross the shock by 2004, and *Voyager 2* by 2010.

CHAPTER EIGHT

The Sun–Earth Connection

Sun image in a tree This cross-section of a young beech tree captures an image of its life-sustaining Sun. The photomicrographer, Jean Rüegger-Deschenaux of Zurich, colored the specimen with chrysoldine and astral blue before shooting it at a magnification of 40-fold. (Winner of the 1994 Nikon International Small World Competition.)

8.1 The Earth's magnetic influence

Invisible magnetic fields emanate from the Earth, as well as the Sun. As early as 1600, William Gilbert, physician to Queen Elizabeth I of England, demonstrated that our planet is itself a great magnet, which explains the orientation of compass needles. It is as if there was a colossal bar magnet at the center of the Earth, with magnetic fields that emerge out of the south geographic polar regions, loop through nearby space, and re-enter at the north polar regions (Fig. 8.1). Since the geographic poles are located near the magnetic ones, a compass needle always points north or south. The magnetic fields are produced by electrically conducting currents in the Earth's molten core, so our planet acts like it has a magnet buried at its center.

There are a few subtle caveats to this picture of a terrestrial

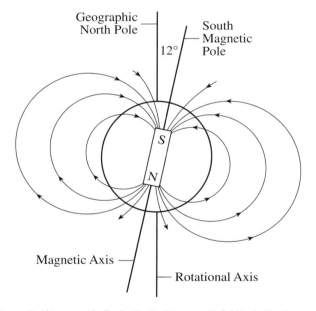

Fig. 8.1 **Earth's magnetic dipole** The Earth's magnetic field looks like that which would be produced by a bar magnet at the center of the Earth, with the North Magnetic Pole corresponding to the South Geographic Pole and *vice versa*. It originates in swirling currents of molten iron deep in the Earth's core, and extends more than 20 Earth radii, or 1.26×10^8 m out into space. Magnetic field lines loop out of the South Geographic Pole and into the North Geographic Pole. A compass needle will always point along a field line. The lines are close together near the magnetic poles where the magnetic force is strong, and spread out where it is weak. The magnetic axis is tilted at an angle of 11.5 degrees with respect to the Earth's rotational axis. Notice that the poles of the magnet are inverted with respect to the geographic poles, following the custom of defining positive, north magnetic polarity as the one in which magnetic fields point out, and negative, south magnetic polarity as the place where magnetic fields point in. This dipolar (two poles) configuration applies near the surface of the Earth, but further out the magnetic field is distorted by the solar wind (see Fig. 8.2).

dipolar magnetic field. The Earth's rotation axis is inclined 12 degrees with respect to its magnetic axis, so a compass needle does not point exactly toward the geographic North Pole, but within 12 degrees of it. Also, since magnetic field lines emerge from a south magnetic pole, the north geographic pole corresponds to the south magnetic pole and *vice versa*.

The surface equatorial field strength of the Earth is 0.000 031 T, or 31 000 nT. That is several times weaker than a typical toy magnet. The terrestrial magnetic fields fall off in strength as they extend to greater distances outside the Earth. Yet, they remain strong enough to shield the Earth from the full force of the solar wind.

Fortunately for life on Earth, the terrestrial magnetic field deflects the Sun's wind away from the Earth and hollows out a protective cavity in the solar wind. This magnetic cocoon is called the magnetosphere. It diverts most of the solar wind around our planet at a distance far above the atmosphere, like a rock in a stream or a windshield that deflects air around a car. The magnetosphere thus protects humans on the ground from possibly lethal energetic solar particles.

Spacecraft have detected magnetospheres around six planets (Table 8.1).

The dipolar (two poles) magnetic configuration applies near the surface of the Earth, but further out the magnetic field is distorted by the Sun's perpetual wind. The energy-laden, electrically-charged solar wind blows out from the Sun in all directions and never stops, carrying with it a magnetic field rooted in the star. Although it is exceedingly thin, far less substantial than a terrestrial breeze or even a whisper, the solar wind is powerful enough to mold the outer edges of the Earth's magnetosphere into a changing asymmetric shape (Fig. 8.2), like a tear drop falling toward the Sun.

Invisible powers collide, sometimes violently, in the space just outside the Earth, where the hot, high-speed, magnetized solar wind meets the Earth's magnetic field. They confront each other with opposing forces, forming a bow shock where the meet (Fig. 8.2). The encounter occurs fairly close to home, usually at a distance of about ten times the Earth's radius (Focus 8.1). The stand-off distance for the six planets with detected magnetic fields is given in Table 8.1.

The solar wind pushes the magnetic field toward the Earth on the day side that faces the Sun, compressing the outer magnetic boundary and forming a shock wave. It is called a bow shock because it is shaped like waves that pile up ahead of the bow of a moving ship. The Sun's wind drags and stretches the terrestrial magnetic field out into a long

Planet	Magnetic field at the equator, B_0 (10^{-4} T)	Tilt[b] of magnetic axis (°)	Offset from planet center (R_P)	Bow shock distance[c], R_{bow} (R_P)	Planet equatorial radius, R_P (10^6 m)
Mercury	0.0033	+ 14	0.05 R_M	1.5 R_M	R_M = 2.439
Earth	0.31	+ 11.5	0.07 R_E	10 R_E	R_E = 6.378
Jupiter	4.28	− 9.6	0.14 R_J	42 R_J	R_J = 71.492
Saturn	0.22	< 1.0	0.04 R_S	19 R_S	R_S = 60.268
Uranus	0.23	− 58.6	0.3 R_U	25 R_U	R_U = 25.559
Neptune	0.14	− 40.8	0.55 R_N	24 R_N	R_N = 24.764

[a] The magnetic field strengths are given at the surface of Mercury and the Earth and at the cloud tops for the giant planets. Venus and Mars have no detected global, dipolar magnetic field, with respective upper limits of 2×10^{-9} and 10^{-8} T. Planetary magnetism is also characterized by the magnetic dipole moment, $M_P = B_0 R_P^3 / 2$, which can be computed from the equatorial magnetic field strengths, B_0, and the equatorial radii, R_P, given in this table.

[b] The tilt is the angle between the magnetic axis and the rotation axis.

[c] The bow-shock distance, R_{bow}, is the distance at which the pressure of the planet's magnetic field just balances the solar-wind pressure (see Focus 8.1).

Table 8.1 **Planetary magnetospheres** [a]

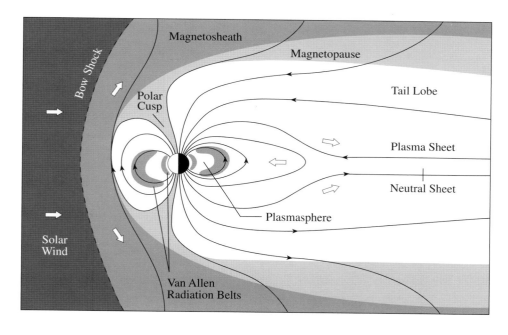

Fig. 8.2 **Magnetosphere** The Earth's magnetic field carves out a hollow in the solar wind, creating a protective cavity, called the magnetosphere. A bow shock forms at about ten Earth radii on the sunlit side of our planet. The location of the bow shock is highly variable since it is pushed in and out by the gusty solar wind. The magnetopause marks the outer boundary of the magnetosphere, at the place where the solar wind takes control of the motions of charged particles. The solar wind is deflected around the Earth, pulling the terrestrial magnetic field into a long magnetotail on the night side. Plasma in the solar wind is deflected at the bow shock (left), flows along the magnetopause into the magnetic tail (right), and is then injected back toward the Earth and Sun within the plasma sheet (center). The Earth, its auroras, atmosphere and ionosphere, and the two Van Allen radiation belts all lie within this magnetic cocoon.

magnetotail on the night side of Earth. The magnetic field points roughly toward the Earth in the northern half of the tail and away in the southern. The field strength drops to nearly zero at the center of the tail where the opposite magnetic orientations lie next to each other and currents can flow (Fig. 8.2).

Thus, the Earth's magnetosphere is not precisely spherical. It has a bow shock facing the Sun and a magnetotail in the opposite direction. The term magnetosphere therefore does not refer to form or shape, but instead implies a sphere of influence. The magnetosphere of the Earth, or any other planet, is that region surrounding the planet in which its magnetic field dominates the motions of energetic charged particles such as electrons, protons and other ions. It is also the volume of space from which the main thrust of the solar wind is excluded.

Yet, some of the energetic particles in space do manage to penetrate the Earth's magnetic defense. Between 0.1 million and 10 million protons and electrons can be found in every cubic meter of the magnetosphere and at its sheath-like boundary. The protons have temperatures of a few million degrees kelvin, and the electrons are at a few hundred thousand degrees. That corresponds to a thermal energy of 0.1 to 1 keV. The magnetic field strength out there is about 50 nT, or 50×10^{-9} T.

The merging between the magnetic fields of the solar wind and the Earth is most effective if they are pointing in

Focus 8.1 Planetary magnetospheres

Six planets are known to have magnetospheres. Their magnetic fields are generated by dynamo action in their interiors, where electrically conducting material is undergoing sufficiently vigorous motions. Mercury and Earth have cores of molten iron alloys. At the high pressures in the interiors of the giant planets Jupiter and Saturn, hydrogen behaves like a liquid metal. For Uranus and Neptune, a water–ammonia–methane mixture forms a deep conduction "ocean".

The equatorial magnetic field strength, denoted by B_0, and the tilt of the dipole axis with respect to the planet's spin axis are given in Table 8.1. The size of the magnetosphere is determined by the distance, R_{bow}, along the planet–Sun line at which the pressure of the planetary magnetic field balances the dynamic ram pressure of the solar wind. The magnetic pressure at the surface of a planet is given by $B_0^2/(2\mu_0)$, where $\mu_0 = 4\pi \times 10^{-7}$ is the permeability of free space. Since the dipole's magnetic field strength falls off as the cube of the distance from the planet, the magnetic pressure decreases as the sixth power of that distance. This means that the stand-off point where the two pressures are equal occurs when:

$$\text{magnetic pressure} = \frac{R_P^6 B_0^2}{2\mu_0 R_{bow}^6} = m_p N V^2 = \text{wind pressure},$$

where the planet's radius is R, the proton mass $m_p = 1.67 \times 10^{-27}$ kg, N is the number density of the protons in the solar wind at the planet's distance from the Sun, and V is the solar-wind velocity at that distance. Solving for R_{bow} we have

$$R_{bow} = \left(\frac{B_0^2}{2\mu_0 m_p N V^2}\right)^{1/6} R_p.$$

At the Earth's distance from the Sun, the number density of the protons in the solar wind is about N = 5 million protons m^{-3} and the wind velocity about $V = 4 \times 10^5$ m s^{-1}. With these numbers our equation gives $R_{bow} = 10\ R_E$, so the bow shock of the Earth is out at about ten times the Earth's radius. The extra ram pressure caused by a powerful mass ejection hitting the Earth can reduce the bow-shock distance to half this value. Unusual drops in the wind's pressure can inflate the leading edge of the magnetosphere five or six times out farther in space, until it engulfs the Moon.

The values of R_{bow} for the other planets can be inferred by noting that the solar-wind number density, N, falls off with the inverse cube of the distance of the planet from the Sun, while the solar-wind velocity remains relatively constant. Values of R_{bow} are given in Table 8.1 for the six planets with detected dipolar magnetic fields.

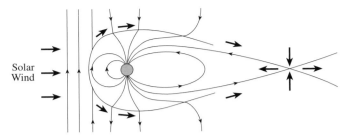

Fig. 8.3 **Magnetic connection at the night side** The Sun's wind brings solar and terrestrial magnetic fields together on the night side of Earth's magnetosphere, in its magnetotail. Magnetic fields that point in opposite directions (*thin arrows*), or roughly toward and away from the Earth, are brought together and merge, reconnecting and pinching off the magnetotail close to Earth. Material in the plasma sheet is accelerated away from this disturbance (*thick arrows*). Some of the plasma is ejected down the magnetotail and away from the Earth, while other charged particles follow magnetic field lines back toward Earth.

back door entry that funnels some of the wind into the magnetosphere. The passing solar wind is slowed down by the connected fields and decelerates in the vicinity of the tail. Energy is extracted from the solar wind and drives a large-scale circulation, or convection, of charged particles within the magnetosphere (Fig. 8.2).

When the solar and terrestrial magnetic fields touch each other in the magnetotail, it can snap like a rubber band that has been stretched too far. The snap catapults the outer part of the tail downstream and propels the inner part back toward Earth. The solar wind is then plugged into the Earth's "electrical socket", and our planet becomes "wired" to the Sun.

Protons and electrons from the solar wind can also leak through the boundary between the solar wind and the magnetosphere. This can happen at the turbulent bow shock, or at the polar cusps that open out to the solar wind like a funnel.

The ionosphere is another important source of particles within the magnetosphere. Solar ultraviolet and X-ray radiation create the ionosphere in our upper atmosphere, which can vary dramatically with solar activity (Section 8.4). Although most of the ionosphere is gravitationally bound to the Earth, some if its particles have sufficient energy to escape. The Earth's magnetic field traps oxygen ions, protons and electrons derived from the ionosphere, creating a plasmasphere that rotates with the Earth's magnetic field and extends outward to between 25 million and 40 million meters. Farther out, the magnetosphere is dominated by its interaction with the solar wind.

When solar-wind electrons and protons enter the Earth's domain, they also become trapped within it and cannot easily get out. In fact, the inner magnetosphere is always filled with a veritable shooting gallery of electrons and protons, trapped within two torus-shaped belts that encircle the Earth's equator but do not touch it (Fig. 8.4). These regions are often called the inner and outer Van Allen radiation belts, named after James A. Van Allen who

opposite directions. With this orientation, the two fields become linked, just as the opposite poles of two toy magnets stick together, and the solar-wind particles can enter the magnetosphere.

The wind's magnetic field will be dragged by the flow of the wind behind the Earth into its magnetotail, wrapping and clinging around the magnetosphere (Fig. 8.3). The magnetosphere can then be punctured in the tail, providing a

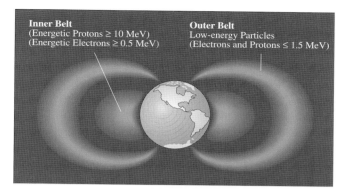

Inner Belt
(Energetic Protons ≥ 10 MeV)
(Energetic Electrons ≥ 0.5 MeV)

Outer Belt
Low-energy Particles
(Electrons and Protons ≤ 1.5 MeV)

Fig. 8.4 **Radiation belts** Electrons and protons encircle the Earth within two donut-shaped, or torus-shaped, regions near the equator, trapped by the terrestrial magnetic field. These regions are now called the inner and outer Van Allen radiation belts, named after James A. Van Allen who first observed them with the *Explorer 1* and *3* satellites in 1958. The inner belt's charged particles tend to have higher energies than those in the outer belt. The trapped particles can damage the microcircuits, solar arrays and other materials of spacecraft that pass through them.

discovered them in 1958. Van Allen used the term "radiation belt" because the charged particles were then known as corpuscular radiation; the nomenclature is still used today, but it does not imply either electromagnetic radiation or radioactivity.

The inner belt is about 1.5 Earth radii from planet center, and the outer belt is located at about 4.5 Earth radii. The Earth's radius is 6.378 million meters, so the radiation belts lie within the plasmasphere.

The inner radiation belt contains protons with energies greater than 10 MeV, with a density of roughly 15 protons per cubic meter. It also contains electrons of lower energy, but exceeding 0.5 MeV. The outer radiation belt also contains protons and electrons, most of which have energies below 1.5 MeV. By way of comparison, a proton in the solar wind usually has an energy of about 1 keV, or 0.001 MeV.

More than half-a-century before the discovery of the radiation belts, Carl Størmer showed how electrons and protons can be trapped and suspended in space by the Earth's dipolar magnetic field. An energetic charged particle moves around the magnetic fields in a spiral path that becomes more tightly coiled in the stronger magnetic fields close to a magnetic pole. The intense polar field acts like a magnetic mirror, turning the particle around so it moves back toward the other pole.

Thus, the electrons and protons bounce back and forth between the north and south magnetic pole (Fig. 8.5). It takes about one minute for an energetic electron to make one trip between the two polar mirror points. The spiraling electrons also drift eastward, completing one trip around the Earth in about half-an-hour. There is a similar drift for protons, but in the westward direction. The bouncing can continue indefinitely for particles trapped in the Earth's radiation belts, until the particles collide with each other or some external force distorts the magnetic fields.

Earth's magnetic dipole is offset from the center of our planet, by about 4.5×10^5 m, as are the dipolar magnetic fields of the other planets - but by different amounts (Table 8.1). Consequently, one side of the Earth's inner radiation belt comes closer to our planet's surface than the other side does. The closest part is known as the South Atlantic Anomaly, because of its anomalous proximity and its location above the east coast of South America. When a satellite goes through the South Atlantic Anomaly, energetic charged particles can penetrate inside it and disrupt its computers or other scientific instruments. Scientists attempt to shield their instruments against the pervasive danger, and try not to use them when passing through the South Atlantic Anomaly.

If the charged particles are trapped within the Earth's magnetism, and cannot cross the magnetic field lines, one wonders how they got into the radiation belts in the first place. They originate from some combination of solar activity, cosmic rays and the ionosphere.

Although the particles in the solar wind are not usually as energetic as the electrons and protons found in the radiation belts, solar flares or coronal mass ejections can accelerate charged particles to much greater energies, as high as 1000 MeV. By way of comparison, a typical proton in the solar wind is one million times less energetic, with a kinetic energy of about 1000 eV. Observations indicate that such solar explosions can disturb the terrestrial magnetic field and inject high-energy charged particles into the magnetosphere. Many of the electrons and protons in the Van Allen radiation belts, particularly the outer one, might therefore originate in solar activity that produces violent gusts in the solar wind.

Cosmic rays may also play a role in feeding the radiation belts, supplying the inner one with its high-energy protons. When cosmic rays bombard the Earth's atmosphere, which lies below the radiation belts, they collide with atoms in our

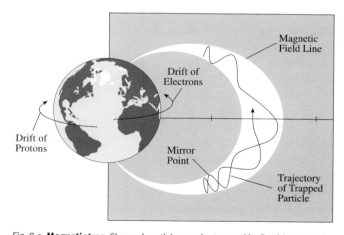

Fig. 8.5 **Magnetic trap** Charged particles can be trapped by Earth's magnetic field. They bounce back and forth between polar mirror points in either hemisphere at intervals of seconds to minutes, and they also drift around the planet on time-scales of hours. As shown by Carl Størmer, in 1907 with the trajectories shown here, the motion is turned around by the stronger magnetic fields near the Earth's magnetic poles. Because of their positive and negative charge, the protons and electrons drift in opposite directions.

air and eject neutrons from the atomic nuclei. These neutrons travel in all directions, unimpeded by magnetic fields since they have no electrical charge.

Once it is liberated from an atomic nucleus, a neutron cannot stand being left alone. A free neutron lasts only 10.25 minutes, on average, before it decays into an electron and proton. A small fraction of the neutrons produced in our atmosphere move out into the inner radiation belt before they disintegrate, producing electrons and protons in places they could not otherwise have reached. These electrically-charged particles are immediately snared by the magnetic fields and remain stored within them, accumulating in substantial numbers over time.

This mechanism might account for the energetic protons in the inner radiation belt. Other electrons and protons can be injected into it from the ionosphere below. The outer belt is mainly fed by the gusty solar wind. Moreover, as next discussed, the terrestrial magnetic fields that hold the radiation-belt particles in place are changing, dynamic entities subject to the vagaries of the Sun's explosive behavior.

8.2 Geomagnetic storms and terrestrial auroras

Large, sporadic geomagnetic storms

We have known about significant variations in the Earth's magnetic field for almost three centuries. They are detected by irregular movements in the direction that compass needles point, with typical fluctuations lasting seconds to days. These variations are caused by invisible geomagnetic storms that rage in the magnetic fields far above our atmosphere. Ground stations throughout the world now monitor such geomagnetic activity, providing a local logarithmic K or linear A index of its strength over three-hour intervals (Table 8.2).

Magnetometers indicate that the great, sporadic geomagnetic storms, that shake the Earth's magnetic field to its very foundations, can produce magnetic fluctuations as large as 1.6 percent at terrestrial mid-latitudes, or 500 nT compared with the Earth's equatorial field strength of 31 000 nT, or 31 μT.

When the Sun shows more sunspots, the terrestrial magnetic field is more frequently disturbed by the most violent storms. If the magnetic measurements are averaged over both a yearly and global scale, the largest storms vary in step with the 11-year sunspot cycle. But it is not the sunspots themselves that produce the geomagnetic storms. They are linked to coronal mass ejections that occur most often when the Sun is more spotted (Section 6.2).

Solar-wind disturbances driven by fast Coronal Mass Ejections, denoted CMEs, are now thought to produce the most intense geomagnetic storms, at least during the maximum in the Sun's activity cycle. Slow CMEs do not produce such events because they lack the strong magnetic fields and high speeds required to stimulate intense magnetic activity on Earth. The Earth intercepts about 70 coronal mass ejections per year when solar activity is at its peak, and less than 10 will have the punch needed to produce large, geomagnetic storms.

The solar wind generally moves slower than a CME, so the ejection plows through the solar wind on its way into interplanetary space, driving a huge shock wave far ahead of it (Fig. 8.6). When directed at the Earth, this shock wave rams into the terrestrial magnetic field and triggers the initial phase, or sudden commencement, of a large geomagnetic storm a few days after the mass ejection leaves the Sun.

Strong interplanetary magnetic fields are also generated by fast CMEs (Fig. 8.6). It is their intense magnetism and high speed that account for the main phase of a powerful magnetic storm, provided that the magnetic alignment is right. The Earth's field is generally directed northward in the

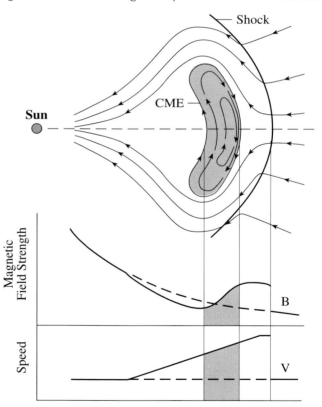

Fig. 8.6 **Interplanetary CME shocks** As it moves away from the Sun (*top left*) a fast coronal mass ejection (CME, *top right*) pushes an interplanetary shock wave before it, amplifying the solar-wind speed, *V*, and magnetic field strength, *B* (*bottom*). The CME produces a speed increase all the way to the shock front, where the wind's motion then slows down precipitously to its steady, unperturbed speed. Compression, resulting from the relative motion between the fast CME and its surroundings, produces strong magnetic fields in a broad region extending sunward from the shock. The strong magnetic fields and high flow speeds commonly associated with interplanetary disturbances driven by fast CMEs are what make such events effective in stimulating geomagnetic activity.

Description	K index	A index
Quiet	0 to 2	less than 8
Unsettled	3	8 to 15
Active	4	16 to 29
Minor storm	5	30 to 49
Major storm	6	50 to 99
Severe storm	7 to 9	100 to 400

Table 8.2 **Classification of geomagnetic disturbances**

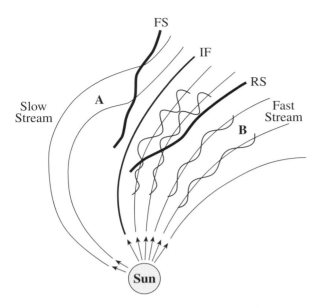

Fig. 8.7 **Co-rotating interaction regions** When fast solar-wind streams, emanating from coronal holes, interact with slow streams, they can produce Co-rotating Interaction Regions (CIRs) in interplanetary space. The magnetic fields of the slow streams in the solar wind are more curved due to the lower speeds, and the fields of the fast streams are more radial because of their higher speeds. Intense magnetic fields can be produced at the interface (IF) between the fast and slow streams in the solar wind. The CIRs are bounded by a forward shock (FS) and a reverse shock (RS).

outer day-side magnetosphere, so a fast CME is more likely to merge and connect with the terrestrial field if it points in the opposite southward direction. The rate of magnetic reconnection, and hence the rate at which energy is transferred from the solar wind to the magnetosphere, increases with the strength and speed of the interplanetary magnetic field. The energy gained drives currents that generate the intense magnetic storm.

Moderate, 27-day recurrent geomagnetic activity

Unlike the great sporadic storms, moderate geomagnetic activity does not exhibit a well-defined connection with sunspots or any other indicator of solar magnetic activity. Indeed, these weaker events, sometimes referred to as substorms, can occur when there are no visible sunspots. They are most noticeable near the minimum of the 11-year solar activity cycle, and produce mid-latitude magnetic fluctuations of about 0.1 percent, or tens of nanotesla, lasting a few hours.

A solar connection is nevertheless indicated by their 27-day repetition period, corresponding to the rotation period of the Sun at low solar latitudes when viewed from the Earth. It means that something on the Sun is triggering moderate geomagnetic activity every time it rotates into alignment with the Earth.

The recurrent activity is linked to long-lived, high-speed streams in the solar wind that emanate from coronal holes

(Section 7.2). When the Sun is near a lull in its 11-year activity cycle, the fast wind streams rushing out of coronal holes can extend to the plane of the solar equator. When this fast wind overtakes the slow-speed, equatorial one, the two wind components interact, like two rivers merging to form a larger one. This produces shock waves and intense magnetic fields that rotate with the Sun (Fig. 8.7). Such Co-rotating Interaction Regions, or CIRs for short, can periodically sweep past the Earth, producing moderate geomagnetic activity every 27 days. Near solar maximum, at the peak of the 11-year activity cycle, coronal mass ejections dominate the interplanetary medium, producing the most intense geomagnetic storms, and the low-level activity is less noticeable.

The auroras

Forceful CMEs can generate exceptionally intense auroras when solar activity is at its peak. When their magnetic fields are swept around and next to the Earth's magnetotail, the interaction can create an opening in the Earth's magnetic barrier, allowing solar-wind particles and energy to pour into the plasma sheet at the center of the magnetotail. The energy gained during this process not only produces intense geomagnetic storms; it also accelerates the infiltrating solar-wind particles and local particles already in the magnetosphere. At such times, the accelerated electrons hurtle along magnetic conduits connected to the upper atmosphere, or ionosphere, in both polar regions, generating spectacular auroras.

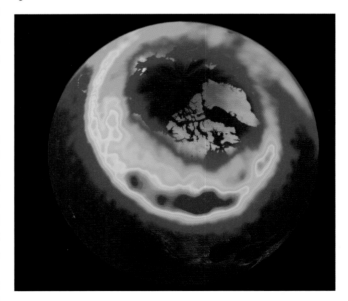

Fig. 8.8 **The auroral oval** *POLAR* looks down on an aurora from high above the Earth's north polar region on 22 October 1999, showing the northern lights in their entirety. The glowing oval, imaged in ultraviolet light, is 4.5 million meters across. The most intense auroral activity appears in bright red or yellow. It is typically produced by magnetic reconnection events in the Earth's magnetotail, on the night side of the Earth. (Courtesy of the Visible Imaging System, University of Iowa and NASA.)

Fig. 8.9 **Aurora Borealis**
Spectacular colored curtains light up the northern sky, like a cosmic neon sign. This photograph of the fluorescent Northern Lights, or Aurora Borealis, was taken in December 1989 at Arctic Valley, Anchorage, Alaska. (Photo © 1989 Cary Anderson.)

Today spacecraft look down on the auroras from high above, showing them in their entirety (Fig. 8.8). They form an oval centered at each magnetic pole, resembling a fiery halo.

The height of the aurora oval above Earth's surface is much smaller than either the radius of the oval, at 2.25 million meters, or the radius of the Earth, 6.38 million meters. An observer on the ground therefore sees only a small, changing piece of the aurora oval, which can resemble a bright, thin, windblown curtain hanging vertically down from the Arctic or Antarctic sky.

The northern or southern lights, named respectively the *aurora borealis* and *aurora australis* in Latin, are one of the most magnificent and earliest-known examples of solar-terrestrial interaction. They illuminate the cold, dark Arctic and Antarctic skies respectively with curtains of green and red light that dance and shimmer across the night sky far above the highest clouds (Fig. 8.9), pulsating and flickering for hours to days.

Brilliant auroras, associated with great magnetic storms, break out as far south as Athens, Rome or Mexico City. They were noted by the ancient Greeks, Plutarch in 467 B.C. and by Aristotle in 349 B.C., but auroras do not extend down to Greece very often, perhaps every 50 or 100 years.

The multi-colored lights have been most frequently observed at far northern latitudes where they have been documented for centuries.

The reason that auroras are usually located near the polar regions is that the Earth's magnetic fields guide energetic electrons there. The auroral lights form when high-speed electrons rain down along the Earth's magnetic field lines into the upper atmosphere in the polar regions, like electricity making the gas in a neon-light shine. The cascade of electrons collides with oxygen and nitrogen in our

Wavelength (nm)	Emitting ion, atom or molecule	Height (km)	Visual color
391.4	N⁺ (nitrogen ion)	1000	violet-purple
427.8	N⁺ (nitrogen ion)	1000	violet-purple
557.7	O (oxygen atom)	90–150	green
630.0	O (oxygen atom)	greater than 150	red
636.4	O (oxygen atom)	greater than 150	red
661.1	N_2 (nitrogen molecule)	65 to 90	red
669.6	N_2 (nitrogen molecule)	65 to 90	red
676.8	N_2 (nitrogen molecule)	65 to 90	red
686.1	N_2 (nitrogen molecule)	65 to 90	red

Table 8.3 **Frequent spectral features in the auroral emission**

atmosphere, boosting them to higher energies and causing them to glow. It is something like the beam of electrons that strikes the screen of your color television set, making it glow in different colors depending on the type of chemicals (phosphors) that coat the screen.

When the electrons slam into the rarefied upper atmosphere, at ionosphere heights of 100 to 400 thousand meters, they excite the abundant atoms to metastable states unattainable in the denser air below. The pumped-up atoms then give up the energy acquired from the electrons, emitting a burst of light.

The precipitating electrons in the auroral ionosphere have energies ranging from 10 eV to hundreds of keV, with characteristic values of about 6 keV and speeds of about 50 million meters per second. At a height of 300 km, their number density is 10^{11} to 10^{13} electrons m^{-3}, and the total auroral particle energy is about 10^{14} J.

The color of the aurora depends on which atoms or molecules are struck by the precipitating electrons, and the atmospheric height at which they are struck (Table 8.3). Excited oxygen atoms radiate both green (557.7 nm) and red (630.0 and 636.4 nm) light. Each color also has a specific altitude range; the green oxygen emission appears at about 100 thousand meters and the red oxygen light at 200 to 400 thousand meters. At these heights, the auroras shine from the ionosphere, an electrically conducting layer in the upper atmosphere (Section 8.4). The bottom edge of the most brilliant green curtains is sometimes fringed with the pink glow of neutral (un-ionized) molecular nitrogen, and rare blue or violet colors are emitted by ionized nitrogen.

Even though the Sun may provide the energy for northern and southern lights, it may not supply the particles that produce them. A popular misconception holds that auroras are caused when electrically-charged particles from the Sun plunge directly into the Earth's atmosphere from the magnetic poles. The electrons that cause the auroras come from the Earth's magnetic tail and are also energized locally within the magnetosphere.

Changing solar-wind conditions can temporarily pinch off the Earth's magnetotail, releasing magnetic energy and pushing electrons up and down the tail. The electrons can also be accelerated within the magnetosphere as they travel down into the upper atmosphere. Currents that flow between the ionosphere and the magnetosphere above it may also play a role, helping to account for persistent low-level auroras when solar activity is low.

Data from NASA's POLAR spacecraft and Japan's Geotail spacecraft has provided the first direct evidence of magnetic reconnection that permits the solar wind to cross into the magnetosphere and produce auroras in the polar regions. During reconnection, the magnetic fields heading in opposite direction – having opposite north and south polarities – break and reconnect (see Section 8.1, Fig. 8.3). The spacecraft have pinpointed the area in the magnetotail where reconnection occurs, at 140 to 160 million meters downwind of Earth on its night side. Moreover, they have shown a clear association between the reconnection and the aurora; the magnetic coupling precedes the brightening of the auroral oval by minutes. As the magnetic field lines on the night side snap and reconnect, they open a valve that lets the solar-wind energy cross into the magnetosphere and additionally shoot energy stored in the magnetic tail back toward the auroral zones near the poles.

The rare, bright, auroras seen at low terrestrial latitudes only become visible during very intense geomagnetic storms. The storms enlarge the magnetotail, spreading the auroral oval down as far as the tropics in both hemispheres. Since these great magnetic storms are produced by solar explosions, it may be the Sun that controls the intensity of the brightest, most extensive auroras, like the dimming switch of a cosmic light.

8.3 Danger blowing in the wind

Fig. 8.10 **Man in space** Astronaut Donald Peterson, on a 50-foot (15-m) tether line during his 4-hour, 3-orbit space walk, moving toward the tail of the *Space Shuttle Challenger* as it glides around the Earth. Hundreds of miles above the Earth, there is no air and an astronaut must wear a space-suit. It supplies the oxygen he needs and insulates his body from extreme heat or cold. However, a space-suit cannot protect an astronaut from energetic particles hurled out from explosions on the Sun. He or she must then be within the protective shielding of a spacecraft or other shelter to avoid the danger. (Courtesy of NASA.)

Sun-driven space weather endangers humans whenever they venture into space. Down here on the ground, we are shielded from the direct onslaught of the raging solar wind by the Earth's atmosphere and magnetic fields, but out in deep space there is no place to hide. Harmful high-energy particles, carried by gusts and squalls in the solar wind, can wipe out unprotected astronauts and destroy satellite electronics. They are of serious concern to future astronauts who might construct the *International Space Station* and explore the Moon or Mars.

Satellites can also be disabled by stormy weather in space. Powerful blasts from coronal mass ejections can compress the Earth's magnetic field and send energetic particles into the magnetosphere, providing threats to Earth-orbiting satellites. Intense radiation from solar flares can change the electrical properties of our atmosphere, disrupting radio navigation or communication systems, and making the atmosphere expand farther into space than usual. Friction can develop between the expanded atmosphere and satellites

traveling in it, slowing down the satellites, altering their orbits, and bringing them to a premature end. Forceful coronal mass ejections can also generate strong currents in our atmosphere, overloading transmission lines on the ground and producing power surges that can blackout entire cities.

Our technological society has become so vulnerable to the potential devastation of these storms in space that national centers employ space-weather forecasters, and continuously monitor the Sun from ground and space to warn of threatening solar activity.

The hazards of space travel

The ultimate vacation, a trip into deep space, is fraught with danger, primarily from energetic particles. Even in the comparative safety of low-Earth orbits, beneath the protection of Earth's magnetic field, astronauts have reported flashing lights inside their eyes. Energetic protons,

Fig. 8.11 **Unprotected from space weather** The first untethered walk in space, on 7 February 1984, where there is no place to hide from inclement Sun-driven storms. Bruce McCandless II, a mission specialist, wears a 300-pound (136-kg) Manned Maneuvering Unit (MMU) with 24 nitrogen gas thrusters and a 35 mm camera. The MMU permits motion in space where the sensation of gravity has vanished, but it does not protect the astronaut from solar flares or coronal mass ejections. (Courtesy of NASA.)

perhaps trapped in the Van Allen radiation belts, pass through the satellite walls and the astronaut's eyelids, striking their retinas and making their eyeballs glow inside.

Once outside the Earth's magnetosphere, astronauts are exposed to the full blast of the ever-flowing solar wind. They could then suffer serious consequences from solar energetic particles even within their spacecraft, resulting in cataracts, skin cancer or even lethal radiation poisoning.

Impulsive solar flares eject protons and electrons into interplanetary space with energies up to a million times greater than those usually present in the solar wind. Flaring protons reach one thousand million electron volts, 1000 MeV or 1 GeV, in energy, and the electrons from solar flares can attain 100 MeV. In comparison, a solar-wind proton, moving at a speed of 5×10^5 m s^{-1} has a kinetic energy of about 1000 eV, or one-millionth the energy of a powerful flare proton.

Energetic protons hurled out from intense solar explosions are especially hazardous. The largest events could inflict serious radiation damage on any astronaut caught in space without adequate shielding (Figs. 8.10, 8.11). Several of these proton events, each lasting 1 to 3 days, occur each year on the average. The high-speed solar protons could even kill an unprotected astronaut who ventures into space. Astronauts walking on the lunar surface in 1972 had at least one close call involving potentially deadly solar-flare events.

These high-speed, charged particles follow a narrow, curved path once they leave the Sun, guided by the spiral structure of the interplanetary magnetic field. They will only threaten humans on a spacecraft, on the Moon or Mars, if

they occur at just the right place on the Sun, at one end of a spiral magnetic field line that connects the flaring region to that part of space. Given the specific circumstances, with a flare near the west limb and the solar equator, the magnetic spiral acts like an interplanetary highway that connects the flaring particles to the space near Earth.

During future space missions, solar astronomers will keep careful watch over the Sun, providing timely warnings of solar flares and coronal mass ejections. The astronauts will then avoid making repairs to their space stations, and curtail any strolls on the Moon or Mars, instead moving inside storm shelters. Current recommendations advise that spacecraft contain metal sanctuaries with walls of aluminum at least 9 cm, or 90 mm thick. Future settlers on the Moon or Mars might want to build underground caves for protection from Sun-driven space weather.

Fast coronal mass ejections plow into the slower-moving solar wind and act like a piston, accelerating electrons and protons as they go, much as ocean waves propel surfers. The mass ejections move straight out of the Sun and flatten everything in their path, like a gigantic falling tree or a car out of control. They energize particles on a grand scale that covers large regions in interplanetary space.

Solar astronomers sometimes designate the two types of solar energetic particle events as impulsive events, accelerated at the Sun during solar flares, and gradual events, linked with coronal mass ejections and mainly accelerated by their shock waves in interplanetary space. The properties of these two types of events are given in Table 8.4.

The crucial information is how strong the storm is and if

Parameter	Impulsive events	Gradual events
Particles: ^{3}He/^{4}He Fe/O H/He	Electron-rich ≈ 1 ≈ 1 ≈ 10	Proton-rich ≈ 0.0005 ≈ 0.1 ≈ 100
Duration of X-ray emission	short (minutes, hard X-rays)	long (hours, soft X-rays)
Duration of particle event	Hours	Days
Radio bursts	Types III and V^a	Types II and IV
Coronagraph	Nothing detected	Coronal mass ejections, 96%
Solar wind	Energetic particles	Very energetic particles
Longitudinal extent	< 30 degrees	≈ 180 degrees
Events per year	≈ 1000	≈ 100

^{a} Impulsive type III and V bursts can be followed by type II and IV.

Table 8.4 **Properties of impulsive and gradual solar energetic particle events**

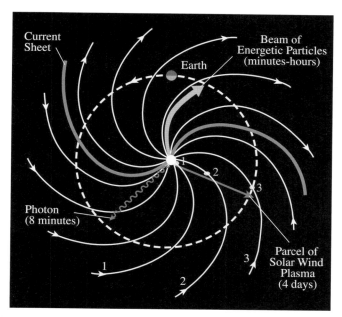

Fig. 8.12 **Travel times** Intense radiation, generated during a solar flare passes right through the interplanetary magnetic field and arrives at the Earth just 8 minutes after being emitted from the Sun. In contrast, a solar flare beams energetic charged particles across a narrow trajectory that follows the interplanetary magnetic spiral, and the time for the particles to reach Earth depends on their energy and velocity, taking roughly an hour for a particle with an energy of about 10 MeV. A coronal mass ejection, or CME, with an average speed of 4.5×10^5 m s^{-1} takes about 4 days to travel from the Sun to Earth's orbit. The CME can energize particles across a wide swath of interplanetary space. The heliospheric current sheet separates magnetic fields of opposite polarities, or directions, denoted by the arrows on the spiral lines. (Courtesy of Frances Bagenal.)

and when it is going to hit us. The exact travel time will depend on the type of solar hazard, since they move with different velocities and on various trajectories in space (Fig. 8.12). Intense radiation from powerful solar flares moves from the Sun to the Earth in just 8 minutes, traveling at the speed of light. Energetic particles, accelerated during the flare process or by the shock waves of coronal mass ejections, can reach the Earth within an hour or less (for energies above 10 MeV). A coronal mass ejection arrives at the Earth as a dense cloud of magnetic fields, electrons and protons one to four days after leaving the Sun.

Humans in space also risk damaging exposure to galactic cosmic rays, the most energetic particles in the interplanetary medium (Section 7.1). Cosmic rays originate outside the heliosphere, and are observed all the time coming from all directions in space, but with a flux that is much lower than Sun-driven particles (Table 8.5). At times of high solar activity, the frequency and intensity of solar explosions is greater, but the enhanced solar winds cut off the flow of cosmic rays into the solar system. The opposite conditions apply near the minimum in solar activity. So, there is some risk whenever someone takes a trip into space.

The dangers posed by cosmic rays are particularly daunting for a visit to Mars. With a projected launch date of

2020, NASA is planning to send astronauts on a three-year mission to the red planet – including six months in transit each way. As an example of the risks involved, iron cosmic-ray particles moving at nearly the speed of light can penetrate deeply into the body, even passing through your skull, ripping up the double helix of DNA molecules in the process. Long exposures to cosmic rays in space also increase the risk of getting cancer, apparently to a forty percent lifetime chance after a voyage to Mars and far above acceptable thresholds of government agencies.

So, there are some real potential hazards to travel in space, depending on the energy and flux of the particles out there, and the time and place of the voyage. Satellites near Earth are in comparable jeopardy, and when they are damaged it could affect large segments of our society.

Satellites in danger

Our networked society, with its computers and global communications systems, has become exceptionally dependent on Earth-orbiting satellites that are vulnerable to space weather. Geosynchronous satellites, that orbit the Earth at the same rate that the planet spins, stay above the same place on Earth to relay and beam down signals used for cellular phones, global positioning systems and internet commerce and data transmission. They can guide automobiles to their destinations, enable aviation and marine navigation, aid in search and rescue missions, and permit nearly instantaneous money exchange or investment choices. Other satellites move in lower orbits and whip around the planet, scanning air, land and sea for environmental change, terrestrial weather forecasting and military reconnaissance.

As many as 1000 satellites are now in operation, providing crucial information to corporations, governments and ordinary citizens. Storms in space can temporarily or permanently wipe out any one of them, affecting the lives of millions of people. Some of these failures are not just

Source	Energy^a (MeV)	Flux (protons m^{-2} s^{-1})
Cosmic rays	1000	6×10^2
Solar flares^b	10	1×10^7
Coronal mass ejections^b	10	3×10^8
Solar wind	0.001	5×10^{12}

^{a} An energy of 1 MeV = 10^6 eV = 1.6×10^{-6} erg = 1.6×10^{-13} J.

^{b} A single-particle event associated with a coronal mass ejection usually has a higher flux of energetic protons than a single-particle event produced by a solar flare, but coronal mass ejections occur about one hundred times less frequently than solar flares.

Table 8.5 **Energy and flux of protons arriving at Earth**

inconveniences, but can have major economic impacts and potentially result in the loss of lives.

Energetic charged particles from solar explosions can seriously damage satellites. When an energetic flaring proton, above 10 MeV in energy, strikes a spacecraft, it can destroy its electronic components. Metal shielding and radiation-hardened computer chips are used to guard against this persistent, ever-present threat to satellites, but nothing can be done to shield solar cells. Since they use sunlight to power spacecraft, solar cells must be exposed to space. Energetic solar protons scour their surface and shorten their lives. They have destroyed the solar cells on at a least one weather satellite.

High-speed electrons can move right through the metallic skin of a spacecraft, sending phantom signals inside and altering the digital bits in its internal data flow. The spurious commands can produce erroneous instrumental data or even send the satellite out of control. When a fast coronal mass ejection scores a direct hit with the Earth, it can produce an intense cascade of high-energy electrons into the Earth's polar regions. Such events have already disabled several communications satellites.

The strong blast of X-rays and ultraviolet radiation from a solar flare alters the Earth's atmosphere, playing havoc with high-frequency radio communications and threatening satellites. The radiation breaks apart atoms in the air, transforming the ionosphere that reflects radio waves to distant locations on Earth. During moderately intense flares, radio communications can be silenced over the Earth's entire sunlit hemisphere, disrupting contact with airplanes flying over oceans or remote countries.

During heightened solar activity, the Earth's atmosphere puffs up like a balloon, causing increased atmospheric drag at orbital altitudes. The enhanced ultraviolet and X-ray radiation from solar flares heats the atmosphere and causes it to expand, and similar or greater effects are caused by coronal mass ejections that produce major geomagnetic storms. The expansion of the terrestrial atmosphere brings higher densities to a given altitude, increasing the drag exerted on a satellite and pulling it to a lower altitude.

Rising solar activity sent both the Skylab Space Station and the Solar Maximum Mission (SMM) satellite into a premature and fatal spiral toward the Earth, and the International Space Station will have to be periodically boosted in altitude to correct for the downward drag. Navigation and global positioning systems, Space Shuttle entry calculations, and accurate monitoring of orbiting objects all depend on accurate knowledge of atmospheric change caused by the intense radiation of solar flares and by coronal mass ejections.

Increases in the dynamic pressure of the Sun's winds during solar activity compresses the magnetosphere and puts high-flying satellites at risk. When a coronal mass ejection slams into the Earth, the force of impact can push the bow shock, at the day side of the magnetosphere, down to half its usual distance of about 10 times Earth's radius. Geostationary spacecraft, that stay over the same spot on Earth, orbit our planet at about 6.6 Earth radii, moving around it once every 24 hours or at the same rate that the planet spins. When the magnetosphere is compressed below their geosynchronous orbits, these satellites are exposed to the full brunt of the gusty solar wind and its charged, energized ingredients.

Turning off the lights

During an intense geomagnetic storm, associated with a colliding coronal mass ejection, strong electric currents flow in the auroral ionosphere. They induce voltage differences in the ground beneath and produce strong currents in any long conductor such as a power line (Fig. 8.13). Up to 100 Amperes of Direct Current, or DC, surge through long-distance power lines designed to carry Alternating Current, or AC, blowing circuit breakers, overheating and melting the windings of transformers, and causing massive failures of electrical distribution systems.

A coronal mass ejection can thereby plunge major urban centers, like New York City or Montreal, into complete darkness, causing social chaos and threatening safety. It is capable of permanently damaging multi-million dollar

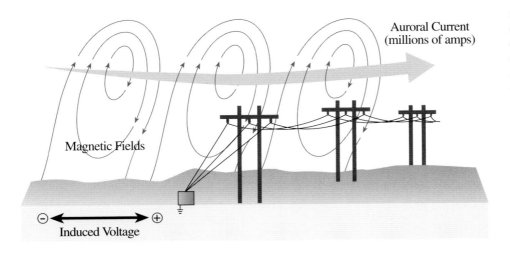

Fig. 8.13 **Power lines in danger** When a coronal mass ejection hits the Earth, electrons in the Earth's magnetosphere cascade into the polar regions, creating a current that flows along the auroral oval at an altitude of about 10^5 m. The magnetic field from this current induces a voltage potential on the surface of the Earth of up to 6 volts per kilometer (V km^{-1}). A strong pulse of direct current enters long conductors like power lines through their ground connection. This can throw circuit breakers, destroy transformers, and shut down power grid systems, sometimes turning off the lights in entire cities.

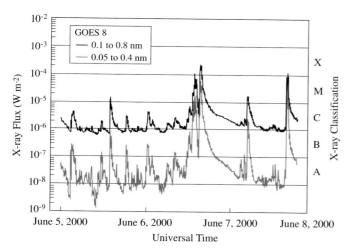

Fig. 8.14 Powerful X-ray flares The X-ray activity of the Sun is monitored by the *Geostationary Operations Environmental Satellites*, or *GOES*. The *GOES* data shown here includes three exceptionally intense, X-class flares emitted from the same active region on 6 and 7 June 2000. The first two flares were also associated with a powerful coronal mass ejection observed with an instrument aboard *SOHO* (Fig. 8.15). The *GOES* data are collected by the National Oceanic and Atmospheric Administration's Space Environment Center and distributed to all interested persons though its space-weather forecasts.

equipment in power-generation plants, and producing hundreds of millions of dollars in losses from unserved power demand or disruption of factories. The threat is greatest in high-latitude regions where the auroral currents are strongest, such as Canada, the northern United States and Scandinavia. In fact, one great magnetic storm in March 1989 put the entire Quebec electric power system out of operation, turning off the lights in a large part of the area for 9 hours.

As electric utility companies rely more and more on enormous power networks that connect widely-separated geographical areas, they become increasingly susceptible to Earth-directed coronal mass ejections and the resulting induced currents. Indeed, some utility companies are now monitoring their power grids for surges produced during geomagnetic storms, and paying close attention to daily space-weather forecasts.

Forecasting space weather

Recognizing our vulnerability to explosions on the Sun, government agencies post forecasts that warn of threatening solar activity. The Space Environment Center (SEC) of the National Oceanic and Atmospheric Administration (NOAA) collects and distributes the relevant data, using satellites and ground-based telescopes to monitor the Sun and relay information about conditions in interplanetary space. Its *Geostationary Operational Environmental Satellites*, or *GOES* for short, monitor threatening activity as it nears the Earth, including

Fig. 8.15 Threatening coronal mass ejection A coronal mass ejection (CME) is observed billowing out from the Sun on 6 June 2000, using the Large Angle Spectrometric COronagraph, or LASCO, on *SOHO*. A central occulting disk blocks out the Sun's intense light to reveal the faint corona, along with background stars and planets. The white circle in the disk denotes the outer edge of the Sun's photosphere. Venus is next to the disk on the right side, seen as a bright line due to an instrumental effect associated with the intense sunlight reflected from the planet, Mars is located at the far left center of the image. This event was a halo mass ejection that grew larger as it expanded, forming a halo around our star, indicating that it was headed toward the Earth. The velocity of the ejected material was at least 9×10^5 m s^{-1}. Although CMEs can occur without a solar flare, this one was accompanied by two intense solar flares (Fig. 8.14). (Courtesy of the *SOHO* LASCO consortium. *SOHO* is a project of international collaboration between ESA and NASA.)

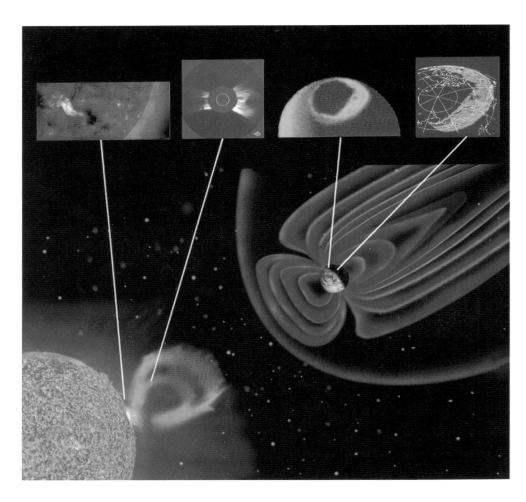

Fig. 8.16 **From Sun to Earth** This illustrative compilation shows how satellites monitor a solar storm from its beginning on the Sun to its interaction with the Earth. *Yohkoh* observes the X-ray emission from tightly coiled magnetic loops, that release their pent-up energy as a coronal mass ejection, detected by *SOHO*'s LASCO. Other satellites track the bubble of magnetized gas on its way to Earth, and then record the collision with our magnetosphere. For instance, a radio experiment on board *WIND* can track the shock waves driven by the mass ejection through the interplanetary medium, and *GEOTAIL* can observe the magnetic connections when the solar ejection collides with the terrestrial magnetic field. The resultant auroras are seen in images obtained with *POLAR* or *IMAGE*.

the powerful X-ray emission of solar flares (Fig. 8.14), and high-speed electrons and protons.

The SEC provides daily space-weather information on the world wide web at the internet address, or Uniform Resource Location (URL) **http://www.sec.noaa.gov**. They have introduced three space-weather scales, one each for geomagnetic storms, solar-radiation storms and radio blackouts. These scales are designated by the letters G, S and R, respectively, followed by a number running from 1 to 5 for minor to extreme hazards.

The worst space weather occurs least often but has the most devastating consequences. An extreme G5 geomagnetic storm occurs about four times each 11-year activity cycle, causing power grid systems to collapse and transformers to experience damage. Severe solar radiation storms, at the S4 level, will produce unavoidable radiation hazard to astronauts on extra-vehicular activity in space, primarily due to protons with energies greater than 10 MeV; such events might happen three times every cycle. For about 8 days during every 11-year activity cycle, a severe R4 high-frequency radio blackout occurs on most of the sunlit side of the Earth, and low-frequency navigation signals produce positioning errors for mariners and aviators.

Predictions about space weather events, based on data from the SEC and NASA satellites are given at **http://www.spaceweather.com**. For instance, the *SOlar and Heliospheric Observatory*, or SOHO, watches solar flares explode on the surface of the Sun at minutes to hours before their particles strike Earth, and it also provides a few days warning of coronal mass ejections that might hit our planet (Fig. 8.15).

The U.S. Air Force operates a global system of ground-based solar telescopes and taps into the output of national, space-borne ones, to continuously monitor solar activity, hoping to forecast events that might severely disrupt military communications and to give operators time to find alternative means of contact.

All of these governmental space-weather forecasts benefit a wide spectrum of people and institutions, including the general public, industries, and government agencies. For example, with an adequate warning, operators can turn off sensitive electronics on satellites, putting them to sleep until the danger passes. Airplane pilots and cellular telephone customers can be warned of potential communication failures. Sensitive navigation and positional systems might be temporarily shut down. The launch of manned space flight missions can be postponed, and walks outside spacecraft or on the Moon or Mars might be delayed. Utility companies can reduce load in anticipation of induced currents on power lines, in that way trading a temporary "brown out" for a potentially disastrous "black out".

Solar astronomers are now looking back at the source of it all, developing methods of predicting solar explosions based on the Sun's magnetic contortions or the growth of active regions on the invisible far side of the Sun (Section 6.4). Space scientists are extending this effort, studying the vital links and dynamic interplay between the Sun and the Earth and viewing them as an interconnected whole. A variety of spacecraft are making coordinated, simultaneous measurements of the Sun, the solar wind, and the Earth's magnetosphere, providing a new global perspective of the intricate coupling between the Sun and Earth under the auspices of an International Solar Terrestrial Physics (ISTP) program. For the first time ever, we can now track every move of possibly destructive events from their beginning on the Sun, to their passage through space, and their ending impact on Earth (Fig. 8.16).

By closely observing the Sun, scientists may someday be able to anticipate changes in the Earth's upper atmosphere. It can be radically transformed by ultraviolet and X-ray radiation during active times on the Sun, and we now turn to this interesting aspect of the Sun–Earth connection.

8.4 The varying Sun and its effect on the Earth's atmosphere

The sun is a magnetic variable star

Our lives depend on the Sun's continued presence and steady output. It illuminates our days, warms our world, and makes life on Earth possible (Section 1.1). The total amount of the Sun's life-sustaining energy is called the "solar constant", perhaps because no variations could be detected in it for a very long time. Yet, as reliable as the Sun appears, it is an inconstant companion. Its luminous output varies in tandem with the Sun's 11-year magnetic-activity cycle.

The solar constant is the average amount of radiant solar energy per second per unit area reaching the top of Earth's atmosphere at a mean distance of one astronomical unit. It can be used with the known distance and radius of the Sun to infer its luminosity (Section 1.3, Focus 1.2). The mean value of the solar constant from 1978 to 1998, was 1366.2 J s^{-1} m^{-2}, with an uncertainty of about ± 1.0 in the same units, whereas the mean value was 1365.6 ±1.0 during the two measured minima in the 11-year solar magnetic-activity cycle.

The discovery that the Sun is a magnetic variable star is relatively recent. Until the early-1980s, it was not known if the Sun was anything but rock-steady because no variations could be reliably detected from the ground. The required measurement precision could not be attained here on Earth because of uncertainty in the changing amount of sunlight absorbed and scattered by our atmosphere.

Stable detectors placed aboard satellites above the Earth's atmosphere have been precisely monitoring the Sun's total irradiance of the Earth since 1978, providing conclusive evidence for small variations in the solar constant (Fig. 8.17). It is almost always changing, in amounts of up to a few-tenths of a percent and on time-scales from 1 second to 20 years. This inconstant behavior can be traced to changing magnetic fields in the solar atmosphere.

The largest downward excursions or dips in the Sun's irradiance of Earth occur when sunspots rotate across the visible solar disk. The magnetic fields in sunspots produce a dimming of the light coming from that part of the Sun, reducing the measured solar constant by a few-tenths of a percent for a few days. The concentrated magnetism in the dark sunspots acts as a valve that blocks the outflow of energy from inside the Sun, producing the observed reductions.

Yet, the Sun becomes brighter and more luminous, rather than fainter, near the maximum in the 11-year cycle of magnetic activity, when sunspots are most numerous. The solar constant increases by about 0.1 percent between activity minimum and maximum. This seems counter-intuitive, since the dark sunspots produce a decrease in the solar output, and sunspots are more numerous near the peak of the activity cycle.

The increase in luminous output is attributed to bright, localized magnetic structures, called *faculae* from the Latin for "little torches" or *plages* after the French word for "beaches". Both faculae and plages mark small, bright regions of concentrated magnetic flux that appear next to sunspots, and also within the magnetic network across the entire solar disk.

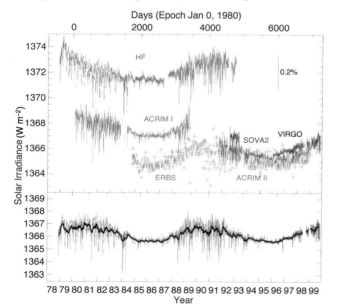

Fig. 8.17 **Variations in the solar constant** Observations with very stable and precise detectors on several Earth-orbiting satellites show that the Sun's total radiative input to the Earth, termed the solar irradiance, is not a constant, but instead varies over time-scales of days and years. Measurements from five independent space-based radiometers since 1978 (*top*) have been combined to produce the composite solar irradiance (*bottom*) over two decades. They show that the Sun's output fluctuates during each 11-year sunspot cycle, changing by about 0.1 percent between maximums (1980 and 1990) and minimums (1987 and 1997) in magnetic activity. Temporary dips of up to 0.3 percent and a few days' duration are due to the presence of large sunspots on the visible disk. The larger number of sunspots near the peak in the 11-year cycle is accompanied by a rise in magnetic activity that creates an increase in luminous output that exceeds the cooling effects of sunspots. The years are 1978 to 1999 and the total irradiance just outside our atmosphere, called the solar constant, is given in units of watts per square meter, W m^{-2}, where 1W is equivalent to 1 J s^{-1}. The capital letters are acronyms for the different radiometers, and offsets among the various data sets are the direct result of uncertainties in their scales. Despite these offsets, each data set clearly shows varying radiation levels that track the overall 11-year solar activity cycle. (Courtesy of Claus Fröhlich.)

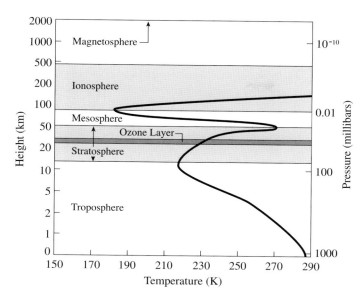

Fig. 8.18 **Sun-layered atmosphere** The pressure of our atmosphere (*right scale*) decreases with height (*left scale*). This is because fewer particles are able to overcome the Earth's gravitational pull and reach greater heights. The temperature (*bottom scale*) also decreases steadily with height in the ground-hugging troposphere, but the temperature increases in two higher layers that are heated by the Sun. They are the stratosphere, with its critical ozone layer, and the ionosphere. The stratosphere is mainly heated by ultraviolet radiation from the Sun, and the ionosphere is created and modulated by the Sun's X-ray and extreme-ultraviolet radiation. A pressure of 1 millibar is equivalent to 0.001 bar and 100 pascal.

The faculae are bright regions in the photosphere that are barely detectable in visible solar light. Excess heating of the solar chromosphere above the facular areas creates emission plages, which are easily detected by their enhanced radiation in violet emission lines of ionized calcium. Designated the H and K lines, they have respective wavelengths of 396.85 and 393.37 nanometers. So, faculae and plages are two words for the same magnetic brightening that is most easily monitored in calcium H and K images of the Sun's chromosphere.

As solar activity increases, there is an increase in the number of sunspots, faculae and plages. The irradiance decrease caused by sunspots is roughly balanced by the increase due to faculae or plages in the active regions near them, and the excess brightness from the magnetic network outside these regions more than compensates for the sunspot deficit. In other words, the total long-term irradiance increase by the bright faculae or plages overwhelms the short-term decrease by dark sunspots as solar activity increases.

The entire spectrum of the Sun's radiation is modulated by varying solar activity. Overwhelmingly the greatest part of solar radiation is emitted in the visible and near-infrared parts of the spectrum where the variations are relatively modest. By contrast, there are enormous changes at the ultraviolet and X-ray wavelengths that contribute only a tiny fraction of the Sun's total luminosity. The ultraviolet emission doubles from activity minimum to maximum, while the X-

ray brightness of the corona increases by a factor of 100 (Section 5.3, Fig. 5.29). Virtually all of this activity goes unseen at visible wavelengths, and most of it contributes negligible amounts to variations in the solar constant.

The Sun's varying ultraviolet and X-ray radiation nevertheless dramatically transforms physical conditions in the Earth's upper atmosphere where it is absorbed. The solar radiation emitted at visible wavelengths passes right through our transparent air, and does not noticeably affect the upper atmosphere.

The Earth's varying Sun-layered atmosphere

Our thin atmosphere is pulled close to the Earth by its gravity, and suspended above the ground by molecular motion. Because air molecules are mainly far apart, our atmosphere is mostly empty space, and it can always be squeezed into a smaller volume.

The atmosphere near the ground is compacted to its greatest density and pressure by the weight of the overlying air. At greater heights there is less air pushing down from above, so the compression is less and the density and pressure of the air falls off into the near vacuum of space. Yet, even at the bottom of our atmosphere the density is only one-thousandth that of liquid water (10^3 kg m^{-3}); an entire liter (10^{-3} m^3) of this air will weigh only one gram.

Not only does the atmospheric pressure decrease as we go upward, the temperature of the air also changes, but it is not a simple fall-off. It falls and rises in two full cycles as we move off into space (Fig. 8.18).

The temperature decreases steadily with increasing height in the lowest layer of our atmosphere, called the troposphere from the Greek *tropo* for "turning". Visible sunlight passes harmlessly through this region to warm the ground below. The temperature above the ground tends to fall at higher altitudes where the air expands in the lower pressure and becomes cooler. The average air temperature drops below the freezing point of water (273 K) about 1000 m (1 km) above the Earth's surface, and decreases progressively until it drops to 220 K at roughly ten times this height.

The temperature then increases at greater heights within the next-lowest atmospheric layer, named the stratosphere, which extends up to about 50 000 m (50 km). The word stratosphere is coined from stratum-sphere, where stratum means "layer". The Sun's invisible ultraviolet radiation is largely absorbed in the stratosphere, where it warms the gas and helps make ozone.

When ultraviolet rays with wavelengths of about 200 nm strike a molecule of ordinary diatomic oxygen that we breathe (O_2), they split it into its two component oxygen atoms (two O). One of the freed oxygen atoms then bumps into, and become attached to, an oxygen molecule, creating an ozone molecule (O_3) that has three oxygen atoms instead of two. The Sun's ultraviolet rays thereby produce a globe-circling layer of ozone in the stratosphere.

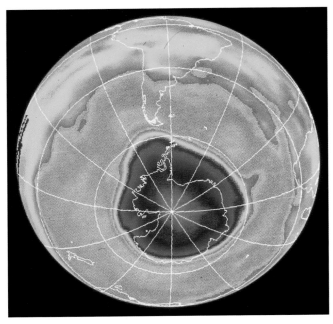

Fig. 8.19. **Hole in the sky** A satellite map showing an exceptionally low concentration of ozone, called the ozone hole, that forms above the South Pole in the local spring. In October 1990 it had an area larger than the Antarctic continent, shown in outline below the hole. Eventually spring warming breaks up the polar vortex and disperses the ozone-poor air over the rest of the planet. (Courtesy of NASA.)

The ozone layer protects us by absorbing most of the solar ultraviolet radiation and keeping its destructive rays from reaching the ground. That is a good thing, for the Sun's virulent ultraviolet radiation is likely to produce eye cataracts and skin cancer in persons exposed to all of it. After all, sunburns are caused by the small amount of ultraviolet radiation that manages to work its way through the ozone layer and down to the Earth's surface.

The threat of dangerous and even lethal ultraviolet rays caused world-wide concern when it was discovered that everyday, man-made chemicals are punching a hole in the ozone layer (Fig. 8.19). The chemicals, called chlorofluorocarbons or CFCs for short, were therefore completely banned by international agreement in 1990. Still, the ozone layer is not expected to regain full strength until well into the latter half of the 21st century.

Since the ozone layer is produced by the Sun's ultraviolet radiation, and the amount of this varies, there are natural fluctuations in the density of the ozone layer. The total global amount of ozone becomes enhanced, depleted and enhanced again by 1 to 2 percent as solar activity goes from its maximum to its minimum and back to its maximum every 11 years. This modulates the protective ozone layer at a level comparable to human-induced ozone depletion by chemicals wafting up from the ground. Monitoring of the expected recovery of the ozone layer from chemicals will therefore require careful watch over how the Sun is changing the layer from above.

The mesosphere, from the Greek *meso* for "intermediate" lies

Focus 8.2 Radio waves are reflected by the changing ionosphere

The ionosphere reflects radio waves up to a maximum frequency that depends on the density of free electrons there. The crucial upper frequency is equal to the plasma frequency, the natural frequency of oscillation for the ionosphere. The plasma frequency, ν_p, varies as the square root of the electron density, N_e, and is given by $\nu_p = 9N_e^{1/2}$ Hz for an electron density in electrons per cubic meter (also see Section 6.1, Focus 6.3).

The ionosphere is transparent to radio radiation above the plasma frequency, so these waves can be used for communication with high-flying satellites or to observe the radio Universe. Cosmic and solar radiation at frequencies below the plasma frequency are reflected back into space from the top of the ionosphere and cannot be observed from the ground. They must be observed from satellites orbiting the Earth above the ionosphere.

The ionosphere is sub-divided into at least three layers, labeled D, E and F from base to top, that reflect waves of different frequencies or wavelengths. The heights of these layers have been inferred from the time between the transmission of a radio pulse and the reception of its reflected echo, and the electron density can be determined from the highest frequency at which a return signal is still received. Radio scientists usually express this frequency in units of megahertz, abbreviated MHz, where one MHz is equivalent to a million hertz, or 1 MHz = 10^6 Hz. The wavelength can be inferred from the frequency by dividing the velocity of light 2.997 924 58 × 10^8 m s^{-1}, by the frequency in Hz.

The uppermost layer of the ionosphere, at a height of roughly 250 km, reflects frequencies between 3 and 30 MHz, or wavelengths from 10 to 100 m, that are used for radio broadcasting and over-the-horizon radar surveillance. Electrons at lower heights, from 95 to 140 km in the E layer, reflect medium frequencies from 0.3 to 3 MHz with wavelengths of 100 to 1000 m. The lowest layer, D, at heights of less than 90 km, reflects the lowest frequencies from 0.003 and 0.3 MHz, or wavelengths of 10^3 to 10^5 m, that are used in some navigation systems such as LORAN and OMEGA.

The maximum plasma frequency of the ionosphere depends on the solar magnetic-activity cycle. At activity maximum, the Sun produces more ultraviolet radiation and X-rays, the ionosphere is more highly ionized by them, and the free-electron density at a given altitude is enhanced (Fig. 8.20). The plasma frequency then increases, so higher radio frequencies are reflected from the ionosphere near the maximum in the 11-year cycle of solar activity. The peak electron density in the ionosphere can reach 100 billion (10^{11}) to a million million (10^{12}) electrons per cubic meter, depending on the level of solar activity (Fig. 8.20). The density of the neutral, un-ionized atoms at a given height also increases, resulting in greater atmospheric drag on satellites (Section 8.3).

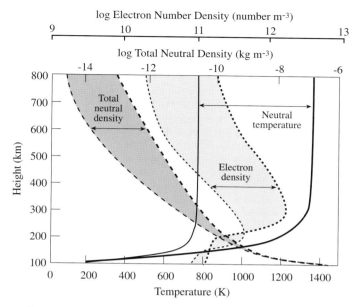

Fig. 8.20 **Varying solar heating of the Earth's upper atmosphere** During the Sun's 11-year activity cycle, the upper-atmosphere temperatures fluctuate by factors of two, and neutral (un-ionized atom) and electron densities by factors of ten. The bold lines register maximum values and the less bold the minimum values. Enhanced magnetic activity on the Sun produces increased ultraviolet and X-ray radiation that heats the Earth's upper atmosphere and causes it to expand, resulting in higher temperatures and greater densities at a given altitude in our atmosphere. (Courtesy of Judith Lean.)

just above the stratosphere. The temperature declines rapidly with increasing height in the mesosphere, reaching its lowest value of ~100 K at the top of this layer. The main reason for the decreasing temperatures is the falling ozone concentration and decreased absorption of solar ultraviolet radiation.

Above the mesosphere, the temperature begins to once more increase with height, attaining values, at upward of about 150 km, that are hotter than those at ground level. This layer is called the ionosphere, a permanent spherical shell of electrons and ions, created and heated by absorbing the extreme-ultraviolet and X-ray portions of the Sun's energy. This radiation tears electrons off the atoms and molecules in the upper atmosphere, thereby creating ions and free electrons that are not attached to atoms.

The ionosphere was postulated in 1902 to explain Guglielmo Marconi's transatlantic radio communications. Since radio waves travel in straight lines, and cannot pass through the solid Earth, they get around the planet's curvature by reflection from electrons in the ionosphere.

Solar X-rays and extreme-ultraviolet radiation both produce and significantly alter the Earth's ionosphere. Their greater intensity near the maximum of the 11-year magnetic-activity cycle produces increased ionization, greater heat, and expansion of our upper atmosphere. At a given height in the ionosphere, the temperature, the density of free electrons, and the density of neutral, un-ionized atoms all rise and fall in synchronism with solar activity over its 11-year cycle (Fig. 8.20). This Sun-induced change in the content and structure of the ionosphere affects its ability to mirror radio waves (Focus 8.2).

It is visible sunlight that passes through to the troposphere where all our weather and climate occur. Global circulation of the air, driven by differential solar heating of the Earth's equatorial and polar surfaces, creates complex, wheeling patterns of weather in this layer. As we shall next see, the variable Sun may have affected the climate in this part of our air during the past millennium.

8.5 The Sun's role in warming and cooling the Earth

The Sun's radiation and global warming

Signs of the Sun's varying activity have been found in records of the temperature at the top of the Sun-warmed oceans. During the past 130 years, the global sea-surface temperature has been swinging up and down by about 0.05 degrees Celsius (°C) in time with the 11-year activity cycle. The temperatures increase or decrease when the solar constant does, at least for the two decades that it has been observed, but by amounts that may be two or three times warmer than can be accounted for by direct solar heating.

The land-surface temperatures have been correlated with the length of the solar cycle. The yearly mean air temperature over land in the northern hemisphere has moved higher or lower, by about 0.2 °C, in close synchronism with the solar-cycle length during the past 130 years (Fig. 8.21). Short cycles are characteristic of greater solar activity that apparently warms our planet, while longer cycles signify decreased activity on the Sun and cooler times at the Earth's surface. These temperature variations might be attributed to solar-driven changes in cloud cover, caused by the Sun's 11-year modulation of the amount of cosmic rays reaching Earth.

At times of enhanced activity on the Sun, the solar wind is pumped up with intense magnetic fields that extend far out into interplanetary space, blocking more cosmic rays that would otherwise arrive at Earth (Section 7.1). The resulting decrease in cosmic rays means that fewer energetic charged particles penetrate to the lower atmosphere where they may help produce clouds, particularly at higher latitudes where the shielding by Earth's magnetic field is less. The reduction in clouds, that reflect sunlight, would explain why the Earth's surface temperature gets hotter when the Sun is more active.

Many of the temperature changes on Earth during the first half of the 20th century might be directly related to brightening and dimming of the Sun. Solar variability provides a reasonable match to the detailed ups and downs of the temperature record during this period (Fig. 8.22).

To fully understand the temperature measurements, scientists have examined historical records of the variable brightness of the Sun and other stars. Their reconstruction of the varying solar irradiance of Earth (Fig. 8.23) shows that the Sun's changing brightness dominated our climate for two centuries, from 1600 to 1800. Cooling by hazy emission from volcanoes next played an important role, but the Sun noticeably warmed the climate for another century, from 1870 to 1970. After that, heat-trapping gases apparently took control of our climate.

The Earth is now hotter than it has been any time during the previous 1000 years (Fig. 8.24). Global warming by the greenhouse effect is probably responsible for this recent, unprecedented rise in temperature. Minor ingredients of the atmosphere, such as carbon dioxide and water vapor, absorb the ground's infrared radiation, holding it close to the planet's surface and elevating the temperature there. Methane and nitrous oxide also act as greenhouse gases, but they are less abundant than carbon dioxide and water vapor.

The greenhouse effect is literally a matter of life and death. Without its atmosphere, the Earth is heated by the Sun to only −18°C, which is well below the freezing point of water (at 0°C). Fortunately for life on Earth, the greenhouse gases in the air warm the planet by as much as 33°C, to an average temperature of about 15°C, and this extra heat keeps the oceans from becoming frozen over. Most of this "natural" greenhouse warming comes from water molecules (60 to 70 percent) and carbon dioxide provides only a few degrees of the temperature increase.

You can nevertheless have too much of a good thing. Since the industrial revolution, humans have released heat-trapping gases into the atmosphere at an ever-increasing rate, creating an "unnatural" greenhouse effect. The amount of carbon dioxide in our air has, for example, grown by more than 13 percent since 1958, when direct monitoring started, and it is still steadily accumulating as the result of rapid growth in the world population and increased burning of fossil fuels like coal and oil.

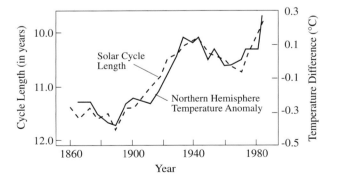

Fig. 8.21 **Global surface temperatures and sunspot cycle length** Variations in the air temperature over land in the Northern Hemisphere (*solid line*) closely fit changes in the length of the sunspot cycle (*dashed line*). Shorter sunspot cycles are associated with increased temperatures and more intense solar activity. This suggests that solar activity is at least partly responsible for the rise in global temperatures over the last century, and that the Sun can substantially moderate or enhance global warming brought about by human increases of carbon dioxide and other greenhouse gases in the atmosphere.

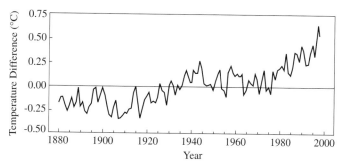

Fig. 8.22 Global temperature changes in the 20th century The annual global mean temperature near the Earth's surface is plotted as a temperature difference from the long-term mean value. At more than 0.5 °C above the mean, recent global temperatures maintain a warming trend for the past two decades. Natural temperature fluctuations prohibit a clear detection of human-induced warming in the early part of the 20th century, but human increases of greenhouse gases in the atmosphere may be noticeably contributing to the recent rise in temperatures. The global mean temperature for the period 1880 to 1997 is 13.8 °C (56.9 °F). (Courtesy of Michael Changery, the National Climatic Data Center (NCDC) of NOAO)

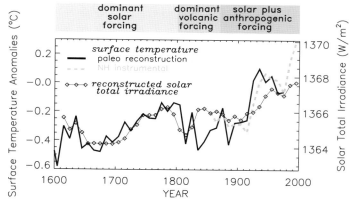

Fig. 8.23 Sources of global temperature variations A reconstruction of the Sun's brightness over the past 400 years indicates that the global temperature fluctuations from 1600 to 1800 were mainly due to variations in solar activity. Cooling by haze from volcanoes played a role during the next 100 years. The global temperature increases after the mid-1970s, are caused mainly by heat-trapping greenhouse gases emitted by industrial economies, the term "anthropogenic" being used here to describe human-induced warming of the atmosphere. Here the decade-averaged values of differences (anomalies) in Northern Hemisphere summer surface temperatures (*scale at left*) are compared with the reconstructed solar total irradiance (*symbols, scale at right*). The dark solid line is paleoclimate temperature data, primarily tree rings, and the gray dashed line is instrumental temperature data. (Courtesy of Judith Lean.)

If current emissions of carbon dioxide and other greenhouse gases go unchecked, their concentrations in the atmosphere are likely to be double pre-industrial levels sometime in the 21st century. If that happens, many experts forecast, the average surface temperature of the globe will rise by about 2 °C, making the Earth hotter than it has been in millions of years, creating widespread drought. Sea levels could rise enough to turn parts of Florida into Atlantis, inundate Venice, and submerge island nations. Although the true threat remains difficult to gauge, it is likely that the increased heat and violent weather will drastically change the climate we are used to, and some of us will feel like the world is melting down in a pool of sweat.

The continued accelerated burning of fossil fuels will someday cause great damage to the environment, so both the developing and industrial nations should now do more to stop it. The Sun's activity can nevertheless substantially enhance or moderate this warming, and there isn't very much we can do about the Sun's changing temperament except monitor it. Moreover, there may be relief on its way when the next ice age begins.

Cooling the Earth down

Spacecraft observations of the varying solar brightness over the past two decades indicate that it has varied by about 0.1 percent. The observed brightness and magnetic variations of other stars with masses and ages close to those of the Sun, indicate that more substantial variations of the Sun's luminosity are possible. They may be associated with dramatic changes in the Earth's climate on time-scales of hundreds, thousands, and hundreds of thousands of years.

Profound Sun-driven transformations in climate are suggested by past solar activity recorded in sunspot observations, tree rings and ice cores. An example is the period from 1645 to 1715, now known as the Maunder Minimum, when sunspot activity dropped to unusually low levels and the world experienced one of the coldest periods of the Little Ice Age in Europe. Comparisons with the brightness variations of Sun-like stars indicate that the Sun was approximately 0.25 percent dimmer at the time of the Maunder Minimum than currently, and was therefore capable of explaining the estimated drop of about 0.5 °C in global mean temperature.

Radioactive isotopes record solar magnetic activity and related climate change for thousands of years. During periods of increased activity on the Sun, when the Earth was presumably warmer, the magnetic fields in the solar wind had a larger shielding effect on cosmic rays. This prevented the energetic charged particles from entering the Earth's atmosphere and producing radioactive isotopes. In contrast, high amounts of the radioactive elements were produced when the Sun was inactive and the climate was cold.

Carbon-14, dubbed radiocarbon and designated ^{14}C, is the first radioactive isotope to be used to reconstruct past solar activity. Radiocarbon is produced in the Earth's atmosphere by a nuclear reaction in which energetic neutrons interact with nitrogen, the most-abundant substance in our air. The neutrons are themselves the products of interactions between cosmic rays and the nuclei of air molecules.

Radiocarbon can be found in annual tree rings dating back to eight thousand years ago. Each radiocarbon atom, ^{14}C, joins with an oxygen molecule, O_2, in the air to produce

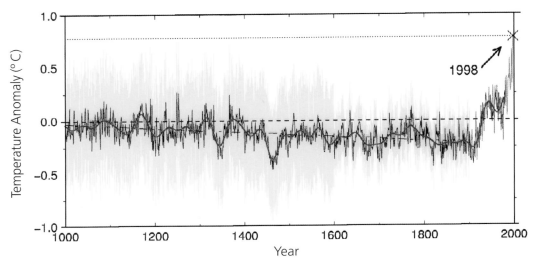

Fig. 8.24 **Unusual heat** This reconstruction of the hemisphere temperature record suggests that the Northern Hemisphere has been warmer in the 20th century than in any other century of the last thousand years. The sharp upward jump in the temperature during the last 100 years was recorded by thermometers at and near the Earth's surface. Earlier fluctuations were reconstructed from "proxy" evidence of climatic change contained in tree rings, lake and ocean sediments, and ancient ice and coral reefs. The heavy solid line is the reconstruction smoothed over 40-year intervals; the long dashes depict the linear trend from A.D. 1000 to 1850, and the short dashes provide the reference zero level, at the mean for the calibration period 1902 to 1980. The farther back in time the reconstruction is carried, the larger the range of possible error, denoted by the yellow shading. (Courtesy of Michael E. Mann.)

a form of carbon dioxide, designated by $^{14}CO_2$, that is assimilated by live trees during photosynthesis, and deposited in their outer rings. The time of assimilation can be determined at any later date from the age of the annual tree ring. Just count the number of tree rings that have been subsequently formed at the rate of one ring per year.

The radiocarbon records confirm that the Maunder Minimum corresponded to a dramatic reduction in solar activity, and show that such prolonged periods of inactivity are a fairly common aspect of the Sun's behavior. During the past two thousand years, the Sun has spent nearly a third of that time in a relatively inactive state (Fig. 8.25). Extended periods of solar inactivity must therefore be considered to be a permanent feature of the Sun, and can be expected to occur again in the future.

The changing Sun has been drastically altering the climate for thousands of years. The Little Ice Age (1400–1800), for example, overlaps the Spörer Minimum (1420–1500) and Maunder Minimum (1645–1715) in solar activity. During this long period of unusual cold weather, alpine glaciers expanded, the river Thames, England and the canals of Venice, Italy, regularly froze over, and painters depicted unusually harsh winters in Europe (Fig. 8.26).

Cores of ice taken from the polar ice caps complement and extend the tree-ring evidence for past Sun–climate connections. The ice contains the radioactive isotope of beryllium, ^{10}Be, that has been deposited there by snows, the later snows compressing earlier ones into ice. Like

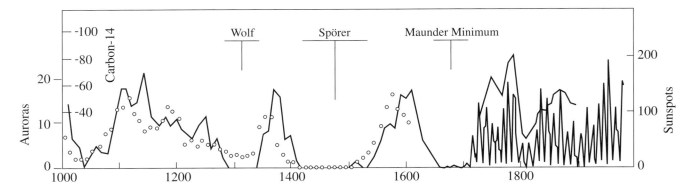

Fig. 8.25 **Long periods of solar inactivity** Three independent indices demonstrate the existence of prolonged decreases in the level of solar activity, such as the Maunder and Spörer minima. The observed annual mean sunspot numbers (*scale at right*) also follows the 11-year solar activity cycle after 1700. The curve extending from A.D. 1000 to 1900 is a proxy sunspot number index derived from measurements of carbon-14 in tree rings. Increased carbon-14 is plotted downward (*scale at left-inside*), so increased solar activity and larger proxy sunspot numbers correspond to reduced amounts of radiocarbon in the Earth's atmosphere. Open circles are an index of the numbers of auroras occuring in the Northern Hemisphere (*scale at left-outside*). (Courtesy of John A. Eddy.)

Fig. 8.26 **Hunters in the snow** This section of a painting by Pieter Bruegel the Elder depicts a time when the average temperatures in Northern Europe were much colder than they are today. Severe cold occurred during the Maunder minimum, from 1645 to 1715, when there was a conspicuous absence of sunspots and other signs of solar activity. This picture was painted in 1565, near the end of another dearth of sunspots, called the Spörer minimum. (Courtesy of the Kunsthistorisches Museum, Vienna.)

radiocarbon, the ^{10}Be is produced by nuclear reactions between energetic neutrons and molecules in the air, a consequence of cosmic rays entering the atmosphere. When solar activity is more pronounced, the cosmic rays arriving at Earth are less numerous and the amount of ^{10}Be is reduced, and *vice versa*. The radioactive isotopes found in both tree rings and ice cores indicate that the Sun's activity has fallen to unusually low levels at least three times during the past one thousand years, each drop apparently corresponding to a long, cold spell of roughly a century in duration.

Further back in time, during the past one million years, our climate has been dominated by the recurrent ice ages, each lasting about 100 thousand years. At the height of each long ice age, the great polar ice sheets advance down to lower latitudes. These glaciations are punctuated every 100 thousand years or so by a relatively short interval of unusual warmth, called an interglacial, lasting 10 or 20 thousand years, when the glaciers retreat. We now live in such a warm interglacial interval, called the Holocene period, in which human civilization has flowered. Still, the die is cast for the next glaciation, and the ice will come again.

The rhythmic alteration of glacial and interglacial intervals is related to periodic alterations in the amount and distribution of sunlight received by Earth over tens of thousands of years. When less sunlight is received in far northern latitudes, the summer temperatures are colder there. So, less polar ice melts in the summer and, over time, the winter snows are compressed into more ice to make the glaciers grow.

Three astronomical cycles combine to alter the angles and distance at which sunlight strikes the far northern latitudes of Earth, triggering the ice ages. This explanation was fully developed by Milutin Milankovitch from 1920 to 1941, so the astronomical cycles are now sometimes called the Milankovitch cycles. They involve periodic wobbles in the Earth's rotation and changes in the tilt of its axis and the shape of its orbit, occurring over tens of thousands of years (Fig.8.27).

In the longest cycle, the shape of the Earth's orbit stretches slightly and its eccentricity changes, from more circular to more elliptical and back again, over a period of 100 thousand years. As its path becomes more elongated, the

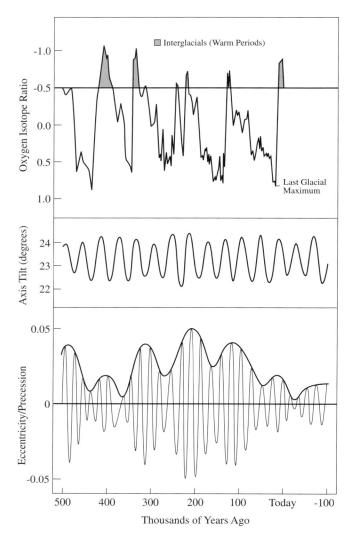

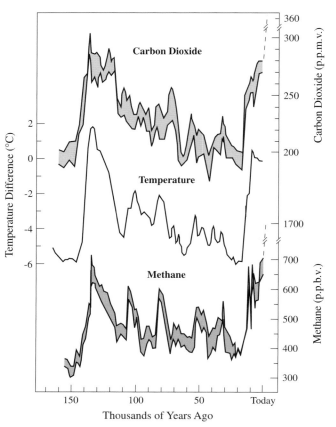

Fig. 8.28 **Ice age temperatures and greenhouse gases** Ice-core data indicate that changes in the atmospheric temperature over Antarctica closely parallel variations in the atmospheric concentrations of two greenhouse gases, carbon dioxide and methane, for the past 160 000 years. When the temperature rises, so does the amount of these two greenhouse gases, and *vice versa*. This strong correlation has been extended by a deeper Vostok ice core, to 3623 meters in depth and the past 420 000 years. The carbon dioxide (parts per million by volume) and methane (parts per billion per volume) increases may have contributed to the glacial-interglacial changes by amplifying orbital forcing of climate change. The ice-core data do not include the past 200 years, shown as broken dashed lines at the right. The present-day levels of carbon dioxide and methane are unprecedented during the past four 100 000-year glacial-interglacial cycles.

Fig. 8.27 **Astronomical cycles cause the ice ages** The advance and retreat of glaciers are controlled by changes in the Earth's orbital shape or eccentricity, and variations in its axial tilt and wobble. They alter the angles and distances from which solar radiation reaches Earth, and therefore change the amount and distribution of sunlight on our planet. The global ebb and flow of ice is inferred from the presence of lighter and heavier forms of oxygen, called isotopes, in the fossilized shells of tiny marine animals found in deep-sea sediments. During glaciations, the shells are enriched with oxygen-18 because oxygen-16, a lighter form, is trapped in glacial ice. The relative abundance of oxygen-18 and oxygen-16 (*top*) is compared with periodic 41 000-year variations in the tilt of the Earth's axis (*middle*) and in the shape, or eccentricity – longer 100 000-year variation, and wobble, or precession– shorter 23 000-year variation, of the Earth's orbit (*bottom*).

Earth's distance from the Sun varies more during each year, intensifying the seasons in one hemisphere and moderating them in the other. Shorter cycles include the periodic wobble of the Earth's axis over a period of about 23 thousand years, and a 41-thousand-year periodic variation in the axial tilt. The greater the tilt, the more intense the seasons in both hemispheres, with hotter summers and colder winters.

The astronomical theory of the ice ages was first documented from the study of sediments that have accumulated at the bottom of the oceans. Scientists investigated the relative amounts of the heavier and lighter forms, or isotopes, of oxygen locked up in the fossilized shells of tiny marine creatures that are found in the deep-sea sediments. The changing ratios of the two forms tell how much ice was present when the sea animals lived, and therefore record the advance and retreat of the glacial ice.

The oxygen-isotope ratio in cores of deep-sea sediments indicate that the greatest switch from cold to warm periods, and back to cold again, occurs roughly every 100 thousand years, at least during the past 500 thousand years.

Cores extracted from the glacial ice in Greenland and Antarctica provide the longest natural archive of the Earth's past climate. They strongly support the idea that changes in the Earth's orbit and spin axis cause variations in the intensity and distribution of sunlight arriving at Earth, which in turn initiate natural climate changes and trigger the ebb and flow of glacial ice.

Air trapped in the polar ice cores indicates that the Antarctica air-temperature changes are associated with varying concentrations of atmospheric carbon dioxide and methane. The temperatures increase whenever the levels of carbon dioxide and methane do, and they decrease together as well (Fig. 8.28). Scientists cannot however, yet agree whether the increase in greenhouse gases preceded or followed the rising temperatures.

The increase in carbon dioxide apparently resolves a difficulty in explaining the ice ages by the astronomical cycles. Although the largest climate variations occur every 100 thousand years, the corresponding rhythmic stretching of the Earth's orbit is far too small to directly create the observed temperature changes. The build up of greenhouse gas apparently amplifies effects triggered and timed by the astronomical rhythm.

The current Holocene glaciation, which has already lasted 11 thousand years, may not last more than a few thousand years more, and we could then enter an ice age. The next time it happens, the advancing glaciers will bury Copenhagen, Detroit and Montreal under mountains of ice, and because of the drop in sea level people might then walk from England to France, from Siberia to Alaska, and from New Guinea to Australia.

Still, we should not discount recent global warming. The concentrations of carbon dioxide and methane have now risen to unprecedented levels in our air, vastly exceeding those at any time during the past 420 thousand years. The warming produced by their greenhouse effect might counteract the cold of the next ice age.

CHAPTER NINE

Observing the Sun

The entire Sun is a giant mass of incandescent gas, unlike anything we know on Earth. The Sun has no surface; its gas just becomes more tenuous the farther out you go. Although we cannot see it with our eyes, very diAstronomers could not obtain scientific understanding of the Sun until they developed techniques to observe the Sun indirectly with telescopes. Important new insights are obtained each time a new part of the electromagnetic spectrum is observed in this way. Indeed, solar astronomy is primarily an observational science, that has both led to, and been driven by, revolutionary scientific discoveries. Early indirect observations of the Sun, using optical telescopes at visible wavelengths, allowed scientific study of the Sun to begin, showing that our star is a dynamic changing body. The development of optical spectroscopy permitted investigation of the magnetic fields, atmospheric motions, and composition of the Sun, as well as a new understanding of the internal structure of the atom and the chemistry of the cosmos. Radio telescopes provided a unique, high-resolution perspective of the changing, million-degree solar atmosphere, powerful explosions on the Sun, and violent activity that characterizes much of the Universe. The development of artificial satellites and other spacecraft then allowed scientists to study the Sun from above the Earth's atmosphere, permitting a full and continuous view of the Sun's ultraviolet and X-ray radiation, and direct sampling of energetic particles and magnetic fields flowing from it.

ludes, from its deepest part outward, the photosphere, chromosphere and corona (see Fig. 4.1). The Sun's magnetism plays an important role in molding, shaping and heating the coronal gas.

The visible photosphere, or "sphere of light", is the part of the Sun we can watch each day. It is the level of the solar atmosphere from which we get our light and heat. Thephotosphere contains sunspots, thousands of times more magnetic than the Earth, and the number and position of sunspots varies over an 11-year cycle of solar magnetic activity.

The visible sharp edge of the photosphere is something of an illusion. It is merely the level beyond which the gas in the solar atmosphere becomes thin enough to be transparent. The chromosphere and corona are so rarefied that we look right through them, just as we see through the Earth's clear air.

The chromosphere is very thin, but the Sun does not stop there. Its atmosphere extends way out in the corona, to the edge of the solar system. The corona's temperature is a searing million degrees kelvin, so hot that the corona is forever expanding into space.

The entire solar atmosphere is permeated by magnetic fields generated inside the Sun, rooted in the photosphere, and extending into the chromosphere and corona.

Observing the expanding Sun Pink, feathery extensions of the solar atmosphere are seen around the black disk of the Moon, observed during a total solar eclipse in 1991. The million-degree, electrically-charged gas is streaming away from the Sun. (Courtesy of the High Altitude Observatory, National Center for Atmospheric Research.)

9.1 Ground-based optical observing

The Sun has been observed with optical telescopes for centuries, gathering all the visible colors of sunlight. These telescopes are used to resolve spatial details that we cannot detect with the unaided eye. Since the Sun is a quarter of a million times closer to us than the next-nearest star, it permits a level of detailed examination that is not possible on any other star.

There are two types of optical telescopes, the refractor and the reflector (Fig. 9.1), and both of them are used to observe the Sun. Never look directly at the Sun through either kind of telescope; it can cause permanent damage to your eyesight. The safest way to observe the Sun with a telescope is to project the Sun's image on a surface and view it there.

In a refractor, sunlight is bent by refraction at the curved surface of a lens, called an objective, toward a focal point where the different rays of light meet (Fig. 9.1). If we place a detector at the focal point, in the plane parallel to the lens, we can record an image of the Sun. The distance from the

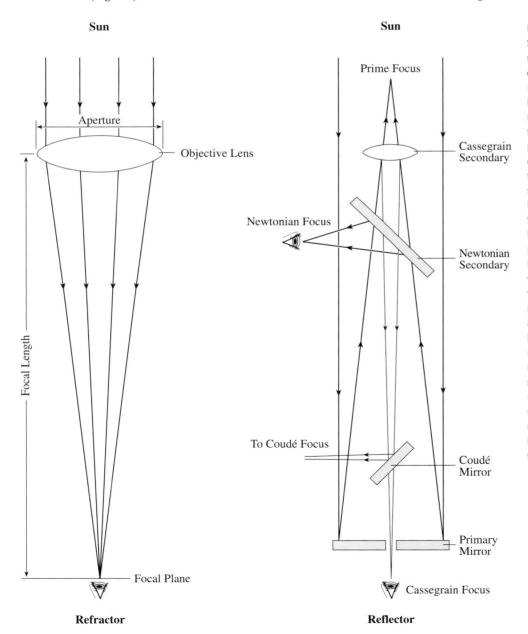

Refractor   **Reflector**

Fig. 9.1 **Refractor and reflector** The Sun can be observed with both refractors and reflectors. The earliest telescopes were refractors (*left*). The curved surfaces of the lens bend the incoming parallel solar light rays and bring them to a focus at the center of the focal plane, where an image of the Sun is created. When using the eye to look through a refracting telescope, a second lens, the eyepiece, is placed behind the focal plane to magnify the image formed by the objective lens and to render the light rays parallel again, but you should never look directly at the Sun, with or without a telescope. The reflecting telescope (*right*) uses a large, parabolic primary mirror to collect and focus light. The prime focus can be used for direct images by removing anything that intervenes in the light path. Large equipment, such as a spectrograph, requires a more mechanically stable platform, so the Cassegrain focus is often used. For massive equipment, the light might be further reflected to the coudé focus.

lens to the focal point is called the focal length, which determines the overall size of the solar image. The critical thing is the diameter, or aperture, of the light-gathering lens. The larger the aperture, the more light is gathered and the finer the detail that can be seen.

The other type of telescope, the reflector, uses a concave primary mirror with a parabolic shape to gather and focus the sunlight. The diameter of this primary mirror determines the telescope's resolution and light-gathering ability. The prime focus is back in the path of the incoming sunlight, so secondary mirrors are sometimes used to reflect the light to another place of observation. There are three types of secondary mirrors called the Cassegrain, coudé, and Newtonian mirrors, that can focus light to different locations (Fig. 9.1).

Larger telescopes permit us to see finer detail on the Sun. This ability to resolve details is called the resolving power of a telescope. It is specified by the angular resolution, a quantity that depends on the size of the telescope lens or mirror and the wavelength of observation. A bigger lens or mirror provides better angular resolution at a given wavelength, and longer wavelengths require larger

---

Focus 9.1. Angular Resolution

The angular resolution, θ, of a lens or mirror of diameter, D, at a wavelength, λ, is given by:

$$\theta = \frac{\lambda}{D} \text{ radians} = 2.062\,648 \times 10^5 \frac{\lambda}{D} \text{ seconds of arc,}$$

where one radian is equivalent to 57.295 78 degrees and 206 264.8 seconds of arc, and there are 3600 seconds of arc in a degree. This equations tells us that a bigger lens or mirror provides finer angular resolution at a given wavelength.

A telescope with a mirror or lens that is only 0.12 m across provides an angular resolution of about 1 second of arc at a visual wavelength of 555 m. Atmospheric turbulence limits the resolution of a bigger optical telescope to about this amount.

Our equation also applies at radio wavelengths where vastly bigger telescopes are required to achieve the same angular resolution as an optical telescope. At a radio wavelength of 0.1 m, an angular resolution of 1 second of arc requires a telescope with a diameter 2×10^4 m. The components of radio interferometer arrays, such as the Very Large Array, are separated by such distances (Section 9.2).

The smallest linear size, L, that can be resolved on the Sun is given by:

$$L = 725\,000\,\theta \text{ m,}$$

where θ is the angular resolution in seconds of arc. The constant 725 000 is the mean distance of the Sun, 1.496×10^{11} m, divided by the number of seconds of arc in a radian. So, the smallest thing you can see on the Sun with the best optical telescope on the ground is usually about 7.25×10^5 m across.

---

telescopes to achieve the same resolution as a smaller telescope at shorter wavelengths (Focus 9.1).

The resolution of ground-based optical telescopes is limited by turbulence in the Earth's atmosphere. It reduces the clarity of the Sun's image at visible wavelengths, limiting the angular resolution to about 1 second of arc. Similar variations cause the stars to twinkle at night, and produce a loss of resolution when we look across a hot road in a desert. This atmospheric limitation to angular resolution at visible wavelengths is called seeing. The best seeing, of 0.2 seconds of arc in unusual conditions, is found only at a few sites in the world, and optical solar observatories are located at most of them. Better visible images with even finer detail can be obtained from the unique vantage point of outer space, using satellite-borne telescopes unencumbered by the limits of our atmosphere.

The effective angular resolution of a typical ground-based optical telescope is about 1 second of arc. At the Sun's mean distance, this corresponds to structures that are 7.25×10^5 m across, about the distance from Boston to Washington, D.C. and about three-quarters the size of France. That is comparable to seeing the details on a coin from 1000 m away. By way of comparison, the angular resolution of the unaided eye is about 60 seconds of arc.

The irony is that atmospheric turbulence is worse when the Sun is in the sky because the Earth's atmosphere and surface are heated and convection is more severe than at night. Local turbulence develops when the nearby ground is heated, but it can be reduced if the observatory is surrounded by water. At least two solar observatories were built in lakes for this reason, one in Big Bear, California and the other in Udaipur, India. The effects of ground level turbulence can also be reduced by placing the telescope in a tall tower.

The largest solar images are obtained by optical telescopes with the longest focal length. The image diameter, d, is given by $d = 2fR_{\odot}/AU = 0.0093\,f$, where $R_{\odot} = 6.955 \times 10^8$ m, the mean distance between the Earth and the Sun is 1 AU = 1.496×10^{11} m, f is the focal length and f and d are in the same units. So, a telescope with a focal length of 50 m provides an image diameter of 0.465 m, and several solar telescopes now do this (Table 9.1).

The McMath solar telescope is a fine example of a modern high-resolution instrument (Fig. 9.2). It has a mirror at the top, called a heliostat, that follows the Sun, and sends a beam of sunlight into the side of a mountain to another mirror that takes the beam and focuses it, forming an image (Fig. 9.3). With a focal length of 82 m, the Sun's image is almost 0.8 m across. The tube that encases the telescope is kept at a cold temperature by pumping water through external tubes around it. This reduces air turbulence inside the tube, resulting in a sharper image.

Some tower telescopes reduce local turbulence by creating a vacuum inside. They can be fed by moving mirror arrangements that track the Sun, such as a coelostat or a turret.

Telescope	Year built	Reflecting system	Aperture (m)	Focal length (m)
Mt Wilson tower	1911	Coelostat	0.30 Lens	45
McMath solar telescope	1962	Heliostat	1.52 Mirror	82
Meudon	1968	Coelostat	0.60 Mirror	45
Sacramento Peak tower	1969	Turret	0.76 Mirror	55
Crimea	1973	Coelostat	0.90 Mirror	50
Kitt Peak	1973	Coelostat	0.60 Mirror	36
Baikal	1978	Siderostat	0.76 Lens	40
Sayan	1979	Coelostat	0.80 Mirror	20
La Palma	1985	Turret	0.50 Lens	22.5
Tenerife	1987	Coelostat	0.70 Mirror	47

Table 9.1 **Optical solar telescopes with large focal lengths**

Fig. 9.2 **Eyes on the Sun** The McMath solar telescope waits for winter skies to clear after a storm (*left*). In another view, scattered sunlight colors the telescope a stunning red, while stars make streaks across the evening sky (*below*). (Courtesy of Gary Ladd (*left*) and William C. Livingston, NOAO (*below*).)

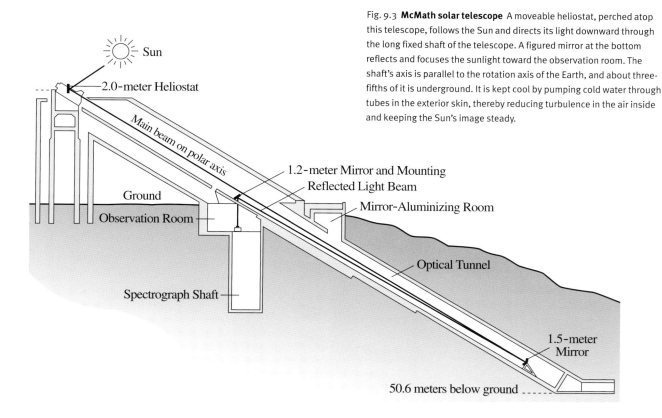

Fig. 9.3 **McMath solar telescope** A moveable heliostat, perched atop this telescope, follows the Sun and directs its light downward through the long fixed shaft of the telescope. A figured mirror at the bottom reflects and focuses the sunlight toward the observation room. The shaft's axis is parallel to the rotation axis of the Earth, and about three-fifths of it is underground. It is kept cool by pumping cold water through tubes in the exterior skin, thereby reducing turbulence in the air inside and keeping the Sun's image steady.

A coelostat is a moving two-mirror system that follows the Sun's apparent motion across the sky and directs sunlight into a fixed direction along the axis of the telescope. A heliostat performs the same function, but with a single mirror, as does the more compact two-mirror system.

Solar astronomy began with more compact telescopes, and some of them are still used by professional solar astronomers today (Table 9.2).

There is another reason for using a large aperture, other than greater resolution. A bigger lens or mirror collects more light. The human eye is severely limited by its inability to gather in light and store the images it sees for more than a few-tenths of a second. The telescope overcame this limitation by collecting light from a large area and by storing it on a photographic plate or electronic chip.

At first sight you would think there is plenty of light coming from the Sun. Because of its closeness, it is a hundred billion times brighter than any other star. But you obtain much less light if just a narrow section of the solar spectrum is isolated and observed for short times, as with a spectroheliograph (Fig. 9.4). Large lenses or mirrors might then be needed to obtain a strong signal, permitting detailed studies of spectral components such as absorption or emission lines.

Telescope	Year built	Aperture (m)	Primary focal length (m)	Effective focal length[a] (m)
Domeless coudé, Capri	1966	0.35 Lens	4.5	16
Pic du Midi	1972	0.50 Lens	6.45	35
Big Bear	1973	0.65 Mirror	3.5	50
Domeless coudé, Hida	1979	0.60 Mirror	3.15	32.2
Gregory coudé, Tenerife	1985	0.45 Mirror	2.48	25
THÉMIS [b], Tenerife	1992	0.90 Mirror	3.15	15

[a] Secondary optical elements are used to form a large image, with the effective focal length given here.
[b] THÉMIS is an acronym for Télescope Héliographique pour l'Étude du Magnétisme et des Instabilitiés Solaires.

Table 9.2 **Optical solar telescopes with short focal lengths**

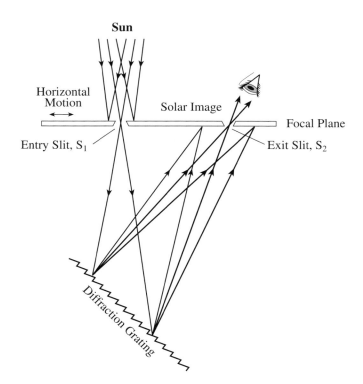

Sun

Horizontal
Motion

Solar Image

Entry Slit, S₁

Focal Plane

Exit Slit, S₂

Diffraction Grating

Fig. 9.4 **Spectroheliograph** A small section of the image at the focal plane of a telescope is selected with a narrow entry slit, S_1, and this light passes to a diffraction grating, producing a spectrum. A second slit, S_2, at the focal plane selects a specific wavelength from the spectrum. If the plate containing the two slits is moved horizontally, then the entry slit passes adjacent strips of the image. The light leaving the moving exit slit then builds up an image of the Sun at a specific wavelength.

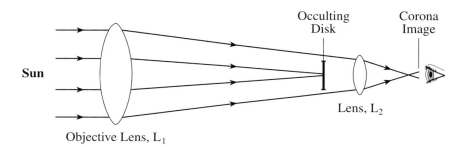

Occulting
Disk

Corona
Image

Sun

Lens, L₂

Objective Lens, L₁

Fig. 9.5 **Coronagraph** Sunlight enters from the left, and is focused by an objective lens, L_1, on an occulting disk, blocking the intense glare of the photosphere. Light from the corona, which is outside the photosphere, bypasses this occulting disk and is focused by a second lens, L_2, forming an image of the corona. Other optical devices are placed along the light path to divert and remove excess light.

A special type of optical telescope is the coronagraph (Fig. 9.5), first developed by the French astronomer Bernard Lyot around 1930. It produces an artificial eclipse of the Sun by means of an occulting disk inside the telescope, blocking out the intense glare of the photosphere. This permits the detection of the faint light of the corona around the perimeter of the occulting disk. Special precautions must be taken to reduce instrumental scattered and diffracted light.

9.2 Ground-based radio observations of the Sun

Because of its proximity, the Sun is the brightest radio object in the sky, and radio telescopes have therefore been used to study our home star for decades. Moreover, the Earth's atmosphere does not distort radio waves that are less than about one meter in wavelength, so we can observe the radio Sun on a cloudy day, just as radio signals are used to communicate with satellites even when it rains or snows outside.

The shapes of most radio telescopes are similar to the reflecting mirror of an optical telescope. The main reflector, called a dish, is a parabolic metal surface that gathers the incoming radio waves, reflecting and focusing them to an electronic receiver at the reflector focus. This receiving system converts the intensity of the incoming radio signal to numbers that are then transmitted to a computer. The data is stored in the computer as a matrix of numbers and is then manipulated to form images.

Some radio telescopes observe the entire Sun, which is about 32 minutes of arc, or 1920 seconds of arc, in angular diameter. At a wavelength of 0.1 m, any radio telescope smaller than 10 m in diameter can monitor the entire Sun, but the bigger ones collect more radiation. Such a single radio telescope can be used to determine the dynamic spectra of solar radio bursts. An example is the computerized Phoenix-2 radio telescope, located in Bleien, Switzerland, that obtains dynamic radio spectra of solar bursts at frequencies from 0.1 to 4 GHz, where 1 GHz is equivalent to a billion Hz, or 10^9 Hz. These measurements can be used to determine the velocity at which electron beams and shock waves are expelled from the Sun (Section 6.1), and to also specify the acceleration height of high-speed flaring electrons.

If you want to examine the radio Sun in fine detail, a bigger telescope is needed. Since radio waves are millions of times longer than those of light, a radio telescope needs to be at least a million times bigger than an optical telescope to obtain the same resolving power (Focus 9.1). For this reason, the first radio telescopes provided a very myopic, out-of-focus view. This perspective has been compared to looking at the Sun through the bottom of a glass bottle.

But the limitation was soon overcome as radio astronomers built successively larger telescopes, culminating in the 100-m, fully-steerable parabolic dish at Effelsberg, West Germany. Its best angular resolution is about 10 seconds of arc. This may be the largest steerable radio dish that can be built, but a novel way of building an even larger dish was to cover the floor of a valley with metal screen, producing a 305-m dish in Arecibo, Puerto Rico. This antenna relies on the rotation of the Earth to bring different regions of the sky into view. Although the Arecibo and

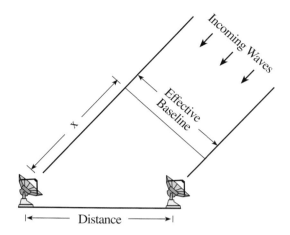

Fig. 9.6 **Two-element interferometer** When an incoming radio wave approaches the Earth at an angle, the crests of the wave will arrive at two separated telescopes at slightly different times. This delay in arrival time is the distance X divided by the velocity of light. If X is an exact multiple of the wavelength, then the waves detected at the two telescopes will be in phase and add up when combined. If not, they will be out of phase and interfere. The angular resolution of the interferometer is equal to the wavelength divided by the effective baseline. When the object being observed is directly overhead, the effective baseline is equal to the distance between the two telescopes.

Effelsberg radio telescopes are sometimes used to observe the Sun, they usually listen to all the rest of the radio Universe.

Nowadays, relatively small radio telescopes, separated by large distances, are combined and coordinated electronically, achieving radio images of the Sun that are as sharp as optical ones. Because it is spread out, an array of small telescopes has the property that is crucial for high resolving power, namely great size relative to wavelength.

The technique is known as interferometry because it analyzes how the waves detected at the telescopes interfere once they are added together. The simplest example is a pair of telescopes with a computer to reconstruct the waves from the combined data (Fig. 9.6).

The sensitivity of an interferometric array is determined by the combined areas of the individual elements, and not by their separations. For example, the two-element interferometer can resolve much finer detail than can its individual component radio telescopes. But the interferometer's collecting area is only twice that of the individual components, so its sensitivity is just twice as great. Many two-element telescope pairs are therefore combined in a full-fledged radio array to gather more energy and obtain a stronger radio signal.

The Very Large Array, abbreviated VLA, is an example of a

Telescope	Location	Frequency [a]	Solar observations
BIMA = Berkeley, Illinois, Maryland Association	Hat Creek, California	100 GHz, 270 GHz	Two-dimensional images, chromosphere
Nancay Radioheliograph [b]	Nancay, France	Five frequencies 150 to 450 MHz	Two-dimensional images, low corona
Nobeyama Radioheliograph [b]	Nobeyama, Japan	17 GHz, 34 GHz	Two-dimensional images, chromosphere
OVRO = Owens Valley Radio Observatory [b]	Owens Valley, California	1 to 18 GHz	Frequency synthesis imaging, radio bursts
RATAN 600	Zhelenchukaya, Russia	1.4 to 30 GHz	Polarimetry, moderate resolution
VLA = Very Large Array	Socorro, New Mexico	Eight frequencies 75 MHz to 23 GHz	Two-dimensional images, low corona

[a] A frequency of 1 MHz = 1 million Hz, or 10^6 Hz, and 1 GHz = 1 billion Hz, or 10^9 Hz.
[b] Telescopes dedicated to observing the Sun.

Table 9.3 **Interferometric radio arrays used to observe the Sun**

modern interferometric radio array that is used to observe the Sun (Fig. 9.7). It consists of 27 radio telescopes placed along the arms of a Y-shaped array at separations of up to 34 thousand m (Fig. 9.8). The telescopes are connected electronically and linked to a central computer, providing a total of 351 pairs of telescopes. When the telescopes are all pointed at the Sun, the received signals are combined to create images of temperature and magnetic structures on the Sun with an angular resolution of up to 0.1 seconds of arc, equal to or better than any ground-based optical telescope.

Interferometric radio arrays that are used to observe the Sun are given in Table 9.3. Some of them are dedicated to solar observing. Others are used to investigate a wide range of cosmic radio sources, but they provide unique information when trained on the Sun. The radio arrays see down to different levels in the solar atmosphere, depending on the frequency of observation. Higher radio frequencies generally detect lower levels in the solar corona. The highest frequency in use looks through the corona to the chromosphere. Circular polarization information is often obtained, leading to specification of the strength and structure of the magnetic fields in the corona (Section 5.3, Focus 5.4).

Fig. 9.7 **Rainbow above the VLA** The Very Large Array, abbreviated VLA, operates at radio wavelengths. In this photograph, it is strikingly portrayed against the colors of a double rainbow. The VLA is a collection of radio telescopes interconnected electronically to provide a total of 351 pairs of telescopes. The combined signals obtain two hundred thousand pieces of information every hour, so giant computers are also required to carry out radio investigations of the Sun. (Courtesy of Douglas Johnson, Batelle Observatory, Washington.)

Fig. 9.8 **VLA telescopes** Each component of the Very Large Array is a radio telescope measuring 25 m in diameter and weighing 235 tons (2.35×10^5 kg). Twenty-seven of these telescopes are placed along the arms of Y-shaped array on a desert near Socorro, New Mexico. Each arm of the Y is 20 thousand meters long. The telescopes can be rolled along tracks to change their configuration and create a radio zoom lens. When the telescopes are crowded in towards the center of the Y, they provide a wide-angle view, and when they are pushed to the outer ends of each arm, the telescopes zero in for a closer look. Their output, when combined in a computer, creates a radio telescope with a diameter as large as 34 thousand meters and an angular resolution that can be smaller than 1 second of arc.

9.3 Observing the Sun from space

Radio telescopes do not provide our only window on the invisible Sun. There are the invisible gamma rays, X-rays, and ultraviolet radiation. They are all absorbed in our atmosphere and must be observed with telescopes in satellites that orbit the Earth above its atmosphere. These space telescopes measure the intensity of the incoming signal and convert these measurements into radio transmissions that are sent to radio telescopes on the ground. Spacecraft can also directly sample the particles and magnetic fields flowing from the Sun.

Solar astronomy from space has several advantages over ground-based observations. The weather in space is always perfectly clear, and the images are not blurred by the atmosphere. Moreover, if the space telescope is placed in the right location, it can view the Sun 24 hours a day, every day, for years.

What's more, you do not need a very big telescope to observe the ultraviolet and X-ray wavelengths with high angular resolution. The aperture only has to be 0.002 m across to achieve an angular resolution of 1 second of arc at a soft X-ray wavelength of 10^{-8} m.

Studying the Sun from space has revolutionized solar physics. It all began in 1957 when the first artificial satellite, *Sputnik*, was launched, followed in 1960–61 by measurements of the solar wind by the *Luna 2* spacecraft and in 1962–63 when *Mariner 2* determined the speed and density of the solar wind on its way to Venus.

Since then many missions have been devoted to studying the Sun itself. Among the first were NASA's pioneering series of small satellites known as the *Orbiting Solar Observatories*, or OSOs, launched from 1962 (OSO 1) to 1975 (OSO 8), observing the Sun over an entire 11-year magnetic-activity cycle. NASA's more ambitious orbiting *Skylab* was launched on 14 May 1973, and manned by three-person crews until 8 February 1974. It demonstrated the ubiquitous nature of the X-ray-emitting coronal loops, albeit with somewhat blurred images without fine spatial detail or high temporal resolution.

The Solar Maximum Mission

NASA's *Solar Maximum Mission* (SMM) satellite, launched on 14 February 1980, obtained a new perspective of explosions on the Sun by examining ultraviolet radiation, X-rays and gamma rays emitted during powerful solar flares (Table 9.4). Similar spectroscopic information was obtained at about the same time using a soft X-ray spectrometer aboard the P78-1 spacecraft, launched on 24 February 1979 as part of the U.S. Department of Defense Space Test Program. SMM and P78-1 both also had coronagraphs on board that vastly increased our knowledge of Coronal Mass Ejections, or CMEs.

Three modern spacecraft, *Yohkoh*, *Ulysses* and the *SOlar and Heliospheric Observatory*, abbreviated SOHO, have recently made revolutionary discoveries about the Sun, presented in

Experiment	Energy or Wavelength range [a]
Gamma-Ray Spectrometer (GRS)	0.1 to 17 MeV and 10 to 160 MeV
Hard X-Ray Burst Spectrometer (HXRBS)	25 to 500 keV
Hard X-Ray Imaging Spectrometer (HXIS) (Angular resolution 8 to 32 seconds of arc, effective time resolution ≥ 10 seconds)	3.5 to 30 keV (six channels)
X-Ray Polychromator (XRP, with BCS and FCS) (BCS = Bent Crystal Spectrometer) (FCS = Flat Crystal Spectrometer)	0.176 to 0.323 nm and 0.144 to 2.24 nm
UV Spectrometer and Polarimeter (UVSP)	110 to 330 nm
Coronagraph and Polarimeter (C/P)	443.5 to 658.3 nm
Active Cavity Radiometer Irradiance Monitor (ACRIM)	Ultraviolet to infrared

[a] A wavelength of one nm, or one nanometer, is equivalent to a billionth (10^{-9}) of a meter, and an energy of about 1 keV. Solar astronomers also often use the ångstrom unit of wavelength, where one ångstrom (1 Å) is equal to 0.1 nm, and an energy of about 10 keV. One keV is equivalent to 1.6022×10^{-16} J or 1.6022×10^{-9} erg, and 1 MeV = 1000 keV.

Table 9.4 **Instruments aboard the *Solar Maximum Mission***

previous chapters of this book. Major new instruments aboard these spacecraft have traced the flow of energy and matter from down inside our star to the Earth and beyond, providing insights that are vastly more focused and detailed than those of previous solar missions. Indeed, we may have obtained more essential new information about the Sun from these spacecraft than from the entire previous century of investigations.

Yohkoh

Following its pioneering *Hinotori*, or fire-bird, solar flare mission, launched on 21 February 1981 near a maximum in the 11-year solar activity cycle, the Institute of Space and Astronautical Science (ISAS) in Japan organized a new mission, initially called *Solar-A*. It was designed to investigate solar flares with improved angular and energy resolution during the next activity maximum; quiescent, non-flaring structures and pre-flare activity would also be scrutinized. The 390-kg spacecraft was launched by ISAS from the Kagoshima Space Center on 30 August 1991, into a 96-minute, nearly-circular Earth orbit. After launch, the mission was renamed, *Yohkoh*, which means "sunbeam" in English. So, *Yohkoh* was the first of our revolutionary scientific troika to begin observing the Sun, and it continued to do so right through the turn of the century.

Yohkoh detects high-energy radiation from the Sun, at soft X-ray and hard X-ray wavelengths (Table 9.5). The *Yohkoh* telescopes have full-Sun fields of view with unparalleled angular and spatial resolution. The Hard X-ray Telescope, or HXT, images hard X-rays emitted by high-speed electrons accelerated in impulsive flares. Both the flaring and quiescent,

or non-flaring, Sun are detected at soft X-ray wavelengths with a rapid, uniform rate using the Soft X-ray Telescope, or SXT for short. It routinely images high-temperature gas, above 2 to 3 million degrees kelvin, across the Sun. Solar images from the soft X-ray telescope, which produces most of the data that make *Yohkoh* so widely known, are available at the instrument home page: **http://www.lmsal.com/SXT/** on the World Wide Web; its images may also be obtained at the Solar Data Analysis Center at: **http://umbra.nascom.nasa.gov/**.

Yohkoh's principal telemetry and operation control center is provided by ISAS near Tokyo, Japan, while NASA obtains telemetry data captured during *Yohkoh* passes over its Deep Space Network ground stations in California, Australia and Spain. The operation of the *Yohkoh* mission is the responsibility of the *Yohkoh* science team. It includes scientists from Japan, the United Kingdom and the United States.

The primary scientific objective of *Yohkoh* is to obtain high-resolution spatial and temporal information about high-energy flare phenomena, permitting detailed scrutiny of where and how flare energy is released and particle acceleration takes place. As was discussed in Chapter 6, *Yohkoh* has fulfilled these objectives. The soft X-ray telescope has also led to many important discoveries about the Sun's dynamic million-degree atmosphere (Section 5.3).

Ulysses

In the 1990s the 367-kg *Ulysses* spacecraft traveled over the poles of the Sun for the first time. Before that, interplanetary spacecraft moved close to the ecliptic, the plane of the Earth's orbital motion about the Sun. This is because they often were intended to rendezvous with another planet whose orbit lies

Instrument	Measurement
BCS	The Bragg Crystal Spectrometer measures X-ray spectral lines of highly ionized iron, Fe XXV and Fe XXVI, calcium, Ca XIX, and sulfur, S XV, between 0.18 and 0.51 nm with a time resolution as short as 0.125 s and across the full solar disk.
HXT	The Hard X-ray Telescope images flare radiation with energy from 20 to 80 keV in four channels, with an angular resolution of about 7 seconds of arc, a time resolution of 0.5 s and a field of view that includes the entire visible disk.
SXT	The Soft X-ray Telescope achieves 4 seconds of arc angular resolution and 2 s temporal resolution, detecting radiation with energy between 1 and 4 keV (1.2 to 0.3 nm) with a wide dynamic range (up to a ratio of 200 000 to 1). It renders images taken at a uniform rate across the full solar disk for both the faint, quiescent and intense flaring X-ray structures with temperatures above 2 to 3 million degrees kelvin.
WBS	The Wide Band Spectrometer measures X-rays and gamma-rays from 3 keV to 30 MeV during solar flares, and is also sensitive to neutrons emitted during flares; it observes the full solar disk with a time resolution as short as 0.125 s.

[a] Energetic radiation is often expressed in terms of its energy content, in units of kilo-electron volts, or keV, rather than the more conventional units of wavelength. For conversions, $1 \text{ keV} = 10^3 \text{ eV} = 1.602 \times 10^{-16}$ J and the photon energy, E, associated with a wavelength λ in nanometer is $E = 1.986 \times 10^{-16} / \lambda$ J. The wavelength units of ångstroms are often used, where one ångstrom $= 1 \text{ Å} = 10^{-10}$ m $= 0.1$ nm.

Table 9.5 *Yohkoh*'s instruments arranged alphabetically by acronym [a]

near that plane, and also because their launch vehicles obtain a natural boost by traveling in the same direction as the Earth's spin and in the plane of the Earth's orbit around the Sun.

The thrust and high speed needed to send a probe directly on a trajectory over the poles of the Sun is beyond the capability of today's most powerful rockets. *Ulysses* therefore had to first move outward away from the Sun, not toward it, in order to voyage across the top of our star. By traveling out to Jupiter, the spacecraft could use the planet's powerful gravity to cancel excess momentum acquired from the Earth and to accelerate, propel and re-orient the spacecraft in slingshot fashion into an inclined orbit that sent it under the Sun.

Launched by NASA's *Space Shuttle Discovery* on 6 October 1990, *Ulysses* encountered Jupiter on 8 February 1992, which hurled the spacecraft into a Sun-centered elliptical orbit, with a period of 6.2 years. Its distance from the Sun varies from 5.4 AU, near Jupiter's orbit, to 1.3 AU at its closest approach to the Sun; the distance over the Sun's poles ranges from 2.0 AU (north) to 2.3 AU (south). The astronomical unit, 1.0 AU, is the mean distance from the Sun to the Earth, or about 150 billion (1.5×10^{11}) m. Thus, at the time of polar passage, *Ulysses* was not close to the Sun; it was more than twice as far from the Sun as the Earth.

After traveling about 3 million million (3×10^{12}) m for nearly four years since leaving Earth, *Ulysses* passed beneath the south pole of the Sun on 13 September 1994. In response to the Sun's gravitational pull, *Ulysses* arched up toward the solar equator, and crossed over the solar north pole on 31 July 1995.

The *Ulysses* mission is a joint undertaking of the European Space Agency (ESA) and the United States (U.S.) National Aeronautics and Space Administration (NASA). The spacecraft and its operations team have been provided by ESA, the launch of the spacecraft, radio tracking, and data management operations are provided by NASA. Scientific experiments have been provided by investigation teams both in Europe and the United States.

The primary scientific objective of *Ulysses* is to explore and define the heliosphere in three dimensions. It carries nine instruments (Table 9.6) that make measurements of fundamental parameters as a function of distance from the Sun and solar latitude; the solar latitude is the angular distance from the plane of the Sun's equator. They include

Instrument	Measurement
COSPIN	The COsmic ray and Solar Particle INstrument records ions with energies of 0.3 to 600 MeV per nucleon and electrons with energies of 4 to 2000 MeV.
DUST	Measures interplanetary DUST particles with mass between 10^{-19} and 10^{-10} kg.
EPAC/GAS	The Energetic PArticle Composition and interstellar neutral GAS instrument records the amount of helium atoms and the composition of energetic ions with energies of 80 keV to 15 MeV per nucleon.
GRB	The Gamma-Ray Burst instrument measures solar flare X-rays and cosmic gamma-ray bursts with energies of 15 to 150 keV.
GWE	The Gravitational Wave Experiment records Doppler shifts in the satellite radio signal that might be due to gravitational waves.
HISCALE	The Heliosphere Instrument for Spectra, Composition, and Anisotropy at Low Energies measures ions with energies of 50 keV to 5 MeV and electrons with energies of 30 to 200 keV.
SCE	The Solar Corona Experiment uses the radio signals from the spacecraft to measure the density, velocity and turbulence spectra in the solar corona and solar wind.
SWICS	The Solar-Wind Ion Composition Spectrometer records elemental and ionic-charge composition, temperature, and the mean speed of solar wind ions for speeds of 1.45×10^5 (H$^+$) to 1.35×10^6(Fe^{+8}) m s^{-1}.
SWOOPS	The Solar-Wind Observations Over the Poles of the Sun instrument measures ions from 237 eV to 35 keV per charge and electrons from 1 to 860 eV.
URAP	The Unified Radio And Plasma wave experiment records plasma waves at frequencies from 0 to 60 kHz, remotely senses traveling solar radio bursts exciting plasma frequencies from 1 to 940 kHz, and measures the electron density.
VHM/FGM	A pair of magnetometers, the Vector Helium Magnetometer and the scalar Flux Gate Magnetometer, that measure the magnetic field strength and direction in the heliosphere and the Jovian magnetosphere from 0.01 to 44000 nanoteslas or from 10^{-7} to 0.44 gauss.

[a] Particle energies are often expressed in electron volts, or eV, where 1 eV = 1.6022×10^{-19} J = 1.6022×10^{-12} erg. High-energy particles can have energies expressed in kilo-electron volts, or keV, and Mega-electron volts, or MeV, where 1 keV = 10^3 eV = 1.602×10^{-16} J and 1 MeV = 10^6 eV = 1.6022×10^{-13} J. The magnetic field strength is given in units of tesla, where 1 T = 10^4 G or 1 nT = 10^{-9} T = 10^{-5} G.

Table 9.6 ***Ulysses'* instruments arranged alphabetically by acronym** [a]

instruments to measure the solar-wind speed, a pair of magnetometers that measure magnetic fields, and several instruments that study electrons, protons and heavier ions that come from both the quiescent and active Sun and from interstellar space (cosmic rays). These quantities have now been measured from pole to pole, with striking results given primarily in Chapter 7 of this book.

The data from *Ulysses* are gathered by NASA's Deep Space Network, and spacecraft operations and data analysis are performed at the Jet Propulsion Laboratory (JPL) in Pasadena, California by a joint ESA/JPL team. Fundamental information about the various experiments can be found on the *Ulysses* home page (**http://ulysses.jpl.nasa.gov/**) or (**http://helio.estec.esa.nl/ulysses/**).

This hardy spacecraft will pass beneath the Sun's south pole on 27 November 2000 (80.2 degrees south), and over its north pole on 13 October 2001 (80.2 degrees north). *Ulysses* will then probe the effects of increasing solar activity when the heliosphere has a more complex shape, and is punctuated by frequent powerful solar outbursts. It will be complemented by related observations of solar activity with the *SOHO* and *Yohkoh* spacecraft.

SOlar and Heliospheric Observatory, SOHO

The SOlar and Heliospheric Observatory, or SOHO for short, has stared the Sun down with an unblinking eye for years (Fig. 9.9). This 1.33 ton (1350 kg), one-billion-dollar spacecraft was launched from Cape Canaveral Air Station, Florida aboard a two-stage Atlas/Centaur rocket on 2 December 1995 and reached its permanent position on 14 February 1996.

SOHO is located sunward at about 1.5 billion (1.5×10^9) m out in space, or at about one percent of the way to the Sun. At this place, the spacecraft is balanced between the pull of the Earth's gravity and the Sun's gravity. Such a position is known in astronomy as the inner Lagrangian, or L_1, point after the French mathematician Joseph Louis Lagrange, who first calculated its position near the end of the 19th century. From this strategic vantage point, SOHO orbits the Sun together with Earth, continuously gazing at the Sun 24 hours a day, every day, for years. All previous solar satellites have been in orbit around our planet, so their observations were periodically interrupted when they entered our Earth's shadow.

SOHO is a joint project of the European Space Agency (ESA) and the United States (U.S.) National Aeronautics and Space Administration (NASA). The spacecraft was designed and built in Europe. NASA launched SOHO and operates the satellite from the Experimenters' Operations Facility at the Goddard Space Flight Center in Greenbelt, Maryland. NASA's Deep Space Network is used to track the satellite and retrieve its data.

SOHO has examined the Sun from its deep interior, through its million-degree atmosphere and ceaseless wind, to our home planet, Earth (Table 9.7). Three devices probe the Sun's internal structure and dynamics; six measure the solar atmosphere; and three keep track of the star's far-

Fig. 9.9 **SOHO** An artist's impression of the *SOlar and Heliospheric Observatory*, or *SOHO* for short. It is one of the best solar observatories ever launched into space, at least so far. The 1.33 ton (1350 kg), one-billion-dollar spacecraft was launched from Cape Canaveral Air Station, Florida, on 2 December 1995, and was still in full operation at the turn of the century. *SOHO* is a joint project of the European Space Agency (ESA) and the United States (U.S.) National Aeronautics and Space Administration (NASA). The spacecraft contains 12 instruments that study the Sun 24 hours a day, every day for years, probing the Sun's deep interior, out through its expanding atmosphere and solar wind, to the Earth and beyond.

reaching winds. Some of these instruments examine many different levels in the solar atmosphere simultaneously, including for the first time the locations where the million-degree gases are heated and the Sun's winds are accelerated. SOHO therefore provides a completely new perspective of how agitation inside the Sun, transmitted through the solar atmosphere, directly affects us on Earth.

The SOHO mission has three principal scientific goals: to measure the structure and dynamics of the solar interior; to gain an understanding of the heating mechanisms of the Sun's million-degree atmosphere, or solar corona; and to determine where the solar wind originates and how it is accelerated. As discussed in Chapters 4, 5 and 7, it has accomplished most of these objectives. Many of the unique images and movies that SOHO's instruments obtain can be found at the SOHO home page: **http://sohowww.nascom.nasa.gov**

Observing the Sun–Earth connection

Earth's space environment traditionally has been studied as a set of independent parts – the Sun, the interplanetary region, the magnetosphere, and the Earth's upper atmosphere and ionosphere. Consequently, past missions have understood these phenomena only individually. To understand the Sun–Earth system as a whole, scientists have established the International Solar–Terrestrial Physics (ISTP) program in the 1990s. An armada of at least 25 satellites, coupled with ground-based observations and modeling centers, allow scientists to study the Sun, the Earth and the space between them in unprecedented detail from many perspectives and in different ways. When linked together with each other and the resources on the ground, the satellites perceive the entire Sun–Earth environment.

Key areas of the Earth's magnetosphere and the solar wind are monitored from satellites placed within them. The missions of NASA's Sun–Earth Connection program are described at the internet address: **http://sec.gsfc.nasa.gov/sec_missions.htm**. Recent examples include:

GEOTAIL, launched in 1992 by Japan's Institute of Astronautical Science (ISAS) and by NASA to study the distant reaches of Earth's magnetic tail, or the region downwind of Earth. It skims along the outer edge of the magnetosphere and dives deep into the magnetotail.

Instrument	Measurement
Helioseismology instruments	
GOLF	The Global Oscillations at Low Frequencies device records the velocity of global oscillations within the Sun.
MDI/SOI	The Michelson Doppler Imager/Solar Oscillations Investigation measures the velocity of oscillations, produced by sounds trapped inside the Sun, and obtains high-resolution magnetograms.
VIRGO	The Variability of solar Rradiance and Gravity Oscillations instrument measures fluctuations in the Sun's brightness, as well as its precise energy output.
Coronal instruments	
CDS	The Coronal Diagnostics Spectrometer records the temperature and density of gases in the corona.
EIT	The Extreme-ultraviolet Imaging Telescope provides full-disk images of the chromosphere and the corona.
SUMER	The Solar Ultraviolet Measurements of Emitted Radiation instrument gives data about the temperatures, densities and velocities of various gases in the chromosphere and corona.
LASCO	The Large Angle Spectrometric COronagraph provides images that reveal the corona's activity, mass, momentum and energy.
UVCS	The UltraViolet Coronagraph Spectrometer measures the temperatures and velocities of hydrogen atoms, oxygen ions and other ions in the corona.
SWAN	The Solar Wind ANisotropies device monitors latitudinal and temporal variations in the solar wind.
Solar wind "in-situ" instruments	
CELIAS	The Charge, ELement and Isotope Analysis System quantifies the mass, charge, composition and energy distribution of particles in the solar wind.
COSTEP	The COmprehensive SupraThermal and Energetic Particle analyzer determines the energy distribution of protons, helium ions and electrons.
ERNE	The Energetic and Relativistic Nuclei and Electron experiment measures the energy distribution and isotopic composition of protons, other ions and electrons.

Table 9.7 **SOHO's instruments arranged alphabetically by acronym within three areas of investigation**

WIND, launched by NASA in 1994 into an orbit that placed it on the side of Earth facing the Sun, where the satellite could sample the solar wind.

SOHO, a joint project of the European Space Agency (ESA) and NASA, launched in 1995 and stationed about 1 percent of the distance to the Sun, or at about 200 Earth radii from the Earth in the direction of the Sun. Here the satellite can directly sample the solar wind well above the edge of the magnetosphere, located at about 10 Earth radii from the Earth in the direction of the Sun, and also continuously look back at the source of it all, the Sun.

POLAR, launched by NASA in 1996 in an orbit that swings over Earth's poles and looks down to monitor the auroras and other physical activity.

IMAGE, launched in 2000 by NASA into a polar orbit to obtain images that study the injection of plasma into the magnetosphere during geomagnetic storms and solar-wind changes.

CLUSTER II, launched in 2000, is a set of four identical spacecraft that simultaneously obtain in-situ measurements of the Earth's magnetosphere, permitting the accurate determination of three-dimensional and time-varying phenomena. It can distinguish between spatial and temporal variations of small-scale plasma structures in the solar wind and bow shock, magnetopause, polar cusp, magnetotail and auroral zone

The combined set of measurements, taken at a multitude of points in space, is specifying how energy is generated in the Sun and transferred to the space near Earth, with its complex magnetic and atmospheric environment. They are also telling us how energy is coupled back and forth between the gusty, incident solar wind and the magnetosphere on the one hand and between the magnetosphere and the Earth's upper atmosphere on the other.

Other spacecraft, planned for the early-2000s, are designed to amplify and extend the understanding gained from these investigations of the Sun–Earth connection.

9.4 The next solar missions

In the first decade of the 21st century, scientists intend to launch several spacecraft that will investigate the explosive Sun (Chapter 6) and space weather (Chapter 8). They include NASA's *High Energy Solar Spectroscopic Imager*, or *HESSI* for short, that will be launched in 2001, as well as NASA's *Solar–TErrestrial RElations Observatory*, abbreviated *STEREO* and the Japanese Institute of Space and Astronautical Science (ISAS) *Solar-B* mission, both tentatively scheduled for launch in 2004.

The primary science goals of *HESSI* are to study the solar-flare particle acceleration and flare energy release using imaging spectroscopy of X-rays and gamma rays with an energy ranging from 3 keV to 20 MeV. *HESSI*'s images will determine the frequency, location and evolution of impulsive flare energy release in the corona and locate the sites of particle acceleration and energy deposition at all phases of solar flares. The extraordinary sensitivity of *HESSI* will enable it to detect many more X-ray and gamma-ray flares than any previous spacecraft.

The acceleration, propagation and evolution of solar-flare electrons will be revealed by hard X-ray bremsstrahlung (Section 6.1) which *HESSI* will image with the finest angular resolution (2 to 7 seconds of arc) ever achieved, and over a broad energy range (a few keV to hundreds of keV) at a time resolution as short as tens of milliseconds. *HESSI* will additionally provide the first imaging spectroscopy of hard X-ray solar flares. The hard X-ray images will permit investigations of plasma at tens of millions of degrees kelvin and determine its relationship to particle acceleration.

The acceleration, propagation and evolution of solar-flare protons and heavier ions will be revealed by gamma-ray lines generated when these energetic charged particles are beamed down into the chromosphere, producing nuclear reactions there (Section 6.1, Table 6.3). The energy resolution of the *HESSI* spectra (2 to 5 keV) will be sufficient to resolve the gamma-ray lines and to measure their shape for the first time, thereby determining the abundance of accelerated and ambient ions in flares.

A crucial aspect of the *HESSI* mission will be supporting context observations of the hot plasma, accelerated electrons, and magnetic fields of the solar flares at regions where the hard X-ray and gamma-ray sources are situated. This information will be obtained by imaging and spectroscopy at soft X-ray and ultraviolet wavelengths, using other spacecraft, and at radio and optical wavelengths with telescopes on the ground (Sections 9.1, 9.2).

The *STEREO* mission will study Coronal Mass Ejections, or CMEs, which can hurl as much as fifty billion tons (5×10^{13}

9.10 ***STEREO*** An artist's conception of the *Solar–TErrestrial Relations Observatory*, or *STEREO*, in which two spacecraft will provide the images for a stereo reconstruction of Coronal Mass Ejections, or CMEs. One spacecraft will lead Earth in its orbit and one will be lagging. When simultaneous telescopic images from the two spacecraft are combined with data from observatories on the ground or in low Earth orbit, the buildup of magnetic energy, the lift off, and the trajectory of CMEs can be tracked in three dimensions. (Courtesy of Johns Hopkins, Applied Physics Laboratory.)

kg) of the Sun's atmosphere out into space (Section 6.2). If they encounter the Earth, the CMEs can create intense geomagnetic storms, damage satellites in Earth orbit, and threaten astronauts (Sections 8.2, 8.3).

Previous observations of CMEs have utilized coronagraphs aboard a single spacecraft, such as *SOHO*, that provided an edge-on, two-dimensional view of the corona. *STEREO* consists of two spacecraft, both moving along the Earth's orbit but one moving well ahead of the Earth and one behind (Fig. 9.10). They will simultaneously examine CMEs with perspectives never obtained before.

The instruments aboard the two spacecraft (Table 9.8) will, when combined, provide spectroscopic observations of the CMEs from their onset at the Sun to the orbit of the Earth. They will measure energetic particles generated by the CME

Investigation	Measurement
IMPACT	The *In situ* Measurements of PArticles and Cme Transients investigation is a suite of seven instruments that will sample the three-dimensional distribution of solar-wind electrons, the characteristics of solar-wind electrons, energetic solar ions, and the vector magnetic field in the photosphere.
PLASTIC	The PLAsma and SupraThermal Ion and Composition experiment will provide plasma characteristics of protons, alpha particles and heavy ions.
SECCHI	The Sun–Earth Connection Coronal and Heliospheric Investigation encompasses four instruments: an extreme-ultraviolet imager, two white-light coronagraphs, and a heliospheric imager. They will study the three-dimensional evolution of CMEs from their origin at the Sun through the corona and the interplanetary medium to the Earth's orbit, in some cases recording their eventual impact with Earth.
SWAVES	The Stereo/WAVES is an interplanetary radio-burst detector that will track the generation and evolution of traveling radio disturbances from the Sun to the orbit of Earth.

Table 9.8 *STEREO's* investigations arranged alphabetically by acronym

disturbances as they pass through the interplanetary medium, and sample magnetic fields and particles in these disturbances when they pass near Earth.

STEREO's science objectives are to understand the origin and evolution of CMEs, and to thereby improve space-weather forecast capability. It will investigate solar magnetic field changes that result in CMEs, and then track CME-driven disturbances from the Sun to Earth's orbit. It will determine the three-dimensional structure of the disturbed interplanetary plasma and magnetic fields, and investigate the mechanisms and sites of energetic particle acceleration by CMEs. STEREO's unique perspective will enable accurate measurements of the speed and trajectory of CMEs, providing early warning of those that might hit Earth. The solar dynamo will also be investigated through study of cyclic phenomena in the corona and interplanetary space.

Solar-B is an ISAS mission proposed as a follow-on to the Japan/U.S./U.K. Yohkoh (Solar-A) collaboration. The mission consists of a coordinated set of optical, extreme-ultraviolet and X-ray instruments (Table 9.9) that will investigate the interaction between the Sun's magnetic field and its corona. The result will be an improved understanding of the mechanisms which give rise to solar magnetic variability and

how this variability modulates the total solar output (Section 8.4) and creates the driving force behind space weather.

Solar-B will measure the detailed density, temperature and velocity of structures in the photosphere, transition region and low corona with high spatial, spectral and temporal resolution. It will, for the first time, provide quantitative measurements of the full vector magnetic field in the photosphere on small enough scales to resolve the elemental flux tubes. The field of view and sensitivity allow changes in the magnetic energy to be related to both coronal heating (Section 5.3) and explosive solar flares or coronal mass ejections (Sections 6.3, 6.4).

The science objectives of Solar-B include:

(a) Investigations of the creation and destruction of the Sun's magnetic field, which is continuously being generated within the Sun's interior (Section 5.1) and swept out into space.
(b) Studies of the modulation of the Sun's luminosity, which varies in phase with the Sun's 11-year magnetic-activity cycle (Section 8.4).
(c) Explanations for the generation of solar ultraviolet and X-ray radiation, that also vary in phase with the Sun's magnetic-activity cycle (Section 8.4).

Instrument	Measurement
EIS	The Extreme-ultraviolet Imaging Spectrometer will obtain monochromatic images, and Doppler line widths and shifts, in the wavelength range 25 to 29 nm, observing spectral lines formed at temperatures between 0.1 to 20 million degrees kelvin. It will obtain plasma velocities to an accuracy of 10 km s^{-1} along with temperatures and densities in the transition region and corona with 2 seconds of arc angular resolution.
SOT	The Solar Optical Telescope has a 0.5-m aperture with an angular resolution of 0.25 seconds of arc. It will provide time series measurements of motions and vector magnetic fields in the photosphere, as well as the photospheric intensity, with a time resolution of 5 minutes.
XRT	The X-Ray Telescope will provide solar images at soft X-ray wavelengths of 0.2 to 6.0 nm (0.16 to 5 keV) with an angular resolution of 1.0 to 2.5 seconds of arc, studying the low corona at temperatures from 0.5 to 14 million degrees kelvin

Table 9.9 *Solar-B's* instruments arranged alphabetically by acronym

(d) Investigations of the eruption and expansion of the Sun's atmosphere, which are intimately related to magnetic reconnection (Sections 6.3, 6.4 and 7.2).

Thus, *Solar-B* will build upon recent investigations with *SOHO*, *Yohkoh*, *Ulysses* and ground-based radio and optical observatories to refresh, reshape and enhance our understanding of how the Sun influences our home planet, the Earth.

Appendix 1

FURTHER READING

AN ANNOTATED LIST OF BOOKS PUBLISHED BETWEEN 1989 AND 2000.

Bone, Neil: *The Aurora: Sun–Earth Interactions*, Second Edition. John Wiley and Sons, New York 1996.

A complete review of all aspects of the aurora, describing the mechanism of auroral displays and their observation, including the history of aurora investigations from the earliest ideas to scientific investigations with satellites. Solar activity, Earth's magnetosphere and solar–terrestrial interactions are also reviewed.

Cox, Arthur N., Livingston, William C., and Matthews, Mildred S. (editors): *Solar Interior and Atmosphere*. The University of Arizona Press, Tucson 1991.

A collection of 38 technical articles by acknowledged experts, including the solar interior, solar neutrinos, solar oscillations, magnetic fields, the faint-young-Sun problem, coronal heating, solar activity, and the relation of the Sun to other stars.

Foukal, Peter V.: *Solar Astrophysics*. John Wiley and Sons, New York 1990.

Technical astrophysics applied to the Sun, including radiative transfer, solar spectroscopy, plasma dynamics, internal structure, energy generation, rotation, convection, oscillations, activity, magnetism, the solar atmosphere, the solar wind and solar variability.

Golub, Leon, and Pasachoff, Jay M.: *The Solar Corona*. Cambridge University Press, New York 1997.

An advanced textbook that presents our recent understanding of coronal physics. It is written for graduate and advanced undergraduate students.

Hargreaves, John Keith: *The Solar–Terrestrial Environment*. Cambridge University Press, New York 1992.

An introductory textbook description of physical conditions in the upper atmosphere, ionosphere and magnetosphere of the Earth, together with basic physics, principles of investigation and technological applications.

Hufbauer, Karl: *Exploring the Sun – Solar Science Since Galileo*. Johns Hopkins Press, Baltimore, Maryland 1991.

A historical account that emphasizes the solar wind and the variable Sun. Techniques, theories, scientific communities and patronage are singled out for discussion.

Jokipii, J. Randy, Sonett, Charles P., and Giampapa, Mark S. (editors): *Cosmic Winds and the Heliosphere*. The University of Arizona Press, Tucson 1997.

A collection of 28 technical articles that discuss the solar heliosphere, the solar wind, the physics of wind origins, winds from other stars, the physical properties of the solar wind, and interactions of winds with the surrounding medium.

Kaler, James B.: *Stars. Scientific American Library*. W. H. Freeman, New York 1992.

This elegantly written and well illustrated volume explores the nature of stars, including the Sun, describing our current knowledge of their origin, variety, distribution, composition, and distinctive histories.

Kippenhahn, Rudolf: *Discovering the Secrets of the Sun*. John Wiley and Sons 1994.

A nicely written popular book that surveys our current knowledge of the Sun, including solar energy, sunlight, eclipses, sunspots, magnetic fields, solar plasmas, nuclear fusion and solar influences on Earth.

Lang, Kenneth R.: *Sun, Earth and Sky*. Springer-Verlag, New York 1995.

A lavishly illustrated book that introduces the Sun, its physics and its impact on life here on Earth, written in a light and friendly style with apt metaphors, similes and analogies, poetry, art, history, and vignettes of scientists at work.

Lang, Kenneth R.: *Astrophysical Formulae. Volume I. Radiation, Gas Processes and High Energy Astrophysics. Volume II. Space, Time, Matter and Cosmology*. Springer-Verlag, New York 1999.

The third, enlarged edition of a comprehensive, widely-used reference to the fundamental formulae employed in astronomy, astrophysics and general physics, including 4000 formulae and 5000 references to the original papers. It includes all aspects of astronomy and astrophysics that are relevant to studies of the Sun.

Lang, Kenneth R.: *The Sun From Space*. Springer-Verlag, New York 2000.

A comprehensive account of solar astrophysics and how our perception and knowledge of the Sun have gradually changed. Timelines and hundreds of seminal papers are provided for key discoveries during the past two centuries, but the emphasis is on the last decade which has seen three successful solar spacecraft missions: *SOHO*, *Ulysses* and *Yohkoh*. Together, these have confirmed many aspects of the Sun and its output, and provided new clues to the numerous open questions that remain. This generously illustrated book is written in a clear and concise style, covering all levels from the amateur astronomer to the expert.

Littman, Mark, Wilcox, Ken and Espanak, Fred: *Totality. Eclipses of the Sun*, Second Edition. Oxford University Press, New York 1999.

This book includes everything you might want to know about eclipses of the Sun, including their dates, history, mythology, observations, photographs, scientific implications and visibility.

Phillips, Kenneth J. H.: *Guide to the Sun*. Cambridge University Press, New York 1992.

This introduction to the Sun includes the solar interior, photosphere, chromosphere and corona, the active Sun, the heliosphere, the Sun and other stars, solar energy and observing the Sun.

Schrijver, Carolus J., and Zwaan, Cornelis: *Solar and Stellar Magnetic Activity*. Cambridge University Press, New York 2000.

A comprehensive review and synthesis of our current understanding of the origin, evolution and effects of magnetic fields in the Sun and other cool stars. Topics include solar differential rotation and meridional flow, solar magnetic configurations and structure, properties of the global solar magnetic field, the solar dynamo and the solar atmosphere. It combines studies of the Sun with investigations of other stars, and discusses how stellar activity evolves and depends on the mass, age and chemical composition of stars.

Sonnett, Charles P., Giampapa, Mark S., and Matthews, Mildred S. (editors): *The Sun in Time*. The University of Arizona Press, Tucson 1991.

A collection of 36 technical articles on energetic solar particles, solar isotopes, the Sun and climate, the early Sun, and solar-like stars.

Stix, Michael: *The Sun*. Springer-Verlag, New York 1989.

A thorough and sometimes mathematical account of the Sun's atmosphere, chromosphere, convection, corona, internal structure, luminous output, magnetism, oscillations, rotation and winds.

Tayler, Roger J.: *The Sun as a Star*. Cambridge University Press, New York 1997.

This textbook provides a broad and wide-ranging introduction to the Sun as a star using mathematics appropriate for advanced undergraduate students in physics. Topics include the properties of spectral lines, the theory of stellar oscillations, plasma physics, magnetohydrodynamics and dynamo theory.

Taylor, Peter O.: *Observing the Sun* Cambridge University Press, New York 1991.

An account of how amateur astronomers might observe the Sun, including a historical background, new techniques and equipment, observations of sunspots and eclipses, and reporting observations.

Appendix 2

Glossary

A The sum of the number of protons and neutrons in an atomic nucleus, known as the mass number.

absolute luminosity Denoted by L, the amount of energy radiated per unit time, measured in units of watts. One joule per second is equal to one watt of power, and to ten million ergs per second. The absolute luminosity of the thermal emission from a black-body is given by the Stefan–Boltzmann law in which $L = 4\pi\sigma R^2 T_e^4$, where $\pi = 3.141\ 59$, the Stefan–Boltzmann constant, $\sigma = 5.670\ 51 \times 10^{-8}$ J m^{-2} K^{-4} s^{-1}, the radius is R and the effective temperature is T_e. See luminosity, and Stefan–Boltzmann law.

absolute magnitude The absolute magnitude of a celestial object is the apparent magnitude the object would have at a standard distance of 10 parsecs (pc) from the Earth. The absolute magnitude is denoted by the symbol M, and the apparent magnitude by m. The absolute magnitude may be derived from the apparent magnitude and the parallax by the formula $M = m + 5 + 5\log\pi_t$, where π_t is the annual trigonometric parallax in seconds of arc. The annual parallax is the reciprocal of the distance, D, in units of parsecs, so $M = m + 5 - 5\log D$. The absolute visual magnitude of the Sun is $+4.83$ and its apparent visual magnitude is -26.74. The absolute magnitudes of most stars lie between -5 and $+15$. See magnitude, and visual magnitude.

absolute temperature The temperature as measured on a scale whose zero point is absolute zero, the point at which all motion at the molecular level ceases. The unit of absolute temperature is the degree kelvin, denoted by K. The freezing temperature of water is 273 K and the boiling temperature of water is 373 K. Absolute zero is 0 K. The equivalent temperature in degrees Celsius is $°C = K - 273$, and the equivalent temperature in degrees Fahrenheit is $°F = (9/5)K - 459.4$. See Celsius, Fahrenheit, and kelvin.

absorption The process by which the intensity of radiation decreases as it passes through a material medium. The energy lost by the radiation is transferred to the medium.

absorption line A "dark" line of decreased radiation intensity at a particular wavelength of the electromagnetic spectrum, formed when a cool, tenuous gas located between a hot, radiating source and an observer, absorbs electromagnetic radiation of that wavelength. This feature looks like a line when the radiation intensity is displayed as a function of wavelength; such a display is called a spectrum. Different atoms, ions and molecules produce characteristic patterns of absorption lines, and observations of these lines enable identification of the chemical ingredients of the gas between the radiating source and the observer. Absorption lines are found in the spectra of stars, for the radiation from their hot interiors is absorbed by the cooler, outer stellar layers. The absorption lines in the Sun's spectrum are often called Fraunhofer lines. See Fraunhofer lines, spectroheliograph, and spectrum.

accretion The capture of gas or dust by a single body to form a larger one, like a moon, planet or star. Also, the coalescence of small particles in space as a result of collisions, and the gradual building up of larger bodies from smaller ones by gravitational attraction.

accretion disk A flattened disk of gas and/or dust in orbit around a star in its equatorial plane. Planets are thought to be formed in the accretion disk of young stars. The disk materials can also slowly spiral inward, eventually "accreting", or merging with the star's mass.

ACE Acronym for the *Advanced Composition Explorer*, a NASA spacecraft launched on 25 August 1997. It samples low-energy particles of solar origin and high-energy galactic particles from the inner Lagrangian point of Earth–Sun gravitational equilibrium. *ACE* also provides near-real-time solar-wind information, reporting space weather that can provide an advance warning (about one hour) of geomagnetic storms. See Lagrangian point.

acoustic waves Sound waves that propagate through the interior of the Sun. They are produced when there is a small displacement of the gas, with pressure acting as the dominant restoring force. Convective turbulence in the Sun and some other stars acts as a source of acoustic waves. Sound waves in the Sun produce oscillations of its visible photosphere with periods of about five minutes; observations of these oscillations can be used to infer the interior constitution of the Sun. Acoustic waves also play a role in heating the chromosphere. See five minute oscillations, and helioseismology.

ACRIM Acronym for the Active Cavity Radiometer Irradiance Monitor, an experiment on the *Solar Maximum Mission* satellite that measured the total radiative output of the Sun. See solar constant.

active region A region in the solar atmosphere, from the photosphere to the corona, that develops when strong magnetic fields emerge from inside the Sun. The magnetized realm in, around and above sunspots is called an active region. Radiation from active regions is enhanced, when compared to neighboring areas in the chromosphere and corona, over the whole electromagnetic spectrum, from X-rays to radio waves. Active regions may last from several hours to a few months. They are the sites of intense explosions, called solar flares, that last a few minutes to hours. The number of active regions varies in step with the 11-year sunspot cycle. *See* flare, solar activity cycle, and sunspot cycle.

activity index Any one of a number of indicators of the level of solar activity at a given time. Indices to measure solar activity include the number of sunspots, the total area covered by sunspots on the visible hemisphere, the areas and brightness of plages, and the total radio and X-ray radiation from the Sun.

a **index** A 3-hourly index of local geomagnetic activity; when the *a* index increases from 0 to 400 a similar 3-hourly K index ranges from 0 to 9. *See* K index.

A **index** A daily index of local geomagnetic activity, the average of the eight 3-hourly a indices.

Alfvén velocity The velocity of propagation of disturbances in a magnetized plasma that propagates along the magnetic field lines. The Alfvén velocity is given by $B/(4\pi\rho)^{1/2}$ for a magnetic field strength of B and a gas mass density of ρ. *See* Alfvén wave.

Alfvén wave A wave motion occurring in a magnetized plasma in which the magnetic field oscillates transverse to the direction of propagation without a change in magnetic field strength. The tension in the field lines acts as a restoring force. These magnetohydrodynamic waves propagate at the Alfvén velocity in the direction of the magnetic field lines. *See* Alfvén velocity, and magnetohydrodynamics.

Alpha Centauri The star system closest to Earth, consisting of three co-orbiting stars. The brightest member of the three stars is *Rigil Kentaurus* (Foot of the Centaur); this star is also sometimes called Alpha Centauri. It is the brightest star in the constellation Centaurus, and the third-brightest star in the sky, with a magnitude of −0.27 and a distance of 4.3 light-years, or 1.3 parsecs (pc). The closest star to our solar system is the faint red-dwarf component of the Alpha Centauri system, known as Proxima Centauri, with a parallax of 0.761 seconds of arc.

alpha particle The nucleus of a helium atom, consisting of two protons and two neutrons. The helium nucleus has a charge twice that of a proton and a mass just 0.007, or 0.7 percent, less than the mass of four protons. The alpha particle is a helium ion. *See* helium, and ion.

Andromeda A large constellation of the northern hemisphere, adjoining the Square of Pegasus.

Andromeda galaxy The most-distant object visible to the unaided eye, also known as M31 or NGC 224, located at a distance of 2.6 million light-years, or 0.8 megaparsec (Mpc). The closest spiral galaxy, similar in shape and size to our own Galaxy. It is visible to the naked eye from the northern hemisphere as a fuzzy, luminous patch in the constellation of Andromeda, and is about 1 by 4 degrees in angular extent. The Andromeda galaxy is accompanied by two elliptical galaxies, M32 and NGC 205, both visible in small telescopes.

ångstrom A unit of length equal to 10^{-10} m, or 0.1 nm. An ångstrom is on the order of the size of an atom.

angular momentum The product of angular velocity and mass.

angular resolution *See* resolution.

annual parallax Half the apparent angular displacement of a nearby star observed against the more distant stars at intervals of six months from opposite sides of the Earth's orbit. The annual parallax is also known as the trigonometric parallax. The distance of a nearby star in parsecs is given by the reciprocal of its annual parallax in seconds of arc.

aperture The diameter of the light-gathering lens or mirror of a telescope.

aphelion The point in a planet's orbit that is farthest from the Sun.

Ap **index** A planet-wide index of geomagnetic activity derived as an average of local *A* indices. A severe geomagnetic storm has an *Ap* index of 100 or more, moderate geomagnetic storms have an *Ap* index of between 50 and 100, and minor geomagnetic storms have an *Ap* index between 29 and 50. *See* A index, and Kp index.

apparent magnitude A measure of the relative brightness of a star, or other celestial object, as perceived by an observer on Earth. It is denoted by the symbol m. Apparent magnitude depends on both the absolute amount of light-energy emitted, or reflected, and the distance to the object. The smallest apparent magnitudes correspond to the greatest brightness. The apparent magnitude of the Sun is −26.74, that of the full Moon is −12.6, and the apparent magnitude of the planet Venus at its brightest is −4.7. *See* absolute magnitude, magnitude, parallax, and visual magnitude.

arcade A series of arches formed by magnetic loops that confine hot gas.

arc degree *See* degree.

arc minute *See* minute of arc.

arc second *See* second of arc.

astronomical unit Abbreviated AU, the astronomical unit is the average distance between the Earth and the Sun. It is equal to $1.495\ 978\ 7 \times 10^{11}$ m, or about 150 million kilometers and 215 solar radii. The astronomical unit is also defined as the distance from the Sun at which a massless particle in an unperturbed orbit would have an orbital period of 365.256 898 3 days.

astronomy The scientific study of planets, stars, galaxies and the Universe, and the processes by which they formed and evolve.

ATM Acronym for the Apollo Telescope Mount on the *Skylab* space station.

atmosphere The gases surrounding the surface of a planet or natural satellite, and the outermost gaseous layers of a star, held near them by their gravity. Since gas has a natural tendency to expand into space, only bodies that have a sufficiently strong gravitational pull can retain atmospheres. The ability of a planet or satellite to retain an atmosphere depends on the gas temperature, determined by its distance from the Sun, and its escape velocity, which increases with the object's mass. Hotter, lighter gas molecules move faster, and are more likely to escape from a planet or satellite; less-massive objects such as our Moon are less likely to retain an atmosphere. We use the term atmosphere for the tenuous outer material of the Sun because it is relatively transparent at visible wavelengths. The solar atmosphere includes, from the deepest layers outward, the photosphere, the chromosphere, and the corona. The Earth's atmosphere is transparent and consists mainly of molecular nitrogen (77 percent) and molecular oxygen (21 percent) with trace amounts of carbon dioxide (0.033 percent) and water vapor (variable, a few times 0.0001 percent). *See* escape velocity.

atmospheric window A wavelength band in the electromagnetic spectrum that is able to pass through the Earth's atmosphere with relatively little attenuation through absorption, scattering or reflection. There are two main windows – the optical window and the radio window.

atom The basic unit of matter composed of a central nucleus, containing protons and neutrons, and a cloud of orbiting electrons equal in number to the number of protons. The hydrogen atom has one proton and no neutrons in its nucleus. An atom is the smallest particle of an element that still has the characteristics of that element. *See* atomic number, and element.

atomic mass unit Abbreviated a.m.u., the unit of atomic mass equal to $1.660\ 54 \times 10^{-27}$ kg. A proton and neutron have respective masses of 1.007 276 5 and 1.008 664 9 a.m.u.

atomic number Denoted by Z, the number of protons in the nucleus of an atom. The atomic number of an element indicates its place in the periodic table of elements. Isotopes have the same atomic number, but a different number of neutrons in the nucleus. *See* isotope.

AU Abbreviation for astronomical unit. *See* astronomical unit.

AURA Acronym for Association of Universities for Research in Astronomy. *See* National Optical Astronomy Observatory.

aurora A display of rapidly varying, colored light usually seen from magnetic polar regions of a planet. The light is given off by collisions between charged particles and the atoms, ions and molecules in a planet's upper atmosphere or ionosphere. Auroras are visible on Earth as the aurora borealis, or northern lights, near the North Pole, and the aurora australis, or southern lights, near the South Pole. They include the green and red emission lines from oxygen atoms and nitrogen molecules that have been excited by electrons coming from the Earth's magnetotail and energized by magnetic reconnection with the solar wind. Terrestrial auroras occur at heights of around 100 to 250 thousand meters, which is within the Earth's ionosphere.

auroral oval One of the two oval-shaped zones around the Earth's geomagnetic poles in which the night-time auroras are observed most often. They are located at latitudes of about 67 degrees north and south, and are about 6 degrees wide. Their position and width both vary with geomagnetic and solar activity. The auroral ovals are detected from Earth-orbiting satellites. Ground-based observers detect only a section of an oval, as a shimmering ribbon and curtain of light.

Baily's beads A string of bright lights observed at the extreme edge of the Sun's disk, seen during a total eclipse of the Sun just before or after totality. They are caused by rays of light from the Sun shining through the valleys on the Moon's limb, while other rays are blocked by mountains on the limb. They were first described by the English astronomer Francis Baily in 1836.

Balmer continuum Radiation associated with the recombination of free electrons to the second bound level of the hydrogen atom.

Balmer series Hydrogen lines seen at visible wavelengths. The first Balmer line is called the hydrogen-alpha line, denoted by Hα, at a wavelength of 656.3 nm; the second and third, respectively denoted by Hβ and Hγ, are at wavelengths of 486.1 and 434.2 nm. The formula governing the wavelengths of these lines was specified by the Swiss mathematics teacher Johann Balmer in 1885.

Bartel's rotation number A cumulative sequence of 27-day rotation periods of solar and geophysical parameters, assigned by Julius Bartel to begin in January 1833. A system for numbering 27-day solar rotations.

BCS Acronym for the Bragg Crystal Spectrometer on *Yohkoh*.

beta decay Emission of an electron (or a positron) and a neutrino by a radioactive nucleus. *See* positron, and neutrino.

beta particle An electron or positron emitted from an atomic nucleus as a result of a nuclear reaction or in the course of radioactive decay. Such particles were originally called beta rays. *See* positron, and neutrino.

big bang A theory that explains the current expansion of the Universe in terms of a giant explosion 10 to 20 billion years ago, prior to which all the mass of the known Universe was compressed to a very high density and an extremely high temperature exceeding 10 billion degrees kelvin. The cosmic microwave background radiation is considered to be the residual radiation of the explosive, big-bang fireball. *See* cosmic microwave background.

billion One thousand million, written 1 000 000 000 or 10^9.

bipolar outflow The high-speed jetting of gas in opposite directions from disk-shaped systems, usually in directions perpendicular to the plane of the disk.

bipolar region A region in the photosphere where the magnetic fields show opposite magnetic polarities, directed in and out of the Sun, including groups of sunspots, active regions without sunspots and small, ephemeral regions. *See* coronal loops.

black-body A body that absorbs all the radiation incident on it, a perfect absorber of radiation. The intensity of the radiation emitted by a black-body and the way it varies with wavelength depends only on the temperature of the body and can be predicted by quantum theory.

black-body radiation The radiation that would be emitted by a black-body, a hypothetical object which absorbs all thermal radiation (heat) falling upon it, and is a perfect emitter of thermal radiation. The spectrum of the radiation emitted by a black-body is a continuous spectrum, and the wavelength of greatest emission is inversely proportional to the body's temperature. The total amount of energy emitted by a black-body is proportional to the product of its area and the fourth power of its temperature. *See* continuous spectrum, effective temperature, Stefan–Boltzmann law, and thermal radiation.

black hole An extremely dense object whose gravity is so intense that neither matter nor radiation can escape, or in other words the object's escape velocity exceeds the velocity of light. The radius of this localized region of space is called the Schwarzschild radius. *See* escape velocity, and Schwarzschild radius.

blueshift The Doppler shift in the wavelength of a spectral line toward a shorter wavelength. A blueshift arises when the source of radiation and its observer are moving toward each other. *See* Doppler shift.

Bode's law A numerical sequence announced by Johann Daniel Bode in 1772 which matches the distances from the Sun of the then known planets. It is formed by adding 4 to each number in the sequence 0, 3, 6, 12, 24, 48, 96 and 192, with the Earth's distance being 10. More properly called the Titius–Bode law, owing to its prior statement by Johann Elert Titius in 1766. *See* Titius–Bode law.

bolometric magnitude An apparent or absolute magnitude, respectively denoted by m_{bol} and M_{bol}, which specifies the total amount of radiation emitted by a body at all wavelengths. *See* apparent magnitude, and absolute magnitude.

Boltzmann's constant The constant of proportionality between energy per degree of freedom and temperature, denoted by the symbol k. Boltzmann's constant has the value $k = 1.380\,66 \times 10^{-23}$ J K^{-1}.

bound–bound transition A transition between energy levels of an electron bound to a nucleus, both before and after the transition.

bow shock The boundary on a planet's sunlit side where the solar wind is deflected and there is a sharp decrease in the wind's velocity. The plasma of the solar wind is heated and compressed at the bow shock. The stream of particles in the solar wind is deflected around the planet, like water flowing past the bow of a moving ship.

bremsstrahlung Radiation that is emitted when an energetic electron is deflected by an ion. It is also called a free–free transition because the electron is free both before and after the encounter, remaining in an unbound hyperbolic orbit without being captured by the ion to make an atom. *Bremsstrahlung* is German for "braking radiation".

brightness *See* luminosity.

burst A sudden, transient increase in solar radio radiation during a solar flare, emitted by energetic electrons. They are probably caused by the sudden release of large amounts (up to 1025 J) of magnetic energy in a relatively small volume in the solar corona. Bursts detected at meter wavelengths are divided into several types, depending on their time–frequency behavior. *See* active region, flare, solar activity cycle, solar flare, and type I, II, III and IV radio burst.

butterfly diagram A plot of the heliographic latitudes of sunspots over the course of the 11-year sunspot cycle. At the start of a cycle there are sunspots at solar latitudes of up to 35 to 45 degrees north and south, but very few near the equator. As the cycle progresses, sunspots occur nearer the equator. This graphical representation of sunspot latitudes as a function of time was first plotted in 1922 by Edward Walter Maunder and is also known as the Maunder diagram. The plot resembles butterfly wings, which gives the diagram its popular name. *See* heliographic latitude, and sunspot cycle.

calcium H and K lines Spectral lines of singly ionized calcium, denoted by Ca II, in the violet part of the spectrum at 396.8 nm (H) and 393.4 nm (K). They are conspicuous features in the spectra of many stars, including the Sun. The designations H and K were given by Fraunhofer and are still commonly used. *See* Fraunhofer lines, plages, and spectroheliograph.

calcium network A pattern of emission features seen in the lines of ionized calcium. *See* calcium H and K lines, chromospheric network, and network.

carbon–nitrogen–oxygen cycle A sequence of nuclear reactions, abbreviated as the CNO cycle, which accounts for the production of energy inside main-sequence stars that are more massive and hence hotter than the Sun. The reactions involve the fusion of four hydrogen nuclei into one helium nucleus, with carbon acting as a catalyst and remaining unchanged at the end of the cycle. It occurs at temperatures above 15 million degrees kelvin. *See* hydrogen burning.

Carrington longitude A system of fixed longitudes rotating with the Sun introduced by Richard C. Carrington in the 19th century.

CDS Acronym for the Coronal Diagnostic Spectrometer on *SOHO*.

CELIAS Acronym for the Charge, ELement and Isotope Analysis System instrument on *SOHO*.

Celsius A unit of temperature denoted by °C for degrees Celsius. Absolute zero on the Celsius temperature scale is at −273 °C, the freezing point of water is at 0 °C, and the boiling point of water is at 100°C. The equivalent temperature in degrees kelvin is given by K = °C + 273. The equivalent temperature in degrees Fahrenheit is given by °F = (9/5)°C + 32. *See* absolute temperature, Fahrenheit, and kelvin.

Centigrade An older name for the Celsius unit of temperature, derived from the fact that there are 100 degrees between the freezing and boiling temperatures of water. *See* Celsius.

Cepheid variable star A class of variable stars that periodically expand and contract, producing a regular variation in their luminosity. They take their name from the first of the type to be discovered, Delta Cephei, the variability of which was noted by John Goodricke in 1784. Since the absolute luminosity or magnitude of a Cepheid variable star is related to its variation period, one can determine its distance from observations of its apparent luminosity or magnitude and period. The period–luminosity relation of Cepheid variable stars was discovered by Henrietta Leavitt in 1912. *See* period–luminosity relation.

charged particles Fundamental components of sub-atomic matter, such as protons, and electrons, that have electrical charge. The charged sub-atomic particles are surrounded by electrical force fields that attract particles of opposite charge and repel those of like charge. Charged particles are either negative (electron) or positive (proton) and are responsible for all electrical phenomena.

Cherenkov radiation Light or other forms of electromagnetic radiation emitted when a charged particle has a velocity that exceeds the velocity of light in a medium. A neutrino can be detected by the cone of blue Cherenkov light emitted as the neutrino enters a large underground tank of water and accelerates an electron by collision to a velocity faster than the velocity of light in water. *See* KAMIOKANDE, Sudbury Neutrino Observatory, and SUPER KAMIOKANDE.

chlorine experiment The subterranean neutrino detector located in the Homestake Mine near Lead, South Dakota, in which a neutrino strikes a chlorine nucleus to produce a nucleus of radioactive argon.

chromosphere The layer or region of the solar atmosphere lying above the photosphere and beneath the transition region and the corona. The Sun's temperature rises to about 10 000 K in the chromosphere. The name literally means "sphere of color". The chromosphere is normally invisible because of the glare of the photosphere shining through it, but it is briefly visible near the beginning or end of a total solar eclipse as a spiky red rim, seen beyond the occulting Moon's disk. At other times, the chromosphere can be studied by spectroscopy, observing it across the solar disk in the red light of hydrogen alpha. A thin transition region separates the chromosphere and the corona. Spicules containing chromospheric material penetrate well into the corona (to heights of 10 million meters) at the edges of the supergranulation cells. *See* hydrogen-alpha line, spicules, supergranulation cells, surge, and transition region.

chromospheric network A large-scale cellular pattern visible in spectroheliograms taken in the calcium H and K lines. The network appears at the boundaries of the photospheric supergranulation cells, and contains magnetic fields that have been swept to the edges of the cells by the flow of material in the cell. *See* supergranulation cells.

CIR Acronym for Corotating Interaction Region. *See* corotating interaction region.

climate The average weather conditions of a place over a period of years.

CLUSTER II A set of four identical spacecraft launched on 9 August 2000 to obtain simultaneous in-situ measurements of small-scale plasma structures in the solar wind and bow shock, magnetopause, polar cusp, magnetotail and auroral zones.

cluster of galaxies A group of galaxies containing up to a thousand galaxies. The nearest rich clusters of galaxies are the Virgo Cluster, 65 million-light-years, or 20 megaparsec (Mpc) away, and the Coma Cluster, 326 million light-years, or 100 Mpc, away.

CME Acronym for Coronal Mass Ejection. See coronal mass ejection.

CNO cycle Acronym for Carbon–Nitrogen–Oxygen cycle. See carbon–nitrogen–oxygen cycle.

CNRS Acronym for Centre National de la Recherche Scientifique, France.

color index The difference in brightness of a star at two different wavelengths, used as a measure of the star's color and hence its temperature. When the blue (B) and yellow (V, for visual) wavelengths are used, one obtains the B–V color index, which is positive for cool red stars and negative for hot blue ones. See UBV system.

comet An icy body, or dirty snowball, much smaller than a planet and about ten thousand meters across, orbiting the Sun usually far beyond the planets and often in a highly eccentric orbit. A comet partially vaporizes when it nears the Sun, within about three times the Earth's distance from the Sun. The comet then develops a diffuse, luminous coma of dust and gas, and, normally, one or more tails. These tails can become longer than the distance between the Earth and the Sun. The curved dust tails are pushed away from the Sun by the pressure of solar radiation, and the straight ion tails are swept away by the solar wind. Most comets are stored far from the planetary system in two large reservoirs: the Kuiper belt beyond the orbit of Neptune, and the Oort cloud at near-interstellar distances. See Kuiper belt, Oort cloud, and solar wind.

continuous spectrum An unbroken distribution of radiation over a broad range of wavelengths. See continuum, and spectrum.

continuum That part of a spectrum that has neither absorption nor emission lines, but only a smooth wavelength distribution of radiant intensity. See continuous spectrum, and spectrum.

convection The transfer of energy by material motion. The physical upwelling of hot matter, thus transporting energy from a lower, hotter region to a higher, cooler region. Gas that is hotter than its surroundings expands and rises vertically, resulting in transport and mixing. When it has cooled by passing on its extra heat to its surroundings, the gas sinks again. Convection can occur where there is a substantial decrease in temperature with height as in the Earth's troposphere, the Sun's convective zone or a boiling pot of water. See convective zone, and troposphere.

convective zone A layer in a star in which convection currents, or mass motions, are the main mechanism by which energy is transported outward. In the Sun, a convective zone extends from just below the photosphere down to 0.713 of the solar radius. The opacity in the convective zone is so large that energy cannot be transported by radiation.

core The central region of a planet, star or other celestial object. For the Sun, the central location where energy is generated by nuclear reactions that convert hydrogen into helium. See fusion, hydrogen burning, and proton–proton chain.

corona The outermost, high-temperature region of the solar atmosphere, above the chromosphere and transition region, consisting of almost fully ionized plasma contained in closed magnetic loops, called coronal loops, or expanding out along open magnetic field lines to form the solar wind. The corona is a highly rarefied, low-density gas, with electron densities of less than 10^{16} electrons per cubic meter, heated to temperatures of millions of degrees. The corona is briefly visible to the unaided eye as a white halo during a total eclipse of the Sun, for at most 7.5 minutes; at other times it can be observed in visible white light only by using a special instrument called a coronagraph. The visible, white-light corona may be divided into the inner K or electron corona, with a continuous spectrum, and the outer F or dust corona that displays Fraunhofer absorption lines. The E corona is the emission-line corona. The corona can always be observed across the solar disk at X-ray and radio wavelengths. The shape of the corona is determined by the distribution of solar magnetic fields that varies with the solar activity cycle. See coronagraph, F corona, K corona, and solar activity cycle.

coronagraph An instrument, used in conjunction with a telescope, for observing the faint solar corona in white light, or in all the colors combined, at times other than during a solar eclipse. The bright light of the Sun's photosphere is blocked out by an occulting disk, providing an artificial eclipse, with additional precautions for removing all traces of stray sunlight. The coronagraph was invented by Bernard Lyot in 1930. Even with a coronagraph located at a high site where the sky is very clear, scattering of light by the Earth's atmosphere is a problem. Coronagraphs therefore work best when placed on satellites above the Earth's atmosphere.

coronal bright point See X-ray bright point.

coronal green line An emission line of Fe XIV at 530.3 nm, the strongest visible line in the solar corona.

coronal hole An extended region in the solar corona where the density and temperature are lower than other places in the corona. The weak, diverging and open magnetic field lines in coronal holes extend radially outward and do not immediately return back to the Sun. The high-speed part of the solar wind streams out from coronal holes. The low density of the gas makes these parts of the corona appear dark in extreme-ultraviolet and soft X-ray images of the Sun, as if there were a hole in the corona. Coronal holes are nearly always present near the solar poles, and can also occur at lower latitudes.

coronal loop A magnetic loop that passes through the corona and joins regions, called footpoints, of opposite magnetic polarity in the underlying photosphere. Coronal loops can have exceptionally strong magnetic fields, and they often contain the dense, million-degree coronal gas that emits intense X-ray radiation. See footpoint.

coronal mass ejection Abbreviated CME, the transient ejection of plasma and magnetic fields from the Sun's corona into interplanetary space, detected by white-light coronagraphs. A large body of magnetically confined, coronal material being released from the corona. A coronal mass ejection contains 5 to 50 billion tons (5×10^{12} to 50×10^{12} kg) of gas and can travel through interplanetary space at a high, supersonic speeds up to 1.2×10^{6} m s^{-1}. It is often associated with an eruptive prominence, and sometimes with a strong solar flare. Coronal mass ejections produce intense shock waves, accelerate vast quantities of energetic particles, grow larger than the Sun in a few hours, and when directed at Earth can cause intense geomagnetic storms, disrupt communications, damage satellites and produce power surges on electrical transmission lines. See geomagnetic storm, and prominence.

coronal streamer A magnetically confined, loop-like coronal structure in the low corona straddling a magnetic neutral line on the solar photosphere. These high-density, bright coronal structures have ray-like stalks that extend radially outward to large distances in the outer corona. Near the minimum in the 11-year solar activity cycle, coronal streamers are located mostly near the solar equator and appear to be the source of the slow-speed solar wind. See helmet streamer, solar activity cycle, and solar wind.

coronal transient See coronal mass ejection.

coronium A supposedly unknown chemical element emitting unidentified emission lines in the spectrum of the solar corona. It was later discovered that the mysterious lines are produced by highly ionized forms of known elements, such as iron.

corotating interaction region A region in interplanetary space where there is an interaction between high- and low-speed streams of the solar wind that causes a shock wave of high density, pressure, and magnetic field.

corpuscular radiation Charged particles (mainly protons and electrons) emitted by the Sun. The stream of electrically-charged particles was hypothesized in the 1950s and later renamed the solar wind. See solar wind.

cosmic microwave background The remnant of the radiation from the hot, early Universe following the big bang. It is now observed as black-body radiation at a temperature of 2.73 K and has an almost equal intensity in all directions in space. It is called the microwave background since the radiation was first detected at a microwave wavelength of 0.073 m.

cosmic rays High-energy charged particles that enter the solar system from interstellar space, moving at speeds approaching the speed of light, and attaining energies greater than 1 MeV. Protons are the most-abundant kind of cosmic rays, but they include lesser amounts of heavier atomic nuclei and electrons. Those beyond the Earth's atmosphere are known collectively as primary cosmic rays. When they enter the Earth's atmosphere, cosmic rays collide with atmospheric atoms and produce various atomic and sub-atomic particles, called secondary cosmic rays, which can be detected at ground level. Sometimes the term galactic cosmic ray is used to distinguish those coming from interstellar space from high-energy, charged particles coming from the Sun. The latter have historically been called solar cosmic rays, but we do not use that designation in this book.

COSPIN Acronym for the COsmic ray and Solar Particle INstrument on *Ulysses*.

COSTEP Acronym for the COmprehensive SupraThermal and Energetic Particle analyzer on *SOHO*.

coudé An arrangement of auxiliary mirrors used with a telescope to direct the light path to a fixed focal point. It is used in conjunction with large, immoveable instruments, such as a spectrograph, which are too massive to be mounted on the telescope itself. The word comes from the French *coudé*, meaning "bent like an elbow".

Coulomb barrier The electric field repulsion experienced when two charged nuclear particles approach. When the barrier is overcome by the tunnel effect, nuclear reactions that produce the Sun's energy and radiation can occur. See tunnel effect.

Coulomb interactions Long-range interaction through the Coulomb force that occurs between particles that are electrically charged.

Crab nebula The nebula M1, or NGC 1952, a remnant of a supernova explosion that was noted by Chinese astronomers in July 1054. It is at a distance of 6295 light-years, or 1930 pc, and contains a radio pulsar, or rotating neutron star, at its center, spinning 30 times a second. The nebula was discovered in 1731 by the English astronomer John Bevis, and independently by Charles Messier in 1758. It gained its popular name of Crab nebula after it was sketched by Lord Rosse to resemble a crab. The nebula is an intense source of radio emission known as Taurus A and is also a source of X-rays. Its synchrotron radiation extends from the radio domain into the visible. *See* supernova remnant, and synchrotron radiation.

critical density The average density of matter in the Universe required to just barely continue its expansion forever. If the Universe had a mass density greater than its critical value, then the Universe would stop expanding in the future, and then collapse back on itself. There is no direct observational evidence for such a dense Universe.

current sheet The two-dimensional surface within a magnetosphere that separates magnetic fields of opposite polarities or directions. *See* heliospheric current sheet.

curvature of space The distortion of space in the neighborhood of matter, one of the consequences of Einstein's General Theory of Relativity. This curvature makes rays of light follow curved paths when passing near a massive object, including the Sun and clusters of galaxies.

cyclotron radiation Electromagnetic radiation emitted by electrons traveling in circular paths in a magnetic field.

dark matter Unseen matter that is inferred to surround bright galaxies and to exist in the space between galaxies. Dark matter may be ten times more abundant than luminous matter.

daughter isotope An isotope produced by radioactive decay of a parent isotope. *See* isotope, parent isotope, and radioactivity.

decametric radiation Radio radiation of approximately 10 m in wavelength.

decimetric radiation Radio radiation of approximately 0.1 m in wavelength.

degree The unit of angular measurement, equal to 1/360 of a full circle. Denoted by the symbol °.

density The mass per unit volume of an object. The density of water is 1000 kg m^{-3}.

deuterium The heavy isotope of hydrogen containing both a proton and a neutron in its nucleus.

deuteron The nucleus of a deuterium atom. This nucleus, containing one neutron and one proton, is sometimes denoted as ^2D, but as D in this book.

diamond-ring effect An intense point of light seen during a total eclipse of the Sun, just before or after totality, when just the bright central part of the edge of the Sun's disk is visible, giving the appearance of a diamond ring.

differential rotation The rotation of a gaseous, non-solid body, such as the Sun, at a rate that varies with latitude. The equatorial regions rotate more quickly than those at higher latitudes. Thus, the outer part of the Sun revolves faster at its equator than at its poles. A solid body like the Earth cannot undergo differential rotation; it must rotate so that the angular velocity is the same everywhere.

diffraction grating A metal or glass surface on which a large number of equidistant parallel lines have been ruled, typically at 100 to 1000 per millimeter. Light striking a grating is dispersed by diffraction into a high-quality spectrum, which is reflected by a metal grating and transmitted by a glass one. *See* échelle grating.

dipole Magnetic fields that include both a north and a south magnetic pole.

disk The visible part of the Sun.

D layer The lowest part of the Earth's daytime ionosphere at heights between about 50 and 90 km. This is the layer that reflects radio waves at frequencies less than about 30 MHz or wavelengths longer than about 10 m.

dM star A red-dwarf star at the lower end of the main sequence of the Hertzsprung–Russell diagram. A dMe star is a red dwarf with hydrogen lines in emission. *See* dwarf star, flare star, M star, and red-dwarf star.

Doppler effect The change in the observed wavelength or frequency of sound or electromagnetic radiation due to the relative motion between the observer and the emitter along the observer's line of sight. The change is to longer wavelengths when the source of waves and the observer are moving away from each other, and to shorter wavelengths when they are moving toward each other. The Doppler effect produces the change in pitch of a siren as an ambulance speeds past. In astronomy, the Doppler effect is used to detect and measure relative motion along the line of sight, determining the radial velocity from the Doppler shift in wavelength. The effect is named after the Austrian physicist Christian Doppler who first described it in 1842. *See* blueshift, Doppler shift, radial velocity, and redshift.

Doppler shift A change in the wavelength or frequency of the radiation received from a source due to its relative motion along the line of sight. A Doppler shift in the spectrum of an astronomical object is commonly described as a redshift when it is towards longer wavelengths (object receding) and as a blueshift when it is towards shorter wavelengths (object approaching). The redshifts of distant galaxies provided the first evidence of the expanding Universe. The Doppler shift makes it possible to determine the radial velocity. *See* blueshift, Doppler effect, radial velocity, and redshift.

DSN Acronym for the Deep Space Network of antennas to receive satellite signals.

dust Small solid particles found in interplanetary and interstellar space, with sizes of between 10^{-6} and 5×10^{-7} m that are comparable to the wavelength of visible light. Interstellar dust reflects and reddens visible starlight, and absorbs energetic ultraviolet light that would otherwise destroy interstellar molecules. Interplanetary dust reflects sunlight, accounting for the F corona. *See* F corona, and zodiacal light.

DUST The cosmic dust instrument on *Ulysses*.

dwarf star The most common type of star in the Galaxy, constituting 90 percent of its stars and 60 percent of its mass. A star on the main sequence of the Hertzsprung-Russell diagram, not to be confused with a much smaller white-dwarf star. The Sun is a typical dwarf star. *See* Hertzsprung–Russell diagram, main sequence and white-dwarf star.

dynamo A mechanism for the cyclic generation of magnetic fields through the interaction of the convection of conducting matter with rotation and rotational shear, resulting from convective and global rotational motions. A dynamo converts the kinetic energy of a moving electrical conductor to the energy of electric currents and a magnetic field. It uses the motion of a conducting, convecting fluid to generate or sustain a magnetic field. The terrestrial magnetic field is supposed to be generated by such a dynamo, located in the Earth's molten core. The Sun's magnetic field is also sustained by an internal dynamo, perhaps located at a region of rotational shear just below the convective zone.

eccentricity Measure of the departure of an orbit from a perfect circle. A circular orbit has an eccentricity $e = 0$; an elliptical orbit has an eccentricity less than 1 and greater than 0 $(0 < e < 1)$; a parabolic orbit has $e = 1$; and a hyperbolic orbit has e greater than 1.

échelle grating A diffraction grating with lines ruled to produce spectra of high dispersion and detail over a narrow band of wavelengths at angles of illumination greater than 45 degrees. *See* diffraction grating.

eclipse The partial or total obscuration of the light from a celestial body as it passes behind another body. In a solar eclipse, the Moon passes between the Sun and the Earth, blocking the Sun. In a lunar eclipse, the Earth passes between the Moon and the Sun, blocking the solar light that the Moon reflects. The eclipse of a star by the Moon or by a planet or other body in the solar system is called an occultation. A total solar eclipse can occur only at the new Moon, and it has a maximum duration of 7.5 minutes. During the brief moments of a total solar eclipse, darkness falls, and the outer parts of the Sun, the chromosphere and the corona, are seen. At any given point on Earth's surface, a total solar eclipse occurs, on the average, once every 360 years. *See* solar eclipse.

ecliptic The projection of the Earth's orbit around the Sun on the celestial sphere, marking the Sun's apparent yearly path against the background stars. The ecliptic plane is the plane of the Earth's orbit around the Sun. It is inclined to the plane of the celestial equator by about 23.5 degrees, and intersects the celestial equator at the two equinoxes. The planets, most asteroids, and most of the short-period comets are in orbits with small or moderate inclinations relative to the ecliptic plane. *See* zodiac.

effective temperature The temperature an object would have if it was emitting black-body radiation at the same energy and at the same wavelengths as the object. The effective temperature is denoted by the symbol T_e, and the effective temperature of the Sun is equal to 5780 K. A star's effective temperature is a good approximation to the actual temperature of its visible disk. *See* black-body radiation.

EIT Acronym for the Extreme-ultraviolet Imaging Telescope on *SOHO*.

E layer A daytime layer of the Earth's ionosphere roughly between altitudes of 95 to 140 km.

electromagnetic radiation Radiation that carries energy through vacuous space, moving through it in periodic waves at the speed of light. The velocity of light, usually designated by the letter c, has a value of $2.997\ 924\ 58 \times 10^8$ m s^{-1}. Electromagnetic radiation originates when charged atomic particles are accelerated, and it propagates by the interplay of oscillating electrical and magnetic fields. Electromagnetic radiation, in common with any wave, has a wavelength, denoted by λ, and a frequency denoted by ν; their product is equal to the velocity of light, or $\lambda\nu = c$, so the wavelength decreases when the frequency increases and *vice versa*. The energy associated with the radiation increases in direct proportion to frequency, and the photon energy, denoted by E, is given by $E = h\nu = hc/\lambda$, where Planck's constant $h = 6.6261 \times 10^{-34}$ J s. There is a continuum of electromagnetic radiation – from long-wavelength radio waves of low frequency and low energy, through visible light waves, to short-wavelength X-rays and gamma-rays of high frequency and high energy. *See* electromagnetic spectrum, gamma-ray radiation, radio radiation, ultraviolet radiation, visible radiation, and X-ray radiation.

electromagnetic spectrum The entire range of all the various kinds or wavelengths of radiation including gamma rays, X-rays, ultraviolet radiation, and radio waves. Light (or the visible spectrum) comprises just one small segment of this much broader spectrum. *See* gamma-ray radiation, optical spectrum, radio radiation, ultraviolet radiation, visible radiation, and X-ray radiation.

electron A negatively charged sub-atomic particle that orbits the nucleus of an atom, but can exist in isolation outside an atom. Free electrons have broken away from their atomic bonds and are not bound to atoms. At the hot temperatures inside the Sun, and in the solar corona and solar wind, the electrons have been set free of their atomic bonds. The mass of an electron is 9.1094×10^{-31} kg, and the charge of an electron is -1.6022×10^{-19} C.

electron neutrino The type of neutrino that interacts with the electron. It is the only kind of neutrino produced by the nuclear reactions that make the Sun shine. *See* neutrino and solar neutrino problem.

electron volt A unit of energy, denoted by eV, often used for measuring the energies of particles and electromagnetic radiation. An electron volt is defined as the energy acquired by an electron when it is accelerated through a potential difference of 1 volt in a vacuum. 1 eV = 1.6022×10^{-19} J. A photon with an energy of 1 eV has a wavelength of 1240 nm and a frequency of 2.42×10^{14} Hz. The energies of X-rays are expressed in thousands of electron volts (keV). Millions (MeV) and billions (GeV) of electron volts are used as units for very energetic charged particles, such as cosmic rays. An electron with a kinetic energy of a few MeV is traveling at almost the velocity of light.

element A chemically homogeneous substance consisting of atoms of all the same atomic number, that cannot be decomposed by chemical means. There are 92 naturally occurring elements, with the number of protons ranging from 1 for hydrogen to 92 for uranium. The atoms of a particular element all have the same number of protons in their nucleus, but the number of neutrons can vary, giving rise to different isotopes of the same element. *See* atomic number, and isotope.

ellipse A closed curve in which the sum of the distances of any point on it from two points within it, the foci, is a constant. The closed orbits of planets, satellites and comets can be described by ellipses.

elliptical galaxy A galaxy of ellipsoidal form. *See* galaxy

emerging flux region An area on the photosphere where new magnetic flux is erupting.

emission line A bright spectral feature at a particular wavelength or narrow band of wavelengths, emitted directly by a hot gas and revealing by its wavelength a chemical constituent of that gas.

emission nebula A cloud of gas in space that emits light. Ultraviolet radiation from nearby hot stars ionizes the gas, and the ions emit visible light when they recombine with free electrons. Interstellar gas is mostly hydrogen, and emission nebulae are therefore often called ionized hydrogen regions, denoted as H II regions. *See* H II region, and Orion nebula.

emission spectrum A series of bright emission lines. *See* emission line.

energy The ability to do work.

EPAC Acronym for the Energetic PArticle Composition instrument on *Ulysses*.

ephemeral region A short-lived, tiny active region. *See* active region.

equator An imaginary line around the center of a body where every point on the line is an equal distance from the poles. The equator defines the boundary between the northern and southern hemispheres.

EQUATOR-S A low-cost mission, launched on 2 December 1997, that studies the Earth's equatorial magnetosphere out to distances of 67,000 km.

erg A unit of energy equal to the work done by a force of 1 dyne acting over a distance of 1 centimeter. An energy of one joule is equal to 10^7, or 10 million, ergs.

ERNE Acronym for the Energetic and Relativistic Nuclei and Electron experiment on *SOHO*.

eruptive prominence *See* prominence.

ESA Acronym for the European Space Agency.

escape velocity The minimum upward speed that a small object must attain to overcome the gravitational attraction of a larger body and leave on a trajectory that does not bring it back again. The escape velocity is denoted by V_{esc}. It depends on the mass, denoted by M, and the radius, specified as R, of the larger body, and is given by $V_{esc} = [2GM/R]^{1/2}$, where the gravitational constant G = 6.6726×10^{-11} N m^2 kg^{-2}.

EUV Acronym for Extreme Ultra-Violet, a portion of the electromagnetic spectrum from approximately 10^{-8} to 10^{-7} m, or 100 to 1000 Å.

eV Abbreviation of electron volt. *See* electron volt.

Evershed effect The radial flow of gases within the penumbra of a sunspot, discovered in 1908 by the English astronomer John Evershed from the Doppler shift in the spectrum of a sunspot. The flow is outward at the level of the photosphere, but inward and downward at higher levels in the solar atmosphere.

excitation An atomic process in which an atom or ion is raised to a higher-energy state by taking one of its electrons from one orbit to another further out from the nucleus of the atom.

expanding Universe The radial motion of galaxies away from us and each other, as discovered in 1929 by the American astronomer Edwin Hubble using the Doppler redshift in their light. The galaxies are receding from us at velocities that are proportional to their distance, a relation known as the Hubble law. The expanding Universe is attributed to the big bang, occurring between 10 and 20 billion years ago. *See* big bang, and Hubble law.

extreme ultraviolet Abbreviated EUV, a portion of the electromagnetic spectrum from approximately 10^{-8} to 10^{-7} m, or 100 to 1000 Å.

faculae Bright regions of the photosphere, associated with sunspots, seen in white light and visible only near the limb of the Sun. They are brighter than the surrounding medium due to their higher temperatures and greater densities. Faculae appear some hours before the associated sunspots, in the same place, but can remain for months after the sunspots have gone. The word *facula* is Latin for "little torch".

Fahrenheit A unit of temperature denoted by °F for degrees Fahrenheit. The freezing temperature of water is 32 °F and the boiling temperature of water is 212 °F. The equivalent temperature in degrees kelvin is $K = (5/9)°F + 255.22$, and the equivalent temperature in degrees Celsius is $°C = 5(°F - 32)/9$. *See* Celsius, and kelvin.

faint-young-Sun paradox The discrepancy between the Earth's warm climate record and a dimmer Sun billions of years ago. Geological evidence indicates the presence of liquid water on Earth from 3.8 billion years ago. Yet, the Sun began shining 4.5 billion years ago with about 70 percent of the brightness it has today; assuming an unchanging atmosphere comparable to that of today, the oceans would have been frozen more than 2 billion years ago.

FAST Acronym for the *Fast Auroral SnapshoT* explorer, launched on 21 August 1996 into a polar orbit to study the detailed plasma physics of the Earth's auroral regions, and to measure the magnetic and electric fields in the upper atmosphere and the mass, charge, and velocity of particles.

F corona One of the components of the white-light corona caused by sunlight scattered from solid dust particles in interplanetary space. The "F" stands for Fraunhofer, and the spectrum of the F corona is that of the Sun including Fraunhofer lines. The F corona extends from about two or three solar radii to far beyond the Earth, and it can also be seen as the zodiacal light. *See* corona, K corona, white light, and zodiacal light.

fermion A sub-atomic particle described by an anti-symmetric wave function and satisfying Fermi–Dirac statistics.

FGM: Acronym for the scaler Flux Gate Magnetometer on *Ulysses*.

fibrils Dark elongated features seen in hydrogen-alpha spectroheliograms of the chromosphere, occurring near sunspots, plages or filament channels. Fibrils form a linear pattern thought to delineate the chromospheric magnetic field.

filament channel A broad pattern of fibrils in the chromosphere, marking the place that a filament may soon form or where a filament recently disappeared.

filaments Masses of relatively cool and dense material suspended above the photosphere in the low corona by magnetic fields, generally along a magnetic inversion, or neutral, line separating regions of opposite magnetic polarity in the underlying photosphere. Filaments appear as dark, elongated features when observed on the solar disk in the light of certain spectral lines, particularly hydrogen alpha and those of ionized calcium. A filament above the limb of the Sun, seen in emission against the dark sky, is called a prominence; a dark filament is a prominence seen in projection against the bright solar disk.

filigree Fine structure visible in the photosphere, collectively forming the photospheric network.

FIP effect FIP is an acronym for the First Ionization Potential. The FIP effect operates in the solar chromosphere, enhancing the abundance of elements with a low first ionization potential in the corona and solar wind relative to the high-FIP elements. *See* ionization potential.

first ionization potential *See* FIP effect, and ionization potential.

fission The splitting, or breaking apart, of a heavy nucleus into two or more lighter nuclei. Fission is spontaneous during the decay of a radioactive element, and it can be induced by particle bombardment.

five-minute oscillations Vertical oscillations of the solar photosphere with a well-defined period of five minutes, usually interpreted in terms of trapped sound waves. The five-minute oscillations are detected in the Doppler effect of solar absorption lines formed in the photosphere and in light variations of the photosphere. *See* acoustic waves, Doppler effect, helioseismology, and photosphere.

flare A sudden and violent release of matter and energy within a solar active region in the form of electromagnetic radiation, energetic particles, wave motions and shock waves, lasting minutes to hours. The frequency and intensity of solar flares increase near the maximum of the solar activity cycle. Solar flares accelerate charged particles into interplanetary space. The impulsive or flash phase of flares usually lasts for a few minutes, during which matter can reach temperatures of tens of millions of degrees kelvin. The flare subsequently fades during the gradual or decay phase lasting about an hour. Most of the radiation is emitted as X-rays but flares are also observed at visible hydrogen-alpha wavelengths and radio wavelengths. They are probably caused by the sudden release of large amounts (up to 10^{25} J) of magnetic energy in a relatively small volume in the solar corona. *See* active region, burst, gradual flare, impulsive flare, solar flare, type I, II, III, and IV radio burst, and X-ray flare class.

flare star A variable star, usually a red dwarf, whose optical and radio luminosity can unpredictably increase in less than a second, decreasing to its normal value in about one minute to one hour. See dM star, M star, red-dwarf star, and stellar activity.

flash spectrum The emission-line spectrum of the solar chromosphere that is seen for a few seconds just before and after totality during an eclipse of the Sun.

F layer The upper layer of the ionosphere, approximately 140 to 1500 km in height. The F2 region at heights between 200 and 600 km, is the densest part of the F layer; the F1 region has a lower peak electron density and forms at lower heights in the daytime.

flux The rate of flow through a reference surface; the flux density is the flux measured per unit area.

focal length The distance between a lens or curved mirror and its focus.

focal ratio The ratio of the focal length of a lens or curved mirror to its diameter. A focal ratio of 8 is written as f/8.

focus The point at which is formed the image of a distant source lying on the axis of a lens or mirror.

footpoint Intersection of a magnetic loop with the photosphere. The lowest visible portion of the magnetic loop. See coronal loop.

forbidden lines Emission lines not normally observed under laboratory conditions because they have a low probability of occurrence, often resulting from a transition between a metastable excited state and the ground state. Under typical conditions on Earth, an atom in a metastable state will lose energy through a collision before it is able to decay to the ground state and emit line radiation. Under astrophysical conditions, including the solar corona and highly-rarefied nebulae, the metastable state can last long enough for the "forbidden" lines to be emitted. See ground state.

Forbush decrease An abrupt decrease in the background cosmic-ray intensity observed at the Earth, by at least 10 percent, a day or two after an explosive outburst on the Sun, recovering over a period of days and weeks. This phenomenon is named after Scott Forbush who first noted it in 1954.

Fraunhofer lines The dark absorption features in the solar spectrum, caused by absorption at specific wavelengths in the cooler layers of the Sun's atmosphere, including the photosphere and chromosphere. Although they were first observed in 1802 by William Hyde Wollaston, they were first carefully studied from 1814 by Joseph von Fraunhofer, who also labeled some of the most prominent

with letters of the alphabet. Some of these identifying letters are still commonly used, notably the sodium D lines and the calcium H and K lines. The most prominent ones at visible wavelengths are caused by the presence of neutral hydrogen, the hydrogen-alpha line designated as C by Fraunhofer, and singly-ionized calcium, sodium, and magnesium. See absorption line, hydrogen-alpha line, H and K lines, and spectrum.

free electron An electron that has broken free of its atomic bond and is therefore not bound to an atom.

freezing in The decoupling of ion charge-states from electron collision equilibrium near the coronal temperature maximum, due to decreasing electron density. Charge states remain unaltered in, or frozen in, the solar wind and thus may be used as coronal thermometers.

frequency The number of crests of a wave passing a fixed point each second usually measured in units of Hertz (one oscillation per second). See hertz, Hz, and MHz.

fusion The joining of two or more lighter nuclei to produce a heavier nucleus, releasing energy as the result. Fusion powers hydrogen bombs and makes stars shine. See thermonuclear fusion and hydrogen burning.

galactic cosmic rays Cosmic-ray particles detected in interplanetary space, entering the Earth's atmosphere, or at the Earth's surface, originating from outside our solar system but from within our Galaxy. See cosmic rays, and solar cosmic rays.

galaxy A collection of millions to hundreds of billions of stars and varying amounts of interstellar gas and dust, usually isolated from other galaxies. All of the galaxies are participating in the expansion of the Universe. The Galaxy that we reside in is designated with a capital G to distinguish it from all of the other galaxies. Our galaxy is spiral in shape, and about 100 000 light-years, or 30 kiloparsec (kpc) in diameter. The Sun is located at the edge of one of the spiral arms, about 27 700 light-years from the center of our Galaxy. The distances between galaxies are usually measured in units of megaparsec, abbreviated Mpc. The mean space density of galaxies is roughly 0.055 galaxies per cubic megaparsec (Mpc^{-3}). Elliptical galaxies are spherical or ellipsoidal systems, while spiral galaxies are flattened disk-shaped systems in which young stars and interstellar gas and dust are concentrated in spiral arms coiling out from a central bulge or nucleus. Galaxies can exist singly, in small groups, or in rich clusters of thousands of galaxies. See cluster of galaxies, elliptical galaxy, and spiral galaxy.

GALLEX Acronym for GALLium EXperiment, started in 1991 within the Gran Sasso Underground Laboratories below a peak in the Apennine mountains of Italy, using 30 tons (3×10^4 kg) of gallium and 1000 tons (10^6 kg) of gallium chloride solution. The experiment measures solar neutrinos of low energy from the proton–proton reaction.

gamma A unit of magnetic field strength equal to 10^{-5} gauss, 10^{-9} tesla, and 1 nanotesla.

gamma-ray radiation The most energetic form of electromagnetic radiation, with the highest frequency and the shortest wavelength. Gamma rays have photon energies in excess of 100 keV or 0.1 MeV and wavelengths less than 10^{-11} m. Because Earth's atmosphere absorbs the radiation at that end of the spectrum, gamma-ray studies of the Sun are conducted from space; they can be observed during very energetic solar flares.

GAS Acronym for the interstellar neutral GAS experiment on *Ulysses*.

gas pressure The outward pressure caused by the motions of gas particles and increasing with their temperature. Gas pressure supports main-sequence stars like the Sun against the inward force of their immense gravity.

gauss The c.g.s. (centimeter–gram–second) unit of magnetic field strength, named after Karl Freidrich Gauss who showed in 1838 that the Earth's dipolar magnetic field must originate inside the planet. The SI (Systéme International) unit of magnetic field strength is the tesla (T), where $1\,T = 10\,000\,T = 10^4\,T$. *See* gamma, and tesla.

Gaussian distribution *See* Maxwellian distribution.

geocorona The outermost part of the Earth's atmosphere, consisting of a halo of hydrogen gas that extends out about 15 Earth radii.

geomagnetic activity Disturbances in the magnetized plasma of a magnetosphere associated with fluctuations of the surface field, auroral activity, reconfiguration and changing flows within the magnetosphere, strong ionospheric currents, and particle precipitation into the ionosphere.

geomagnetic field The magnetic field in and around the Earth. The intensity of the magnetic field at the Earth's surface is approximately 3.2×10^{-5} T, or 0.32 G, at the equator and 6.2×10^{-5} T, or 0.62 G, at the North Pole. To a first approximation, the Earth's magnetic field is like that of a bar magnet (dipole) currently displaced about 500 km from the center of the Earth towards the Pacific Ocean and tilted at 11 degrees to the rotation axis.

geomagnetic index *See* A index, Ap index, and K index.

geomagnetic storm A rapid, world-wide disturbance in the Earth's magnetic field, typically of a few hours duration, due to the passage of a high-speed stream in the solar wind or caused by the arrival in the vicinity of the Earth of a coronal mass ejection. A substorm is a magnetic disturbance observed in polar regions only, and is associated with changes in direction of the interplanetary magnetic field as it encounters the Earth. Auroral activity and disruption of radio communications are common during geomagnetic storms. *See* Ap index, magnetic storm, and substorm.

geosynchronous orbit The orbit of a satellite that travels above the Earth's equator from west to east at an altitude where the satellite's orbital velocity is equal to the rotational velocity of the Earth. In this orbit, a satellite remains nearly stationary above a particular point on the planet. Such an orbit has a height of about 35.9 million meters, or 5.6 Earth radii.

GEOTAIL A joint project of ISAS and NASA launched on 24 July 1992 to measure the global energy flow and transformation in the Earth's magnetotail. *GEOTAIL* is a participating mission of the ISTP program.

giant star A star that is larger than a dwaf (main-sequence) star of the same temperature and generally more luminous than dwarf stars. Giant stars evolve from dwarf stars that have used up the hydrogen fuel in their cores. See dwarf star, Hertzsprung–Russell diagram, and main sequence.

Gleissberg period A period of approximately 80 years in the number of sunspots, the frequency of terrestrial auroras, and the terrestrial radiocarbon variability, named after Wolfgang Gleissberg who first noted the effect in the sunspot records in 1958.

globular cluster A densely packed, spherical star cluster, with ten thousand to a million stars and a low abundance of heavy elements. These star clusters are at least ten billion years old. They are found in the spherical galactic halo, distributed about the galactic center at distances as far out as 50 kiloparsecs (kpc). Our Galaxy has about 140 known globular clusters, whereas the largest elliptical galaxies may have thousands. *See* Population II star.

g-modes Gravity oscillation mode of solar acoustic waves. Buoyancy forces serve as the restoring forces that can lead to this mode of oscillation. The g-modes are concentrated near the center of the Sun. The g-modes have been theoretically predicted, but they have not yet been observed.

GOES 1, 2, ..., 6, 7, 8, 9 Acronym for a series of *Geostationary Operational Environmental Satellites* numbered 1 through 9 in order of launch. These satellites have provided, and continue to provide, information about terrestrial cloud cover, temperatures, water-vapor content, and other meteorological data relayed from central weather facilities to regional stations, as well as space environment monitoring systems to measure proton, electron and solar X-ray fluxes and magnetic fields.

GOLF Acronym for the Global Oscillations at Low Frequency instrument on *SOHO*.

GONG Acronym for the Global Oscillation Network Group of six ground-based telescopes that continuously monitor solar oscillations.

gradual flare A relatively rare type of solar flare that can last for several days and accelerate high-energy protons. *See* impulsive flare, flare, and solar flare.

granulation A mottled, cellular pattern visible at high spatial resolution in the white light of the photosphere, due to hot gases rising from the Sun's interior by convection. *See* granule, and supergranulation cells.

granule One of about a million bright regions, or cells, that cover the visible solar disk at any instant, and that comprise the granulation. The bright center of each granule corresponds to hot gases rising to the photosphere, and its dark edges are the site of cooled gases that are descending toward the interior. Individual granules are often polygon-shaped regions, with an average cell size of about one million meters, or 10^6 m. They appear and disappear on time-scales of about ten minutes.

gravity The universal force of attraction between all particles of matter. According to Newton's law of gravitation, the force of attraction, denoted by F, between two masses, denoted M_1 and M_2, and separated by distance, D, is given by $F = GM_1M_2/D^2$, where the gravitational constant $G = 6.6726 \times 10^{-11}$ N m^2 kg^{-2}. Thus, the gravitational force of attraction increases with mass and falls off as the inverse square of the separation. In his General Theory of Relativity, Einstein modified Newton's law to include the space curvature near a large mass. *See* curvature of space.

GRB Acronym for the solar X-ray and cosmic Gamma-Ray Bursts instrument on *Ulysses*.

greenhouse effect The warming of a planet's surface by heat trapped by its atmosphere, owing to its infrared opacity. Incoming sunlight is absorbed at the surface and re-radiated at longer wavelengths, as infrared heat radiation. If the atmosphere contains greenhouse gases, such as carbon dioxide or water vapor, which are not transparent to the infrared, the heat is reflected back to the ground. The natural greenhouse effect warms the surface of the Earth to an average temperature of about 15 degrees Celsius, or 288 K; without an atmosphere the Earth is heated to only −18 degrees Celsius which is well below the freezing point of water. How much heating the greenhouse effect causes depends on how opaque the atmosphere is to infrared radiation. On Venus, the dense carbon dioxide atmosphere has raised the surface temperature to around 750 K, hot enough to melt lead. Concern has mounted that global warming of the Earth will result from increased concentrations of carbon dioxide and other so-called "unnatural" greenhouse gases released by human activity, particularly the burning of fossil fuels such as coal and oil.

green line *See* coronal green line.

ground state The lowest-energy state of the electron in its orbital motion around the nucleus of an atom.

GSFC Acronym for the Goddard Space Flight Center of NASA.

G star A star of spectral type G, with an effective temperature of between 5000 and 6050 K if it is on the main sequence, and between 4600 and 5600 K if it is among the giants. The Sun is a G-type star. The G-type main-sequence, or dwarf, stars have a mass between 0.8 and 1.1 solar masses. *See* spectral classification.

GWE Acronym for the Gravitational Wave Experiment on *Ulysses*.

Gyr Abbreviation for gigayear, equal to a billion years or to 10^9 years.

gyrofrequency Frequency of the circular motion of a charged particle perpendicular to the magnetic field.

gyroradius Radius of the circular orbit of a charged particle gyrating in a magnetic field and traced by its component of motion perpendicular to a magnetic field.

Hale's law The leading, or westernmost, spots of any sunspot group in the northern hemisphere of the Sun have the same magnetic polarity, while the following, or easternmost, spots have the opposite magnetic polarity. In the southern hemisphere, the leading and trailing sunspots of any sunspot group also exhibit opposite polarities, but the direction of any bipolar group in the southern hemisphere is the reverse of those in the northern hemisphere. All of the spots' magnetic polarities reverse each 11-year solar activity cycle, and return to their original polarity every 22 years.

H alpha (Hα) *See* hydrogen-alpha line.

H and K lines The strongest lines in the visible spectrum of ionized calcium, Ca II, lying in the violet at wavelengths of 393.4 (K) and 396.8 (H) nm. They are conspicuous features in the spectra of many stars, including the Sun. The designations H and K were given by Fraunhofer and are still commonly used. *See* Fraunhofer lines, and spectroheliograph.

hard X-rays Electromagnetic radiation with photon energies of between 10 keV and 100 keV and wavelengths between about 10^{-10} and 10^{-11} m.

heavy water A form of water, or H_2O, in which the abundant, light, hydrogen atoms, H, are replaced by less-abundant, deuterium atoms, a heavy form of hydrogen denoted by D, to form D_2O; here O denotes an atom of oxygen.

helicity A measure of the "twist" an object has, such as the degree of coiling of a magnetic field.

heliocentric Centered on the Sun.

heliographic latitude The angle in degrees from the solar equator to an object as measured along a great circle passing through the poles of the Sun.

heliographic longitude The angle in degrees measured along the solar equator from the central meridian of the Sun to the foot of the great circle that passes through an object and the poles of the Sun. The heliographic longitude is measured positive from 0 to 360 degrees, towards the west in the direction of rotation.

heliopause The outer boundary or edge of the heliosphere, marking the interface of the solar wind with the interstellar medium. The pressure of the solar wind equals that of the interstellar medium at the heliopause. This occurs at a distance of about 100 AU from the Sun, or far beyond the orbits of the known outer planets. The exact position of the heliopause depends both on the strength of the solar wind and on the properties of the local interstellar medium. *Ulysses* data, taken during solar activity minimum, indicate that the solar wind from the Sun's poles has a higher speed than the speed in the ecliptic, so the heliopause at a solar activity minimum should be further from the Sun in the polar direction. *See* termination shock.

helioseismology The study of the interior of the Sun by the analysis of sound waves that propagate through the solar interior and manifest themselves by oscillations at the photosphere. These oscillations have periods of around five minutes and are observed spectroscopically as Doppler shifts in the absorption line spectrum. Helioseismology is a hybrid name combining the Greek words *helios* for "the Sun" and *seismos* for "tremor" or "quake".

heliosphere A vast region, cavity, or bubble carved out of interstellar space by the solar wind. The region of interstellar space surrounding the Sun where the Sun's magnetic field and the charged particles of the solar wind control plasma processes. The heliosphere is immersed in the local interstellar medium, and defines the extent of the Sun's influence. The heliosphere contains our solar system, and extends well beyond the planets to the heliopause, an outer boundary that marks the place where the solar-wind pressure balances that in interstellar space. *See* heliopause.

heliospheric current sheet A thin sheet of current that forms at the interface between oppositely-directed magnetic fields in the solar wind from the north and south solar hemispheres. *See* current sheet.

heliostat A moveable flat mirror used to reflect sunlight into a fixed solar telescope.

heliotail The elongated tail of the heliosphere extending downwind of the approaching interstellar gas formed by the solar wind escaping from the heliosphere.

helium After hydrogen, the second most abundant element in the Sun and the Universe. The nucleus of a helium atom is called an alpha particle; it contains two protons and two neutrons. Helium was first observed in the Sun, and subsequently found on Earth. Most of the helium in the Universe was made in the immediate aftermath of the big-bang explosion that gave rise to the expanding Universe, but helium is also now being synthesized from hydrogen in the cores of main-sequence stars. Helium is one of the rare, inert noble gases. *See* big bang, carbon–nitrogen–oxygen cycle, noble gases, nucleosynthesis, and proton–proton chain.

helmet streamer Named after spiked helmets once common in Europe, helmet streamers form in the low corona over the magnetic inversion, or neutral, lines in large active regions with a prominence commonly embedded in the base of the streamer. The footpoints of the streamer are in regions of opposite magnetic polarity and the streamer itself straddles the prominence. Higher up, the magnetic field is drawn out into interplanetary space within long, narrow stalks. Gas flowing out along these open magnetic fields might help to create the slow-speed solar wind. *See* coronal streamer.

hertz A unit of frequency equal to one cycle per second, abbreviated Hz. One kHz is one thousand Hz, 1 MHz is a million Hz, and 1 GHz is one thousand million Hz.

Hertzsprung–Russell diagram A plot of the absolute magnitude or luminosity of stars against their spectral type, color index or effective temperature, abbreviated H–R diagram. Brightness increases from bottom to top, and temperature increases from right to left. Most stars, including the Sun, lie on the main sequence, which extends from the upper left to the lower right in the H–R diagram. Such stars are called dwarf stars. Another region of giant stars lies in a luminous band above the main sequence. The white-dwarf stars occupy the bottom left of the H–R diagram; these tiny stars are comparable to the Earth in size. An early version of this diagram was presented by the American astronomer Henry Norris Russell in 1913; Ejnar Hertzsprung previously plotted such diagrams for the Pleiades and Hyades star clusters in 1911. *See* dwarf star, giant star, main sequence, and white-dwarf star.

HESSI Acronym for the *High-Energy Solar Spectroscopic Imager*, a future NASA mission to provide high-resolution images and spectroscopy of energetic solar flares at hard X-ray and gamma-ray wavelengths. *HESSI* will study the basic physics of particle acceleration and explosive energy release in solar flares.

high-speed stream A stream within the solar wind having speeds of up to 8×10^5 m s^{-1}, thought to originate mainly from coronal holes.

HIPPARCOS Acronym for the HIgh Precision PARallax COllecting *Satellite*, launched by the European Space Agency, in August 1989. It operated until August 1993, measuring the parallax and proper motions of 120 000 stars with a precision of 0.001 to 0.002 seconds of arc, up to 100 times better than possible from the ground. The acronym also alludes to Hipparchus of Nicea, who recorded accurate star positions more than two millennia previously. *See* parallax, and proper motion.

H II region A cloud of ionized hydrogen in interstellar space, pronounced H-two region. It is produced by ultraviolet radiation of nearby hot stars. *See* emission nebula, and Orion nebula.

H I region A cloud of neutral, or un-ionized, atomic hydrogen gas in interstellar space, pronounced H-one region. Such clouds do not emit visible light, and hence are invisible at optical wavelengths, but they are detected at radio wavelengths through the 21-cm line emission of hydrogen atoms.

HISCALE Acronym for the Heliosphere Instrument for Spectra, Composition, and Anisotropy at Low Energies on *Ulysses*.

HOMESTAKE The South Dakota gold mine where the chlorine neutrino detector is located. *See* chlorine experiment.

homologous flares Solar flares that occur repetitively in the same active region, with essentially the same position and pattern of development. *See* flares.

H–R diagram: Acronym for Hertzsprung–Russell diagram. *See* Hertzsprung–Russell diagram.

Hubble constant The expansion rate of the Universe, denoted by the symbol H_0 where the zero subscript denotes the value at the present time. The rate at which the velocity of recession of galaxies increases with distance. The Hubble constant lies between 50 and 100 kilometers per second per megaparsec ($km\ s^{-1}\ Mpc^{-1}$). *See* Hubble law, and Hubble time.

Hubble law A relationship that states that the farther away from us a galaxy is, the faster it is receding from us. The linear increase of galaxy recession velocity, denoted by V_r, with galaxy distance, specified by D, according to the relation $V_r = H_0D$, where H_0 is the Hubble constant. Although it was not specifically stated by Edwin Hubble, this relation is now known as the Hubble law. *See* Hubble constant.

Hubble Space Telescope Abbreviated HST, the largest astronomical telescope ever put into space, launched by the *Space Shuttle Discovery* in April 1990, and installed with corrective optics in December 1993. The main mirror has a diameter of 2.4 m.

Hubble time The reciprocal of the Hubble constant, denoted by $1/H_0$, which gives a maximum age of the Universe on the assumption that there has been no slowing of the expansion of the Universe. The Hubble time lies between 10 and 20 billion years, and it denotes the time since the big-bang origin of the expanding Universe. *See* Hubble constant.

HXT Acronym for the Hard X-ray Telescope on *Yohkoh*.

hydrogen The lightest, simplest element. The most-abundant element in the Sun and in the Universe. A hydrogen atom consists of one proton circled by one electron. Most of the mass of a hydrogen atom is in its nucleus, the proton, whose mass is 1.6726×10^{-27} kg or 1836 times the mass of an electron. Hydrogen is so light that it is found in only small amounts in Earth's atmosphere. *See* electron, and proton.

hydrogen-alpha line The spectral line of neutral hydrogen in the red part of the visible spectrum, denoted by $H\alpha$. Light emitted or absorbed at a wavelength of 656.3 nm during an atomic transition in hydrogen, the lowest energy transition in its Balmer series. It is the dominant emission from the solar chromosphere. The wavelength also corresponds with a dark line produced by hydrogen absorption in the photosphere, designated with the letter C by Joseph von Fraunhofer in the early 1800s. *See* chromosphere, Fraunhofer lines, spectral line, spectroheliograph, and surge.

hydrogen burning The thermonuclear process in which a star shines by converting hydrogen nuclei into helium nuclei. Stars spend most of their lifetime burning hydrogen, and such stars are the most common kind of star. A star with the mass of the Sun spends around 12 billion years in this stage; more massive stars spend less time. *See* carbon-nitrogen–oxygen cycle and proton–proton chain.

hydromagnetic wave A wave in which both the plasma and magnetic field oscillate.

hydrostatic equilibrium The condition of stability in an atmosphere or stellar interior that exists when the inward gravitational force of the overlying material is exactly balanced by the outward force of the gas and radiation pressure. The pressure at each level in the gas supports the weight of all the overlying material.

Hz Abbreviation for hertz, a unit of frequency equivalent to cycles per second. *See* hertz.

IAS Acronym for the Institut d'Astrophysique Spatiale, France.

ice age A period of cool, dry climate causing a long-term buildup of extensive ice sheets far from the poles. The major ice ages last for about 100 000 years; they are separated by warmer interglacial periods that last roughly 10 000 years. The major ice ages may be triggered by rhythmic variations in the global distribution of sunlight with periods of 23, 41 and 100 thousand years. See Little Ice Age, and Milankovitch cycles.

IMAGE Acronym for *Images for Magnetopause-to-Aurora Global Explorer* launched by NASA on 25 March 2000 into a polar orbit to study the injection of plasma into the magnetosphere during geomagnetic storms and solar-wind changes.

IMF Acronym for Interplanetary Magnetic Field. *See* interplanetary magnetic field.

IMP-1, 2, ..., 8 Acronym for a series of Earth-orbiting *Interplanetary Monitoring Platforms*. IMP-8, launched on 26 October 1973, was instrumented to study cosmic rays, energetic solar particles, plasma, and electric and magnetic fields.

impulsive flare The most common type of solar flare, lasting minutes to hours. *See* flare, gradual flare, and solar flare.

inclination The angle between an object's orbital plane and a reference plane, usually the ecliptic for objects orbiting the Sun and a planet's equator for satellites. *See* ecliptic.

inflation A hypothetical, rapid expansion of the Universe immediately after the big bang, requiring that the Universe has a mass density equal to the critical density. *See* critical density.

infrared radiation Electromagnetic radiation with wavelengths in the range between the visible red spectrum and radio waves, roughly between 10^{-6} and 10^{-4} m in wavelength. Infrared radiation is invisible to the human eye and absorbed almost completely in the lower layers of the Earth's atmosphere, primarily by water vapor. For this reason, infrared astronomy observations have to be conducted from the highest mountain sites, or from aircraft or satellites.

insolation The amount of radiative energy received from the Sun per unit area per unit time at any given location on the Earth's surface. It is also known as the solar insolation, and has the units of watts per square meter (W m^{-2}). The solar insolation is greatest when the Sun is overhead, amounting to about 1000 W m^{-2}. The energy of solar radiation per unit area per unit time received just outside the Earth's atmosphere, known as the solar constant, is 1366.2 ± 1.0 W m^{-2}. *See* solar constant, and solar insolation.

interferometry The electronic combination of signals from two or more telescopes, producing an angular resolution of a much larger instrument, comparable in size to the separation between the component telescopes. This technique is used to give seconds of arc resolution, or better, at long radio wavelengths.

interplanetary magnetic field The Sun's magnetic field carried into interplanetary space by the expanding solar wind, abbreviated as IMF. The Sun's rotation winds the

IMF into a spiral shape. The IMF has a magnetic field strength of about 6×10^{-9} tesla, or 6 nanotesla, or 6×10^{-5} gauss, at the Earth's orbital distance of about 1 AU from the Sun. The interplanetary magnetic field is carried radially outward by the solar wind, but since the magnetic field originates at the Sun, where the magnetic footpoints are located, solar rotation twists the interplanetary magnetic field into a spiral structure known as an Archimedean spiral.

interplanetary medium The medium or material in the space between the planets in the solar system. It is composed of electrically-charged particles, mainly electrons and protons, ejected from the Sun via the solar wind, dust particles from comets, and neutral gas from the interstellar medium.

interplanetary scintillation Fluctuations in the signal received from a distant radio source observed along a line of sight close to the Sun. The scintillation is caused by irregularities in the solar wind.

interstellar dust Solid particles in interstellar space with a size comparable to the wavelength of light. Interstellar dust absorbs, reddens and reflects the light from distant stars. *See* dust.

interstellar medium The gas and dust in the space between the stars in our Galaxy. It is mainly composed of hydrogen atoms, molecules, and solid dust particles. Most of the hydrogen atoms are in cool, un-ionized form at a temperature of 10 to 100 degrees kelvin, but the hydrogen atoms can be ionized by the ultraviolet light of nearby bright stars. Cold, dense molecular clouds consist mainly of hydrogen molecules, with a mass of up to a million solar masses, but these clouds also contain hundreds of other molecules including ammonia, carbon monoxide, formaldehyde, and water. *See* H I region, H II region, interstellar dust, and molecular cloud.

inverse-square law A reduction in the intensity of radiation in proportion to the square of the distance from its source. Also, a reduction in gravitational attraction by the same factor.

invisible radiation Those kinds of radiation to which the human eye is not sensitive; for example, radio and ultraviolet waves, as well as X-rays and gamma rays.

ion An atom that has gained or lost (more usually) one or more electrons, thus having a net electrical charge. By contrast, a neutral atom has an equal number of negatively-charged electrons and positively charged protons, giving the atom a zero net electrical charge.

ionization The process in which a neutral atom or molecule is given a net electrical charge. The atomic process in which ions are produced by removing an electron from an atom, ion or molecule, typically by collisions with atoms or electrons (collisional ionization), or by interaction with electromagnetic radiation (photo-ionization).

ionization potential The amount of energy required to remove the least tightly bound electron from a neutral atom or molecule is called the first ionization potential, denoted by I, and is usually measured in electron volts, or eV. The additional energy needed to remove the next electron is the second ionization potential, etc.

ionosphere The upper region of a planet's atmosphere in which there are free electrons and ions produced when solar ultraviolet and X-ray radiation ionizes the constituents of the atmosphere. Most of the Earth's ionosphere lies between heights of about 50 and 300 km above the ground, though the extent varies considerably with time, season and solar activity. The D layer, between 50 and 90 km, has low electron density. The E and F layers, at about 95 to 140 and 200 to 300 km respectively, form the main part of the ionosphere. The reflecting power of the ionosphere makes long-range broadcasting and telecommunication possible at radio frequencies up to about 30 MHz. *See* D, E, and F layers.

irradiance The solar irradiance is the amount of solar energy received at the Earth outside its atmosphere per unit area per unit time, with units of watts per square meter *See* solar constant.

ISAS Acronym for the Institute of Space and Astronautical Science, Japan.

isotope One of two or more forms of the same chemical element, whose atoms all have the same number of protons in their nucleus, but a different number of neutrons and therefore a different mass. An isotope is one of two or more different species of the same chemical element, having the same atomic number Z but different mass number A. The term, proposed by Frederick Soddy in 1913, derives from the Greek *isos* for "equal" and *topos* for place, and reflects the fact that such species occupy the same position in the periodic table of the elements. *See* atomic number, and mass number.

ISTP Acronym for the International Solar–Terrestrial Physics program for exploring the Sun–Earth connection. The major participating missions include, in order of their launch date, *GEOTAIL*, *WIND*, *SOHO*, *POLAR*, *IMAGE* and *CLUSTER II*.

JPL Acronym for the Jet Propulsion Laboratory.

KAMIOKANDE A massive underground neutrino detector in Japan filled with water, replaced by the SUPER KAMIOKANDE detector. *See* SUPER KAMIOKANDE.

K corona The electron-scattered component of the coronal white-light intensity from which the electron density can be inferred. The inner part of the white-light solar corona which shines by light from the photosphere scattered by high-energy electrons, extending to about 700 million meters from the photosphere. The K corona emits a continuous spectrum without absorption lines, on which are superimposed the Fraunhofer absorption lines of the F corona. The "K" comes from the German *Kontinuum* or *Kontinuierlich*. The K component is polarized and decreases rapidly in intensity with distance from the Sun. *See* corona, and F corona.

kelvin A unit of absolute temperature abbreviated K for degrees kelvin. It is named for the British physicist William Thomson, Lord Kelvin. Motion at the molecular level stops at zero degrees kelvin. Water freezes at 273 degrees kelvin, and the boiling point of water is 373 degrees kelvin. The equivalent temperature in degrees Celsius is $^{\circ}C = K - 273$, and the equivalent temperature in degrees Fahrenheit is $^{\circ}F = (9/5)K - 459.4$. See absolute temperature, Celsius, and Fahrenheit.

Kepler's laws The three fundamental laws governing the motions of the planets around the Sun, first worked out by Johannes Kepler between 1605 and 1619, and based on observations made by Tycho Brahe. They are: (1) The orbit of each planet is an ellipse with the Sun at one focus. (2) A line from the Sun to a planet sweeps out equal areas in equal time. (3) The square of a planet's orbital period is proportional to the cube of its average distance from the Sun. The second law means that a planet moves at a slightly faster speed when its orbit is closest to the Sun. This law is known as the law of areas. The third law indicates that more distant planets move around the Sun at slower speeds and take longer times to complete each orbit.

keV Abbreviation for kilo-electron volt, or one thousand electron volts. A unit of energy with $1 \text{ keV} = 1.6022 \times 10^{-16}$ J. The wavelength of radiation with a photon energy in keV is 1.24×10^{-9} meters/(energy in keV). *See* electron volt.

kilometer A unit of distance, abbreviated km. It is equal to one thousand meters and to 0.6214 mile.

K index A 3-hourly, quasi-logarithmic local index of geomagnetic activity ranging from 0 to 9; the 3-hourly local a index of geomagnetic activity ranges from 0 to 400.

kinetic energy The energy that an object possesses as a result of its motion.

Kirchoff's law The ratio of the emission and absorption coefficients of a black-body is equal to its brightness.

kpc Abbreviation for kilo-parsec, or one thousand parsecs, a unit of distance with $1 \text{ kpc} = 3.085\ 678 \times 10^{19}$ m.

Kp index A 3-hourly index of geomagnetic activity based on the *K* index from 12 or 13 stations distributed around the world. *See K* index.

Kuiper belt A collection of 1 to 10 billion (10^9 to 10^{10}) or more icy bodies orbiting the Sun in low-eccentricity, low-inclination orbits beyond Neptune, extending out possibly to about 1000 AU.

LANL Acronym for the Los Alamos National Laboratory.

LASCO Acronym for the Large Angle Spectroscopic COronagraph on *SOHO*.

Lagrangian point Also known as the inner Lagrangian point and designated L_1, the point about one one-hundredth of the way from the Earth to the Sun, where the gravitational pull of the Earth and Sun balance in such a way as to give an orbit of exactly one Earth year. A spacecraft at the Lagrangian point will orbit the Sun with the Earth without revolving about the Earth, and therefore never experiences night. The Lagrangian point is located at a distance from the Earth of about 1.5×10^9 m towards the Sun. The *ACE* and *SOHO* spacecraft are both near this point.

latitude *See* heliographic latitude.

leader spot The leader spot is the western, preceding part of a magnetically bipolar or multi-polar sunspot group. Since the Sun rotates from east to west, the leader spot precedes the other members of the local group as the Sun rotates.

lepton Electrons, muons and tau particles, along with their corresponding anti-particles, neutrinos, and anti-neutrinos. Any fermion that does not participate in the strong interaction. Leptons are all significantly less massive than most other elementary particles. *Lepton* is the Greek word for "slender".

light The kind of radiation to which the human eye is sensitive. *See* optical radiation, velocity of light, and visible radiation.

light-year The distance that light travels through a vacuum in one year. One light-year is equivalent to $9.460\ 53 \times 10^{15}$ m, to 63 240 AU units, and to 0.3066 pc. Proxima Centauri is the nearest star, other than the Sun, with a distance of 4.3 light-years. It only takes 8 minutes for light to travel from the Sun to Earth.

limb The apparent edge of a celestial object which is visible as a disk, such as the Sun or the Moon. Astronomers refer to the left edge of the solar disk as the Sun's east limb and to the right edge as its west limb.

limb darkening Since radiation moving in the vertical direction is more intense than radiation moving at an angle to the vertical, the edge of the visible disk of the Sun appears darker than the center of the Sun.

Little Ice Age A prolonged period of unusually cold weather in Europe, from about 1400 to 1800, overlapping two periods of unusually low sunspot numbers called the Maunder Minimum (1645 to 1715) and the Spörer Minimum (1420 to 1570). During the Little Ice Age alpine glaciers expanded, the River Thames and the canals of Venice regularly froze over and Europe experienced unusually harsh winters.

LMSAL Acronym for Lockheed Martin Solar and Astrophysics Laboratory.

longitude *See* heliographic longitude.

Lorentz force The total electromagnetic force on a charged particle in the presence of an electric field due to its motion across a magnetic field.

luminosity The absolute brightness or luminosity of a glowing body, denoted by L. The amount of energy radiated per unit time by an object, measured in units of watts. One joule per second is equal to one watt of power, and to ten million ergs per second. The absolute luminosity of the thermal emission from a black-body is given by the Stefan–Boltzmann law in which $L = 4\pi\sigma R^2 T_e^4$, where $\pi = 3.141\ 59$, the Stefan–Boltzmann constant, $\sigma = 5.670\ 51 \times 10^{-8}$ J m^{-2} K^{-4} s^{-1}, the radius is R and the effective temperature is T_e. The brightness of radiation falls off with the inverse square of the distance, D, so the apparent luminosity $l = L/D^2$. *See* absolute luminosity, inverse-square law, and Stefan–Boltzmann law.

Magellanic clouds Two small, irregular galaxies that are satellites of our Galaxy, the Milky Way. On Earth they are visible only from the equatorial regions and the southern hemisphere.

magnetic braking A process proposed to account for the slow rotation of the Sun and some other stars, in which angular momentum is transferred from the star to the surrounding plasma through its magnetic field. *See* angular momentum.

magnetic dynamo *See* dynamo.

magnetic field A magnetic force-field around the Sun, the planets, and any other magnetized body, generated by electrical currents. The Sun's large-scale magnetic field, like that of Earth, exhibits a north and south pole linked by lines of magnetic force, but the Sun also contains numerous dipolar sunspots linked by magnetic loops.

magnetic field lines Imaginary lines that indicate the strength and direction of a magnetic field. The lines are drawn closer together where the field is stronger. Charged particles move freely along magnetic field lines, but cannot move across them.

magnetic inversion line The line where the observed photospheric longitudinal (line-of-sight) magnetic field is zero, also known as the magnetic neutral line or region.

magnetic network *See* calcium network, chromospheric network, and network.

magnetic polarity *See* polarity.

magnetic pressure A type of pressure inherent in magnetic fields, acting on a plasma and given by B^2/μ_0 where B is the magnetic field strength in tesla and the permeability of free space is $\mu_0 = 1.2566 \times 10^{-6}$ newtons per ampere squared ($N\,A^{-2}$).

magnetic reconnection A change in the topology of the magnetic field where the magnetic field lines re-orient themselves by new connections. A process by which magnetic field lines are broken and then rejoined in a new configuration. This can occur when two oppositely-directed magnetic fields, embedded in plasma, move toward each other and reconnect at the place that they touch. Magnetic reconnection is an important mechanism for releasing magnetic energy to heat the Sun's corona and to power explosive phenomenon on the Sun, such as coronal mass ejections and solar flares.

magnetic storm A disturbance in the Earth's magnetic field observed all over the Earth due to the passage of a high-speed stream and/or a coronal mass ejection in the solar wind. A substorm is a magnetic disturbance observed in polar regions only, and is associated with changes in direction of the interplanetary magnetic field as it encounters the Earth. See geomagnetic activity, and geomagnetic storm.

magnetism One aspect of electromagnetism, a fundamental force of nature, whereby a magnetized object can force a charged particle into a new direction of motion and attract or repel other magnetized objects.

magnetogram A computer image, picture or map of the strength, direction and distribution of magnetic fields across the solar photosphere, based on Zeeman-effect measurements. See magnetograph, and Zeeman effect.

magnetograph An instrument used to map the strength, direction and distribution of magnetic fields across the solar photosphere using the Zeeman effect. Normally, the longitudinal (line-of-sight) magnetic field is measured, but the transverse component is also measured with a vector magnetograph. A magnetogram is a map of the measured magnetic field. See magnetogram, and Zeeman effect.

magnetohydrodynamics The study of the behavior of a plasma in a magnetic field, abbreviated as MHD. A description of a plasma as a single conducting fluid. See Alfvén velocity, Alfvén wave, and plasma.

magnetopause The outer boundary layer of a planetary magnetosphere, where it joins the solar wind. The surface which separates the magnetosphere from the solar wind. At the magnetopause, the magnetic pressure of the planet's internal magnetic field deflects the flow of the solar wind around the magnetosphere. A magnetopause only forms if the planet has a sufficiently strong internal magnetic field.

magnetosheath The turbulent region of solar-wind flow between the bow shock and the planet's magnetopause. Here the plasma and magnetic fields are of solar origin, but have been heated and deflected. See bow shock, magnetopause, magnetosphere, and solar wind.

magnetosphere The region of space surrounding a planet in which the planet's magnetic field predominates over the solar wind, and controls the motions of charged particles in it. The magnetosphere is shaped by interactions between a planet's magnetic field and the solar wind. The magnetosphere shields the planet from the solar wind, preventing or impeding the direct entry of the solar-wind plasma into the magnetic cavity. See bow shock, magnetopause, magnetosheath, magnetotail, and Van Allen belts.

magnetotail The portion of a planet's magnetosphere formed on the planet's night side by the pulling action of the solar wind. An elongated extension of the planet's magnetic fields on the side of the planet opposite to the Sun. Magnetotails can extend hundreds of planetary radii.

magnitude A measure of the brightness of a star or other celestial object. On the magnitude scale, the lowest magnitudes refer to objects of greatest brightness. In the 2nd century B.C. the Greek astronomer Hipparchus of Nicea classified the stars into six magnitudes; the brightest stars are of the first magnitude while those of sixth magnitude are just visible to the unaided human eye. In the 19th century a scale of magnitudes was adopted in which a star of first magnitude is exactly 100 times as bright as one of sixth magnitude. The ratio of brightness between one magnitude and the next is thus the fifth root of 100, or 2.512. The apparent magnitude is the brightness of an object as seen from Earth. Absolute magnitude is the true brightness of an object, taking into account its distance from Earth; it is defined as the apparent magnitude an object would have at the arbitrary distance of ten parsecs. The brightness of a star as observed from the Earth, and hence its apparent magnitude, depends on both its absolute magnitude and its distance. The absolute visual magnitude of the Sun is +4.83 magnitudes and its apparent visual magnitude is −26.74 magnitudes. See absolute magnitude, apparent magnitude, and visual magnitude.

main sequence The region in the Hertzsprung–Russell diagram containing most of the stars in our Galaxy, including the Sun. Such stars are known as dwarf stars, but also sometimes called main-sequence stars. They constitute 90 percent of the stars in our Galaxy and 60 percent of its mass. The main sequence extends diagonally from the upper left to the lower right, or from the high-luminosity, high-temperature stars to the low-luminosity, low-temperature stars. Stars spend the majority of their lifetimes on the main sequence, during which they produce energy from the fusion of hydrogen into helium in their cores. The zero-age main sequence is where stars lie when they first start to burn hydrogen. The position of a star on the main sequence depends on its mass, the most-massive stars being the brightest. The more massive a star, the sooner it evolves off the main sequence. *See* dwarf star, and Hertzsprung–Russell diagram.

mass A measure of the quantity of material contained within an object. The total amount of matter in a body. *See* solar mass.

mass–luminosity relation The relation between the absolute luminosity, denoted by L, and the mass, specified by M, for stars on the main sequence of the Hertzsprung–Russell diagram, first derived by Arthur Eddington in 1924. It is expressed as $L = constant \times M^y$, where the exponent y lies between 3 and 4.

mass number Denoted by the capital letter A, the number of nucleons in an atomic nucleus, the sum of the number of protons and neutrons.

Maunder diagram *See* butterfly diagram.

Maunder Minimum The period roughly between 1645 and 1715 when few sunspots were observed. It is named after the British solar physicist Edward Walter Maunder (1851–1928). *See* Little Ice Age.

Maxwellian distribution Distribution of particle velocities for a gas in thermal equilibrium, characterized by a well-defined temperature, T, and root-mean-square velocity, $V_{rms} = (3kT/m)^{1/2}$ for a particle of mass m, where Boltzmann's constant $k = 1.380\,66 \times 10^{-23}$ J K^{-1}.

MDI Acronym for the Michelson Doppler Imager on *SOHO*.

M dwarf A main-sequence star of spectral type M. *See* M star.

megahertz A unit of frequency, abbreviated MHz. It is equal to one million hertz. *See* hertz.

megaparsec A unit of intergalactic distances. One megaparsec, or 1 Mpc, is equivalent to a million parsecs and 3.0857×10^{22} meters.

megaton The energy released in the explosion of one million tons (one billion kilograms) of TNT, equal to 4.2×10^{15} J, or 4.2×10^{22} erg.

MeV A unit of energy equal to one million electron volts and 1.6022×10^{-13} J. This unit is used to describe the total energy of high-velocity particles or energetic radiation photons. *See* cosmic rays, electron volt, and gamma-ray radiation.

MHD Abbreviation for magnetohydrodynamics. *See* magnetohydrodynamics.

MHz Abbreviation for megahertz, or a million hertz. *See* hertz, and megahertz.

microflares Also called nanoflares, these are ubiquitous small brightenings on the Sun, each lasting a few minutes, which can be observed at extreme-ultraviolet, radio, ultraviolet and X-ray wavelengths. Microflares are formed in the chromosphere or low corona, and they may contribute to coronal heating. *See* nanoflares.

microwaves Electromagnetic radiation with wavelengths between 0.001 and 0.06 m, or frequencies from 5 to 300 GHz, at the short-wavelength end of the radio spectrum.

Milankovitch cycles Three rhythmic fluctuations in the wobble and tilt of the Earth's rotational axis and the shape of the Earth's orbit that set the major ice ages in motion by altering the seasonal distribution of the Sun's light and heat on Earth. They are the 23 000-year precessional wobble of the Earth's rotational axis, the 41 000-year variation in the Earth's axial tilt, and the 100 000-year change in the eccentricity or shape of the Earth's orbit. These cycles are named after the Yugoslavian astronomer Milutin Milankovitch who developed the relevant mathematical theory between 1920 and 1941. *See* ice age.

Milky Way The band of light across our night sky, produced by the light of numerous stars as we look out through the disk of our Galaxy. The starlight is obscured in places by clouds of interstellar gas and dust.

million One thousand thousand, written as 1 000 000 or 10^6.

minute of arc A unit of angular measure. One minute of arc is equal to 1/60 degree and 60 seconds of arc. a minute of arc is denoted by the symbol ′.

mirror An optical component of an instrument which redirects an incoming beam of light.

missing mass *See* dark matter.

molecular cloud An unusually dense and large interstellar nebula of gas and dust containing hydrogen molecules with a combined mass of up to a million solar masses. Regions of star formation are usually located near or within molecular clouds. *See* protostar.

molecule A tightly knit group of two or more atoms, bound together by electromagnetic forces among the electrons and nuclei of the atoms.

Moreton wave A wave-like disturbance in the chromosphere initiated by a solar flare. A Moreton wave is detected as a circular hydrogen-alpha brightening that expands away from a flare and across the solar disk to distances of about 10^9 m at velocities of 10^6 m s^{-1}. It is named after Gail E. Moreton who first observed them in 1960.

Mount Wilson magnetic classifications
α – Alpha denotes a unipolar sunspot group.
β – Beta describes a bipolar sunspot group with a distinct division between the polarities.

$\beta\gamma$ – Beta-gamma describes a bipolar sunspot group without a distinct, continuous polarity division.
δ – Delta denotes a complex configuration of sunspots with opposite polarity.
γ – Gamma describes a complex active region, with both positive and negative magnetic polarities, which cannot be described as dipolar.

M star A star of spectral type M, characterized by absorption bands of titanium oxide molecules. M-type dwarf stars, also known as red dwarfs, define the lower end of the main sequence. M stars are smaller, less-massive and cooler than the Sun, and they are the most common class of star in our Galaxy. They have masses less than 0.5 solar mass, burn hydrogen in their core by the proton–proton chain, and have convective zones that can extend throughout the entire star for the lowest mass. *See* convective zone, dM star, flare star, proton–proton chain, red dwarf star, and spectral classification.

MSW effect The transformation of a neutrino of one type or "flavor" into a neutrino of another kind while traveling through matter, named after Lincoln Wolfenstein who originated the theory in the late-1980s and S. P. Mikheyev and Alexis Y. Smirnov who further developed it about a decade later. The MSW effect could explain the solar neutrino problem if some of the electron neutrinos produced in the core of the Sun oscillate into muon or tau neutrinos on their way out of the Sun, thereby becoming invisible to most neutrino detectors. *See* neutrino oscillation, and solar neutrino problem.

muon An elementary particle and a lepton, denoted by the symbol μ. A muon is similar to an electron except for its much greater mass, about 207 times that of an electron, and brief life. A muon decays into neutrinos and an electron in just 2.197×10^{-6} seconds. Muons may have a positive or a negative charge. Muon neutrinos interact with muons, and the anti-muon is the anti-particle of the muon.

nanoflares Ubiquitous low-level flares that might heat the corona, also called microflares. They may be detected at extreme ultraviolet, radio, ultraviolet and X-ray wavelengths. A billion of them would be required to release the same amount of energy as a normal flare. *See* microflares.

nanotesla A unit of magnetic field strength, abbreviated nT. It is equal to 10^{-9} T, to 10^{-5} G, and to 1 gamma.

NASA: Acronym for National Aeronautics and Space Administration, United States.

National Aeronautics and Space Administration Abbreviated NASA, the United States government agency responsible for civilian manned and robotic activities in space, including launch vehicles, scientific satellites, and space probes.

National Optical Astronomy Observatory Abbreviated NOAO, it was formed in 1982 to consolidate under one director several ground-based astronomical optical observatories: the Kitt Peak National Observatory, Cerro Tololo Inter-American Observatory, and the National Solar Observatory with facilities at Sacramento Peak, New Mexico and Kitt Peak, Arizona. NOAO has its headquarters in Tucson, Arizona, is funded by the U.S. National Science Foundation and operated by the Association of Universities for Research in Astronomy, abbreviated AURA.

National Radio Astronomy Observatory Abbreviated NRAO, with its headquarters at Green Bank, West Virginia, it operates the Very Large Array in New Mexico, and other major radio telescope facilities at Green Bank and Tucson, Arizona. NRAO is operated for the U.S. National Science Foundation by Associated Universities, Inc., under a cooperative agreement.

nebula A cloud of gas and dust seen in interstellar space. The remnants of exploding stars, called supernova remnants, are also known as nebulae. Galaxies were also once called spiral or elliptical nebulae, before their enormous sizes and distances were discovered. *See* Crab nebula, emission nebula, H II region, Orion nebula and planetary nebula.

network Chromosphere and photosphere features arranged in a cellular structure, associated with supergranular cells and magnetic fields. *See* calcium network, and super-granulation cells.

neutral line or region A line that separates longitudinal magnetic fields of opposite magnetic polarity. A region where the longitudinal magnetic field strength approaches zero. Generally, neutral regions occur between regions of opposite polarity. See magnetic inversion line.

neutrino A sub-atomic particle with no electric charge and very little mass. Neutrinos travel at nearly the velocity of light and interact very weakly with other matter. There are three types, or flavors of neutrinos, named the electron, muon and tau neutrinos after the particles they interact with. Vast quantities of electron neutrinos are created as the result of nuclear reactions in the Sun's core. Other types of neutrinos are produced in man-made particle accelerators, nuclear reactors and by cosmic rays entering the Earth's atmosphere. *See* beta decay, neutrino astronomy, proton–proton chain, and solar neutrino problem.

neutrino astronomy The attempt to detect neutrinos from cosmic sources, especially the Sun. Because they hardly interact with matter at all, neutrinos are very difficult to detect. Solar neutrinos have been observed with massive detectors placed deep underground. *See* chlorine experiment, GALLEX, HOMESTAKE, KAMIOKANDE, SAGE, SUPER KAMIOKANDE, and Sudbury Neutrino Observatory.

neutrino oscillation The change of one type or flavor of neutrino to another while traveling through matter or a vacuum. Neutrinos come in three flavors, the electron, muon and tau neutrinos. *See* MSW effect, neutrino, and solar neutrino problem.

neutron A sub-atomic particle with no electric charge found in all atomic nuclei except that of hydrogen. The neutron has slightly more mass than a proton. The mass of the neutron is 1.008 665 a.m.u. and 1.6749×10^{-27} kg. The neutron is 1839 times heavier than an electron. When set free from an atomic nucleus, or created outside one, a neutron decays into a proton, an electron and an anti-electron neutrino with a half-life of only 615 seconds or 10.25 minutes, and a mean life of just 887 seconds. *See* atomic mass unit, mass number, and nucleon.

neutron star A very dense star with a radius of about ten thousand meters, consisting of neutrons and produced in the final evolutionary stages of some stars more massive than the Sun. Rotating neutron stars are observed as radio pulsars, and are also thought to be components of X-ray binaries. *See* Crab nebula, and pulsar.

NOAA Acronym for the National Oceanic and Atmospheric Administration, United States.

NOAO Acronym for the National Optical Astronomy Observatory, United States. *See* National Optical Astronomy Observatory.

noble gases The rare and inert gases, or elements, helium, neon, argon, krypton, xenon, and radon, which are almost always mono-atomic and rarely undergo chemical reactions.

non-thermal particle A particle that is not part of a thermal gas. These particles cannot be described by a conventional temperature. *See* thermal gas.

non-thermal radiation The electromagnetic radiation produced by a non-thermal electron traveling at a speed close to that of light in the presence of a magnetic field. Such radiation is called synchrotron radiation after the man-made particle accelerator where it was first seen. More generally, the term non-thermal radiation denotes any electromagnetic radiation from an astronomical body that is not thermal in origin. *See* black-body radiation and thermal radiation.

northern lights A popular name for an aurora when observed from northern latitudes.

NRAO Acronym for the National Radio Astronomy Observatory, United States. *See* National Radio Astronomy Observatory.

NRL Acronym for the Naval Research Laboratory, United States.

NSF Acronym for the National Science Foundation, United States.

nuclear energy The energy obtained by nuclear reactions, the source of the Sun's luminosity.

nuclear fission A reaction involving an atomic nucleus in which the nucleus splits into two or more simpler and lighter nuclei. *See* fission.

nuclear force The force that binds protons and neutrons within atomic nuclei, and which is effective only at distances less than 10^{-15} m.

nuclear fusion The combination of two or more atomic nuclei into a more complex, heavier nucleus. *See* fusion.

nucleon A particle in or from the nucleus of an atom, either a proton or a neutron. *See* mass number.

nucleosynthesis The production of chemical elements from other chemical elements by naturally occurring nuclear reactions. Heavier elements are built up from lighter ones by fusion reactions; fission reactions break up heavy elements into lighter ones. Fusion reactions inside stars have created elements heavier than helium. Although helium is also synthesized from hydrogen inside stars, most of the helium in the Universe was created by nucleosynthesis during the first few minutes following the big bang.

nucleus The small, massive center of an atom, made up of protons and (except for hydrogen) neutrons, bound together by the nuclear force. The nucleus of an atom is positively charged, and contains nearly all of the atom's mass. An atom's electrons orbit the nucleus of the atom.

obliquity The angle between an object's axis of rotation and the pole of its orbit.

ohmic dissipation Conversion of an electrical current to heat because of the resistance of the medium in which it travels.

Oort cloud Spherical cloud of a million million, or 10^{12}, to ten million million, or 10^{13}, comets surrounding the planetary system and extending out about 100 000 AU, or 0.5 pc, from the Sun. The nearest star to the Sun is at a distance of 1.3 pc. *See* parsec.

opacity A measure of the ability of a gaseous atmosphere to absorb radiation and become opaque to it. A transparent gas has little or no opacity. The mean absorption of radiation that is used with a diffusion equation to calculate the photon luminosity flow in the Sun.

optical astronomy The study of visible light from objects in space.

optical depth A measure of the radiation absorbed as it passes through a medium. A transparent medium has an optical depth of zero. The medium is optically thin when the optical depth is less than unity, and optically thick when greater than unity.

optical radiation Electromagnetic radiation that is visible to the human eye, with wavelengths of approximately 385 to 700 nm. See optical spectrum, photosphere, and visible radiation.

optical spectrum Spectrum of a source that spans the visible wavelength range, approximately from 385 to 700 nm. See visible radiation.

Orion nebula A gaseous emission nebula visible to the naked eye as a diffuse glow marking Orion's sword. It is an H II region illuminated by the ultraviolet light of bright young stars in the Trapezium. The Orion nebula is located at a distance of 15 light-years or 500 pc. Star formation is still underway in the Orion nebula and in the dense molecular cloud in which it is embedded.

O star A star of spectral type O, characterized by absorption lines of ionized helium in its spectrum. O stars are the hottest, brightest and most-massive stars on the main sequence, with an effective temperature of up to 50 thousand degrees kelvin, an absolute luminosity of up to a million times that of the Sun, and a mass as large as 50 solar masses. See spectral classification.

ozone A form of molecular oxygen containing three atoms (O_3) instead of the normal two. It is created by the action of ultraviolet sunlight on the Earth's atmosphere.

ozone layer A region in the lower part of Earth's stratosphere (about 20 to 60 km above sea level) where the greatest concentration of ozone (O_3) appears. The ozone layer shields the Earth's surface from the Sun's energetic ultraviolet rays.

pair annihilation Mutual destruction of an electron and positron with the formation of radiation (511-keV gamma rays).

parallax An angular displacement of a nearby star with respect to distant ones, denoted by the symbol π. See annual parallax, solar parallax, and HIPPARCOS.

parent isotope Radioactive isotope from which a daughter isotope is produced by decay. See daughter isotope, and radioactivity.

parsec A unit of stellar distances. One parsec, or 1 pc, is equivalent to 3.0857×10^{16} m, 3.2616 light-years, and 206 265 AU. The parsec is the distance at which one astronomical unit subtends an angle of one second of arc, so the distance of a nearby star in parsecs is given by the reciprocal of its annual parallax in seconds of arc. The nearest star other than the Sun is Proxima Centauri with a distance of 1.3 pc. Distances comparable to the extent of our Galaxy are measured in units of kiloparsec, abbreviated kpc, where 1 kpc = one thousand parsecs = 1000 pc, and intergalactic distances are specified in units of megaparsec, abbreviated Mpc, with 1 Mpc = one million parsecs = 1 000 000 pc. See annual parallax.

PCA Acronym for Polar Cap Absorption. See polar cap absorption.

penumbra The lighter periphery of a sunspot seen in white light, surrounding the darker umbra. The penumbra consists of linear bright and dark elements extending radially from the sunspot umbra.

perihelion For a planet, comet or other object orbiting the Sun, the point in the orbit that is closest to the Sun. See aphelion.

period–luminosity relation A relationship between the period of light variation and the absolute luminosity of a variable star. The star's distance can be determined using this relation with its observed apparent luminosity or magnitude and its variation period. See Cepheid variable star.

photodissociation The breakdown of molecules due to the absorption of light, especially ultraviolet sunlight.

photoionization The ionization of an atom by the absorption of a photon of electromagnetic radiation. Ionization can take place only if the photon carries at least the energy corresponding to the ionization potential of the atom, that is, the minimum energy required to overcome the force binding the electron within the atom. See ionization potential.

photon A discrete unit or quantity of electromagnetic energy. A photon can be described as a particle, or quantum, of light. Photons have no electric charge and travel at the speed of light. See photon energy.

photon energy The energy of radiation of a particular frequency or wavelength. Short-wavelength, or high-frequency, photons have more energy than long-wavelength, or low-frequency, photons. The amount of photon energy is equal to the product of the frequency and Planck's constant $h = 6.6261 \times 10^{-34}$ J s.

photosphere That part of the Sun from which visible light originates. The intensely bright, visible portion of the Sun where most of the Sun's energy escapes into space. The lowest layer of the Sun's atmosphere viewed in white light. The region of any star which gives rise to its visible continuum radiation. Less than 3.5×10^5 m thick, the solar photosphere is a zone where the gaseous layers change from being completely opaque to radiation to being transparent. It is the layer from which the light we actually see is emitted and where most of the Sun's energy escapes to space. The effective temperature of the Sun's photosphere is 5780 K. The continuum radiation of the photosphere is absorbed at certain wavelengths by slightly cooler gas just above it, producing the dark Fraunhofer lines. Sunspots and faculae are observed in the photosphere, and magnetograms describe the magnetic field in the photosphere. The in and out heaving motions, or oscillations, of the photosphere are used to decipher the internal dynamics and structure of the Sun. See faculae, Fraunhofer lines, helioseismology, magnetogram, sunspots, and white light.

photosynthesis The chemical process whereby green plants use sunlight to manufacture carbohydrates from carbon dioxide and water and release oxygen as a by-product.

PI Abbreviation for Principal Investigator.

plages From the French word for "beaches", plages are bright, dense regions in the chromosphere found above sunspots or other active areas of the solar photosphere, and they always accompany and outlive sunspot groups. Plages appear much brighter in the monochromatic light of the hydrogen-alpha line or the calcium H and K lines than the surrounding parts of the chromosphere. Plages are associated with faculae, which occur just below them in the photosphere, and are found in regions of enhanced magnetic field. *See* calcium H and K lines, faculae, hydrogen-alpha line, and H and K lines.

planet A large body orbiting the Sun or another star, but not large enough to generate energy through nuclear fusion at its core. Some definitions demand that a planet should have an atmosphere, and/or a satellite, and/or be large enough to form itself into a sphere by self-gravity.

planetary nebula A nebula of roughly spherical shape, usually ejected from a star located at its center. Because of its round shape, it may resemble the disk of a planet as seen in a telescope, but there is no physical connection with planets. A planetary nebula can be formed when stellar winds are blown out from a red giant or supergiant star, leaving a hot core to excite the gas. The central stars of planetary nebulae are hot and blue, in transition between a giant star and a white dwarf.

planetesimal A small body formed in the early solar system by accretion of dust and ice (if present) in the central plane of the solar nebula. *See* solar nebula.

plasma An ionized gas, consisting of electrons that have been pulled free of atoms and ions, in which the temperature is too high for neutral, un-ionized atoms to exist. The high temperatures result in atoms losing their normal complement of electrons to leave positively-charged ions, or atomic nuclei, behind, and it is too hot for the free electrons and ions to join together and form permanent atoms. Plasma has been called the fourth state of matter, in addition to solid, liquid, and gas. Since the total negative electrical charge of the free electrons is equal to the total positive charge of the atomic nuclei, or ions, a plasma is electrically neutral over a sufficiently large volume. Most of the matter in the Universe is in the plasma state.

plasma beta The ratio of gas pressure to the pressure of the magnetic field within a plasma. The gas pressure increases with temperature, and the magnetic pressure is proportional to the square of the magnetic field strength. *See* gas pressure, and magnetic pressure.

plasmasphere A region of high-density, cold plasma, consisting of protons and electrons, that surrounds the Earth above the ionosphere, at altitudes greater than 1000 km (10^6 m) . The plasmasphere extends out to between three and seven Earth radii (2 to 4.5) $\times$ 10^7 m.

p-mode An acoustic mode of oscillation of the Sun in which pressure is the restoring force. *See* acoustic waves.

POLAR A NASA spacecraft launched on 24 February 1996 to measure plasma, energetic particle and fields in the Earth's high-latitude polar regions, and energy input through the Earth's dayside polar cusp. It also provides global, multi-spectral aurora images and determines the characteristics of the aurora plasma acceleration outflow. POLAR is a participating mission of the ISTP program.

polar cap absorption Abbreviated PCA, an anomalous condition of the polar ionosphere whereby high-frequency and very-high-frequency (3 to 300 MHz) radio waves are absorbed, and low-frequency to very-low-frequency (3 to 300 kHz) radio waves are reflected at lower heights than normal. The absorption is inferred from the proton flux at energies greater than 10 MeV, so PCAs and proton events are simultaneous.

polarity The directionality of a magnet or magnetic field, being north- or south-seeking. According to one convention, magnetic lines of force emerge from regions of positive north polarity and re-enter regions of negative south polarity. However, magnets point towards the Earth's magnetic North Pole.

polarization The process of affecting radiation so that the electromagnetic vibrations are not randomly oriented, but instead have a preferred direction. Synchrotron radiation is polarized, as are the Zeeman components.

Population I star A relatively young star, containing a high abundance of metals that tends to be located in the arms and disk of spiral galaxies. The Sun is a Population I star.

Population II star A relatively old star whose abundance of heavy elements is much less than that of the Sun and other Population I stars, consisting almost solely of hydrogen and helium. Such stars tend to be located in elliptical galaxies, the central regions of spiral galaxies and globular star clusters. *See* globular cluster.

pore Small, short-lived dark area in the photosphere out of which a sunspot may develop.

positron A positively-charged anti-particle of the electron. A sub-atomic particle with the mass of the electron but an equal positive electric charge. *See* beta decay.

post-flare loop An arcade of loops or a loop prominence system, often seen after a major two-ribbon flare, which bridges the ribbons.

potential energy The energy that an object possesses as a result of its position.

p–p chain Abbreviation for proton–proton chain. *See* proton–proton chain.

p–p reaction Abbreviation for proton–proton reaction. *See* proton–proton reaction.

precession The slow, periodic conical motion of the rotation of a spinning body, like a wobbling top or the rotating Earth. The precessional motion of the Earth is caused by the tidal action of the Moon and Sun on the spinning Earth. As a result, the Earth's axis of rotation sweeps out a cone in space, centered around the axis of the Earth's orbit, and completing one circuit in about 26 000 years. This cone has an angular radius, or opening angle, equal to the obliquity of the ecliptic, with a value of 23 degrees 26.36 minutes of arc.

primary cosmic rays The cosmic rays that arrive at Earth's upper atmosphere from outer space.

prime focus The point at which the objective lens or primary mirror of a telescope brings light to a focus in the absence of any other optical component, such as a secondary mirror.

prominence A region of cool (10^4 K), high-density gas embedded in the lower part of the hot (10^6 K), low-density solar corona. A prominence is apparently suspended above the photosphere by magnetic fields. A prominence is a filament viewed on the limb of the Sun in the light of the hydrogen-alpha line, or as bright protrusions at the limb seen during total eclipses or with a coronagraph. A quiescent prominence occurs away from active regions and can last for weeks or many months. It may extend upwards for 10^7 to 10^8 m. An active prominence is a short-lived, high-speed eruption associated with active regions, sunspots and flares. It can appear as a surge, spray or loop, and can reach heights of up to 7×10^8 m in just one hour. An eruptive prominence may arch 10^8 m outward before bursting apart; it is often associated with a coronal mass ejection. *See* coronal mass ejection, filaments, and hydrogen-alpha line.

proper motion The apparent motion of a star on the celestial sphere, as a result of its movement relative to the Sun. *See* HIPPARCOS.

proton A positively-charged, sub-atomic particle located in the nucleus of an atom or set free from it. The nucleus of a hydrogen atom is a proton. The atomic number, denoted by Z, specifies the number of protons in the nucleus of any atom. The proton has a mass of $1.672\ 623 \times 10^{-27}$ kg, and a proton is 1836 times more massive than an electron. *See* atomic number, and nucleon.

proton event By definition the measurement of at least 10^5 protons m^{-2} s^{-1} steradian^{-1} at energies greater than 10 MeV.

proton flare Any flare producing significant fluxes of energetic protons, with energies greater than 10 MeV, in the vicinity of the Earth.

proton–proton chain Abbreviated p–p chain, a series of thermonuclear reactions in which hydrogen nuclei, or protons, are transformed into helium nuclei. It is the main source of energy in the Sun and of all main-sequence stars with a mass comparable to, or less than, the Sun. In the first stage of the proton–proton chain, two protons combine to form a deuterium nucleus, releasing a positron, a neutrino and radiation. In the second stage the deuterium nucleus combines with a proton to form an isotope of helium, denoted ^{3}He, again releasing radiation. In the last stage, two ^{3}He nuclei combine to form the normal helium nucleus, denoted ^{4}He, releasing two protons and radiation. Overall, four protons are converted into one helium nucleus, with the mass difference released as energy.

proton–proton reaction Abbreviated p–p reaction, the first reaction in the proton–proton chain in which two protons combine to form a deuterium nucleus, releasing a positron, a neutrino and radiation. This reaction produces the largest number of neutrinos from the Sun, and these low-energy neutrinos are detected by gallium experiments such as GALLEX and SAGE. *See* GALLEX, proton–proton chain, and SAGE.

protostar An embryonic star in the process of formation, which is luminous owing to the release of gravitational potential energy from the infall of nebula material. A protostar has not yet begun to shine by nuclear fusion in its core. Protostars are formed when dense clumps of gas and dust collapse within massive molecular clouds, and they can be observed at infrared wavelengths. *See* molecular cloud.

pulsar An object emitting periodic pulses of radio radiation, believed to be a rotating neutron star formed as the aftermath of a supernova explosion. Pulsars were first noticed in 1967 by Jocelyn Bell, a graduate student working under the direction of Antony Hewish at the Mullard Radio Astronomy Observatory in Cambridge, England. They were studying rapid fluctuations of extra-galactic radio sources; this twinkling or scintillation is produced by the solar wind. *See* Crab nebula, and neutron star.

quantum mechanics The theory of atomic and sub-atomic systems based on the notion of quantized energy in radiation photons and in the angular momentum and energy levels of atoms, arising from the failure of classical concepts of waves or particles individually to describe such systems. *See* uncertainty principle.

quasar The most luminous objects in the Universe, emitting radiation over a wide range of wavelengths from X-ray to optical and radio wavelengths. The small angular extent of these objects at first suggested that they were stars, hence the name quasi-stellar object shortened to quasar, but they are actually very compact and distant extra-galactic objects with very large redshifts.

quiescent prominence A prominence occurring away from active regions that can last for weeks or months. *See* active region, and prominence.

quiet Sun The Sun when it is at the minimum level of activity in the 11-year solar cycle.

radar A form of radio observation in which a pulsed radio signal is emitted and the reflected signal is received and studied. Acronym for RAdio Detection And Ranging.

radial velocity The speed at which an object is moving either toward or away from an observer. The velocity of an object relative to an observer along the line of sight; it is measured by the Doppler effect. To determine an object's true velocity in space, it is necessary also to know the transverse velocity, which is across the line of sight. *See* Doppler effect, and Doppler shift.

radian A dimensionless unit of angular measures equal to $2.062\,648 \times 10^5$ seconds of arc. There are 2π radians in a full circle of 360 degrees, where $\pi = 3.141\,59$.

radiation A process that carries energy through space. *See* electromagnetic radiation.

radiation belt A ring-shaped region around a planet in which electrically-charged particles, electrons, protons, and other ions, are trapped, following spiral trajectories around the direction of the magnetic field of the planet. The main radiation belts surrounding the Earth are known as the Van Allen belts, containing electrons or protons from the Sun. Similar regions exist around other planets with magnetic fields, such as Jupiter. The Earth also has an inner radiation belt containing ions of material from interstellar space. *See* Van Allen belts, and South Atlantic Anomaly.

radiation pressure The pressure exerted by electromagnetic radiation. The pressure generated by photons during the process of radiation. Radiation pressure can compete with gas pressure in supporting giant stars, and it blows the dust tails of comets away from the Sun. *See* gas pressure.

radiative zone An interior layer of the Sun, lying between the energy-generating core and the convective zone, where energy travels outward by radiation. *See* convective zone.

radioactive decay *See* radioactivity.

radioactivity The spontaneous decay of certain rare, unstable, heavy nuclei into more stable lighter nuclei, with the release of energy. The process by which certain kinds of atomic nuclei naturally decompose or decay with the spontaneous emission of sub-atomic particles and gamma rays. *See* beta decay, daughter isotope, isotope, and parent isotope.

radio burst Radio emission from the Sun during a solar flare. *See* flare, radio radiation, solar flare, and type I, II, III and IV radio bursts.

radio galaxy A galaxy that emits intense radio radiation, often from two lobes apparently hurled out from a central elliptical galaxy. The biggest radio galaxies have lobes around 15 million light-years across, comparable in size to a typical cluster of galaxies. A radio galaxy's enormous energy output comes from the synchrotron radiation of energetic electrons spiraling about a magnetic field and traveling at nearly the speed of light. *See* synchrotron radiation.

radioheliograph A radio telescope designed for mapping the distribution of radio emission from the Sun.

radio interferometer Two or more radio telescopes connected electronically to produce an angular resolution comparable to the telescope separation, but a collecting area comparable to those of the component telescopes. *See* interferometry.

radionuclide A radioactive isotope of an element. An example is carbon-14, called radio carbon, that is the radioactive form of carbon-12.

radio radiation The part of the electromagnetic spectrum whose radiation has the longest wavelengths and smallest frequencies of all types, with wavelengths ranging from about 0.001 m to 30 m and frequencies ranging between 10 MHz and 300 GHz. It includes millimeter waves, from 0.001 to 0.010 m in wavelength, and microwaves with wavelengths from 0.01 to 0.06 m. Solar flares exhibit bursts of radio radiation designated by different types. *See* burst, microwaves, and type I, II, III, and IV radio burst.

radio telescope A large radio antenna designed to concentrate radio waves and permit detection of faint radio signals reaching us from the Sun, stars and distant galaxies. A large single dish or parabolic reflector sends the radio waves to a focus where they are converted into electronic signals. A larger radio telescope collects more energy, detecting fainter sources, and has greater angular resolution, permitting the observation of finer detail. Better resolution can be obtained by linking the electronic signals of two or more dishes to form a radio interferometer with an angular resolution comparable to a single antenna as big as the largest separation of the dishes. An example is the Very Large Array. *See* interferometry, radio interferometer, and Very Large Array.

RAL Acronym for the Rutherford Appleton Laboratory, United Kingdom.

recombination The capture of an electron by a positive ion. It is the opposite process to ionization.

reconnection *See* magnetic reconnection.

red-dwarf star A cool, low-mass star near the lower end of the main sequence, typically of spectral class M, with a mass less than 0.5 solar masses, a radius less than half the radius of the Sun, and an effective temperature less than 4000 K. The red-dwarf, or dwarf M stars are the most common class of star in our Galaxy. *See* dM star, M star, and spectral classification.

red-giant star A star with a vast, expanded low-density atmosphere, often a hundred times as large as the Sun, with a relatively cool temperature when compared with other giant stars. *See* giant star.

redshift The increase in wavelength of electromagnetic radiation caused by the Doppler effect, when the source of radiation is moving away from the observer along the line of sight. At optical wavelengths, the Doppler shift is toward longer, redder wavelengths. *See* Doppler effect, Doppler shift, and radial velocity.

reflecting telescope Also known as a reflector, a telescope that gathers radiation and forms an image by the reflection of light from a primary concave mirror, usually parabolic.

refracting telescope Also known as a refractor, a telescope that gathers radiation and forms an image by the refraction of light from a lens, called the objective.

relative sunspot number *See* sunspot number.

rem A measurement of exposure to radiation used in the United Stares. Internationally, the Sievert scale is used, where 1 rem = 0.01 sieverts. The average annual radiation exposure in the United States is 0.003 sieverts, an exposure of 5 sieverts would likely prove fatal.

resolution The degree to which the fine details in an image are separated and detected. Also known as angular resolution, the smallest size distinguishable by a telescope. The ability of a telescope to discern fine detail on a celestial object. The angular resolution, denoted by θ, in radians is given by the ratio of the wavelength, λ, to the diameter, D, of the primary mirror or reflector, with $\theta = \lambda/D$ radians. To convert to seconds of arc, 1 radian = $2.062\ 648 \times 10^5$ seconds of arc. *See* scintillation, and seeing.

respiration The process by which living things take in oxygen and use it to produce energy.

revolution The orbital motion of one object around another.

ring current Current carried by energetic particles that flow at radial distances beyond a few planetary radii in the near-equatorial regions of a planetary magnetosphere.

rotation The spin of an object about its own axis.

Russell–Vogt theorem The structure of a star is uniquely determined by the mass and the chemical composition of the star. Stars of the same composition, with therefore the same laws of opacity and energy generation, have radii, luminosities, effective temperatures, and mean densities that are determined solely by the stars' masses. This theorem was first derived by Heinrich Vogt in 1926, and included in the following year in a textbook written by Henry Norris Russell, with R. S. Dugan and J. W. Stewart, it apparently having been derived independently, and hence it is generally called the Russell–Vogt theorem.

SAGE Acronym for the Soviet–American Gallium Experiment begun in 1990, an underground neutrino detector in the northern Caucasus that uses 60 tons (6×10^4 kg) of gallium. It measures the low energy neutrinos for the proton–proton reactions in the Sun.

Sagittarius A A strong radio source at the center of our Galaxy.

satellite A natural or artificial body orbiting another of larger size.

Schwabe cycle Historical term for the 11-year sunspot cycle discovered by the amateur German astronomer Samuel Heinrich Schwabe in the early-1840s. *See* solar activity cycle, and sunspot cycle.

Schwarzschild radius The critical radius at which a very massive body under the influence of its own gravitation becomes a black hole. The Schwarzschild radius, $R_S = 2GM/c^2$, where G is the gravitational constant, M is the mass and c is the velocity of light.

SCE Acronym for the Solar Corona Experiment on *Ulysses*.

scintillation The twinkling of the stars caused by wind-blown clouds or other non-uniformity in our atmosphere, and the fluctuations of distant radio sources caused by the solar wind or winds in interstellar space. Scintillation in the Earth's atmosphere limits the angular resolution of even the best ground-based optical telescope to about one second of arc. *See* resolution, and seeing.

SCR Acronym for Solar Cosmic Ray. *See* solar cosmic ray.

SEC Acronym for the Sun-Earth Connection, a division of NASA.

second of arc A unit of angular measure. There are 60 seconds of arc in one minute of arc, and therefore 3600 seconds of arc in one degree. One second of arc is equal to 7.27×10^5 m on the visible disk of the Sun. A second of arc is denoted by the symbol ′′, and it is also called an arc second. There are $2.062\ 648 \times 10^5$ seconds of arc in one radian, and 2π radians in a full circle of 360 degrees, where the constant $\pi = 3.141\ 59$.

secondary cosmic rays Sub-atomic particles produced by collisions between primary cosmic rays and the molecules in the Earth's atmosphere.

sector boundary A place in the solar wind where the predominant direction of the interplanetary magnetic field changes direction, from toward the Sun to away from the Sun or *vice versa*. *See* solar sector.

seeing Fluctuations in a visible-light image due to refractive inhomogenieties in the Earth's atmosphere. The effect of random turbulent motion in the Earth's atmosphere on the quality of the image of an astronomical object at optical wavelengths. In conditions of good seeing, images are sharp and steady; in poor seeing, they are extended and blurred and appear to be in constant motion. Seeing usually limits the angular resolution of ground-based optical telescopes to about one second of arc. *See* scintillation.

SEL Acronym for the Space Environment Laboratory of the National Oceanic and Atmospheric Administration (NOAA), United States.

semi-major axis The average distance between an orbiting object and the object around which it orbits. It is equal to half the major axis of the elliptical orbit.

SEP Acronym for Solar Energetic Particle. *See* solar energetic particle.

shock wave A sudden discontinuous change in density and pressure propagating in a gas or plasma at supersonic velocity. There are associated changes in particle flow speed and in the magnetic and electric field strengths.

sidereal Measured or determined with reference to the stars.

sidereal period The orbital or rotation period of a planet, the Sun, or other celestial body with respect to the background stars. *See* synodic period.

Skylab An American space station, launched into Earth orbit on 14 May 1973. Three crews, each of three men, were sent to the station for periods of several weeks between 1973 and 1974. The station burnt up on re-entering the atmosphere in 1979.

SMM Acronym for the *Solar Maximum Mission*. *See Solar Maximum Mission*.

SNO Acronym for the Sudbury Neutrino Observatory, Canada. *See* Sudbury Neutrino Observatory.

SNR Acronym for SuperNova Remnant. *See* supernova remnant.

SNU Abbreviation for the Solar Neutrino Unit equal to 10^{-36} neutrino captures per target atom per second. *See* solar neutrino unit.

soft X-rays Electromagnetic radiation with photon energies of 1 to 10 keV and wavelengths between about 10^{-9} and 10^{-10} m. *See* SXT and *Yohkoh*.

SOHO Acronym for the *SOlar and Heliospheric Observatory*. *See Solar and Heliospheric Observatory*.

SOI/MDI Acronym for the Solar Oscillations Investigation / Michelson Doppler Imager instrument on *SOHO*.

Solar-A A mission of the Japanese Institue of Space and Astronautical Science (ISAS) launched on 30 August 1991 to study the Sun, particularly solar flares, at X-ray and gamma-ray wavelengths. *Solar-A* was renamed the *Yohkoh*, or "sunbeam" mission after its successful launch. *See Yohkoh*.

solar activity cycle A cyclical variation in solar activity with a period of about 11 years between maxima (or minima) of solar activity. It is characterized by waxing and waning of various forms of solar activity, such as sunspots, flares, and coronal mass ejections. Over the course of an 11-year cycle, sunspots vary both in number and latitude. The complete cycle of the solar magnetic field is 22 years. The solar cycle may be maintained by a dynamo driven by differential rotation and convection. Activity cycles similar to the solar cycle are apparently typical of stars with convective zones. *See* butterfly diagram, Hale's law, solar maximum, solar minimum, Spörer's law, and sunspot cycle.

SOlar and Heliospheric Observatory Abbreviated SOHO, a joint project of ESA and NASA, that was launched on 2 December 1995, and reached its permanent position on 14 February 1996. *SOHO* orbits the Sun at the Lagrangian point where the gravitational forces of the Earth and Sun are equal. It carries twelve instruments designed to investigate the solar atmosphere and how it is heated, solar oscillations, how the Sun expels material into space, the structure of the Sun and processes operating within it. *SOHO* is a participating mission of the ISTP program. *See* Lagrangian point.

solar atmosphere The outer layers of the Sun, from the photosphere through the chromosphere, transition region and corona. The term atmosphere is used to describe the outermost gaseous layers of the Sun because they are relatively transparent at visible wavelengths. *See* atmosphere.

Solar-B A mission of the Japanese Institute of Space and Astronautical Science (ISAS) that is tentatively scheduled for launch in 2004. It will include a coordinated set of optical, extreme-ultraviolet and X-ray telescopes that will investigate the interaction between the Sun's magnetic field and its corona.

solar burst *See* burst, and type I, II, III, and IV radio burst.

solar constant The total amount of solar energy, integrated over all wavelengths, received per unit time and unit area at the mean Sun–Earth distance outside the Earth's atmosphere. Its value is 1366.2 J s^{-1} m^{-2}, which is equivalent to 1366.2 W m^{-2}, with an uncertainty of $\pm$ 1.0 in the same units. *See* insolation, irradiance, and solar insolation.

solar core The region at the center of the Sun where nuclear reactions release vast quantities of energy.

solar corona *See* corona.

solar cosmic rays Abbreviated SCR, a historical name for energetic charged particles, mainly protons and electrons, accelerated to energies greater than 1 MeV by explosive processes on the Sun. We prefer to use the term solar energetic particles to avoid confusion with cosmic rays that enter the solar system from interstellar space. *See* solar energetic particle.

solar cycle The approximately 11-year variation in solar activity, as well as the number and position of sunspots. Taking Hale's law of sunspot magnetic polarity into account leads to a 22-year magnetic activity cycle. *See* Hale's law, solar activity cycle, and sunspot cycle.

solar dynamo *See* dynamo.

solar eclipse A blockage of light from the Sun when the Moon is positioned precisely between the Sun and the Earth. Since the Moon's orbital plane is inclined to the plane of the ecliptic, a solar eclipse can occur only when the Moon is at conjunction, at the phase of the new Moon, and at the same time at or near one of its nodes. A total solar eclipse is seen at places where the umbra of the Moon's shadow cone falls on and moves over the Earth's surface. Although the total duration of a solar eclipse can be as much as four hours, the Sun is completely covered by the Moon, at totality, for at most 7.5 minutes. During the brief moments of a total solar eclipse, darkness falls, and the outer parts of the Sun, the chromosphere and the corona, are seen. At any given point on Earth's surface, a total solar eclipse occurs, on the average, once every 360 years. *See* Baily's beads, chromosphere, corona, diamond effect, and prominence.

solar energetic particle Abbreviated SEP, charged particles, mainly protons and electrons, accelerated to energies greater than 1 MeV by explosive processes on the Sun and detected by spacecraft and on Earth.

solar flare A sudden and violent release of matter and energy within a solar active region in the form of electromagnetic radiation, energetic particles, wave motions and shock waves, lasting minutes to hours. A sudden brightening in an active region observed in chromospheric and coronal emissions which typically lasts tens of minutes. The frequency and intensity of solar flares increase near the maximum of the solar activity cycle. Solar flares accelerate charged particles into interplanetary space. The impulsive or flash phase of flares usually lasts for a few minutes, during which matter can reach temperatures of tens of millions of degrees kelvin. The flare subsequently fades during the gradual or decay phase lasting about an hour. Most of the radiation is emitted as X-rays but flares are also observed at visible hydrogen-alpha wavelengths and radio wavelengths. They are probably caused by the sudden release of large amounts (up to 10^{25} J) of magnetic energy in a relatively small volume in the solar corona. *See* active region, burst, flare, gradual flare, impulsive flare, radio burst, solar activity cycle, and type I, II, III, and IV radio burst.

solar insolation The amount of radiative energy received from the Sun per unit area per unit time at any given location on the Earth's surface. It is also known as the insolation, and has the units of watts per square meter (W m^{-2}). When the Sun is overhead, the solar insolation is greatest, amounting to about 1000 W m^{-2}, when compared to the amount received just outside the Earth's atmosphere, the solar constant of 1366.2 W m^{-2}. *See* insolation, and solar constant.

solar limb The apparent edge of the Sun as it is seen in the sky.

solar mass The amount of mass in the Sun, equal to 1.989×10^{30} kg.

solar maximum The peak of the sunspot cycle when the numbers of sunspots is greatest, and the output of particles and radiation is maximized. The highest level of solar activity, sometimes defined as the month(s) during the solar activity cycle when the 12-month mean of the monthly average of sunspot numbers reaches a maximum, such as July 1989.

Solar Maximum Mission Abbreviated SMM, a NASA satellite, launched on 14 February 1980 for studying the Sun during a period of maximum solar activity. It failed after nine months, but repairs were successfully done by a *Space Shuttle* crew in 1984. SMM re-entered the Earth's atmosphere in 1989.

solar minimum The beginning or end of a sunspot cycle, marked by the near absence of sunspots, and a relatively low output of radiation and energetic particles. The lowest level of solar activity, sometimes defined as the month(s) during the solar activity cycle when the 12-month mean of the monthly average of sunspot numbers reaches a minimum, such as September 1986 or November 1996.

solar nebula The disk-shaped cloud of gas and dust out of which the Sun and planetary system formed.

solar neutrino problem Massive, subterranean neutrino detectors find only one-third to one-half the number of solar neutrinos that theoretical calculations predict. There are two possible explanations for the solar neutrino problem. Either we do not understand exactly how nuclear reactions energize the Sun, or we do not fully understand neutrinos. The latter explanation is the most likely, perhaps because neutrinos are changing into an undetectable form on their way out of the Sun. *See* MSW effect, and neutrino oscillation.

solar neutrino unit Abbreviated SNU, a unit of solar neutrino capture rate by subterranean detectors, with 1 SNU = 10^{-36} solar neutrino captures per second per target atom. *See* SNU.

solar parallax The angular size of the radius of the Earth at a distance of one astronomical unit, amounting to 8.794 148 seconds of arc. *See* astronomical unit.

solar probe A planned NASA spacecraft, the first to fly through the atmosphere of the Sun, taking *in-situ* measurements down to 3 solar radii above the solar low corona where the radiation temperatures exceed 2×10^6 K. *Solar probe* will study the heating of the solar corona and the origin and acceleration of the solar wind.

solar sectors A region in the solar wind that has predominantly one magnetic polarity, pointed away from or toward the Sun. *See* sector boundary.

solar system The Sun, its planets and the smaller bodies orbiting the Sun, including asteroids and comets.

solar wind The expansion of the solar corona to form supersonic plasma streaming in all directions away from the Sun with speed of 3×10^5 to 1×10^6 m s^{-1}. A steady flow of energetic charged particles, mainly electrons and protons, and entrained magnetic field lines moving out from the Sun in all directions into interplanetary space at supersonic and super-Alfvénic speed. The solar wind has a fast component, or high-speed stream, with a velocity of about 8×10^5 m s^{-1}, and a slow-speed component moving at about half this speed. The solar wind is essentially the hot solar corona expanding into interplanetary space. The solar wind carries away about 10^{-13} of the Sun's mass per year. Although the solar wind is diverted around the Earth's magnetosphere, some of its particles enter the magnetosphere. *See* Alfvén velocity, aurora, heliosphere, high-speed stream, magnetosphere, and Van Allen belts.

sound speed *See* velocity of sound.

sound waves *See* acoustic waves.

South Atlantic anomaly A region over the South Atlantic Ocean where the lower Van Allen belt of energetic, electrically-charged particles is particularly close to the Earth's surface, presenting a hazard for artificial satellites.

space weather Changing conditions in interplanetary space and the Earth's magnetosphere controlled by the variable solar wind.

Spartan 201 An autonomous, detached *Space Shuttle* payload that observes the solar corona at far ultraviolet and visible wavelengths. The two instruments on *Spartan 201* were an UltraViolet Coronal Spectrometer (UVCS) and a White-Light Coronagraph (WLC). There is a similar UVCS instrument on *SOHO*.

spectral classification The sequence of stellar spectral types arranged according to the prominence or absence of certain lines in the spectra, designated as O, B, A, F, G, K and M. It is a temperature sequence, from the hot O and B stars to the cool M ones. *See* G star, M star, and O star.

spectral line A radiative feature observed in emission (bright) or absorption (dark) at a specific frequency or wavelength. A spectral line looks like a line in a display of radiation intensity as a function of wavelength or frequency. This spectral feature is produced by atoms or ions as they absorb or emit light, and it can be used to determine the chemical ingredients of the radiating source. Spectral lines are also used to infer the radial velocity and magnetic field of the radiating source. *See* absorption line, Doppler effect, emission line, Fraunhofer lines, H and K lines, hydrogen-alpha line, radial velocity, spectroheliograph, and Zeeman effect.

spectral type *See* spectral classification.

spectrogram A record of the spectrum of an object, produced by a spectrograph.

spectrograph Also known as a spectrometer. An instrument that separates light or other electromagnetic radiation into its component wavelengths, collectively known as a spectrum, and records the result electronically or photographically. Spectra are now often recorded with Charge-Coupled Devices, or CCDs, from which information in digital form can be analyzed by computer. High-dispersion instruments such as the spectro-heliograph separate the spectral lines widely so that a particular wavelength can be studied in detail.

spectroheliogram A monochromatic image of the Sun produced by means of a spectroheliograph or by the use of a narrow-band filter. *See* calcium H and K lines, H and K lines, hydrogen-alpha line, and spectroheliograph.

spectroheliograph A type of spectrograph used to image the Sun in the light of one particular wavelength only, thus permitting researchers to observe features normally lost in the total spectrum of radiation emitted. A second slit placed in front of the photographic plate or detector images a narrow strip of the Sun's disk at a chosen wavelength. By moving the entrance slit and the second slit in tandem, the whole solar disk can be scanned and a monochromatic image of all or part of the Sun is obtained. Such an image is called a spectroheliogram. The chromosphere is imaged when the strong lines of hydrogen alpha or calcium H and K are observed. *See* calcium H and K lines, H and K lines, hydrogen-alpha line, and spectroheliogram.

spectrohelioscope An instrument for viewing the Sun's radiative output in a narrow band of wavelengths by the eye. A spectroheliograph adapted for visual use. *See* spectroheliograph.

spectrometer *See* spectrograph.

spectroscopy The study of a spectrum, including the wavelength and intensity of emission and absorption lines, to determine the composition, motion or magnetic field of the radiating source. *See* spectrum.

spectrum The distribution of intensity of electromagnetic radiation with wavelength. Electromagnetic radiation, arranged in order of wavelength from long-wavelength, low-frequency, radio emissions to short-wavelength, high-frequency, gamma rays; also, a narrower band of wavelengths, called the visible spectrum, as when light dispersed by a prism or rainbow shows its component colors. A continuous spectrum is an unbroken distribution of radiation over a broad range of wavelengths, such as the separation of white light into its component colors from red to violet. Continuous spectra are often punctuated with emission or absorption lines, called line spectra, which can be examined to reveal the composition, motion and magnetic field of the radiating source. *See* absorption line, Doppler effect, Doppler shift, emission line, Fraunhofer lines, spectroheliograph, and Zeeman effect

speed of light *See* velocity of light.

speed of sound *See* velocity of sound.

spicules Narrow, predominantly radial, spike-like structures extending from the solar chromosphere into the corona, observed in hydrogen-alpha lines either at the limb or in the spectroheliograms taken especially in the red wing of the hydrogen-alpha line. They change rapidly, having a lifetime of five to fifteen minutes and velocities of about 2.5×10^4 m s^{-1}. Typically, spicules are 10^3 m thick and more than 10^6 m long. They are not distributed uniformly on the Sun but concentrated along the cell boundaries of the supergranulation pattern.

spiral galaxy A galaxy containing spiral arms coiling and winding out from a central bulge or nucleus, forming a flattened, disk-shaped region. The arms contain gas, dust, and young stars, while the nucleus contains old stars. Our Galaxy, the Milky Way, and the Andromeda galaxy are spiral galaxies. *See* Andromeda galaxy, Galaxy, and Milky Way.

Spörer Minimum A period of low sunspot activity in the 15th century (about A.D. 1420–1500), named after the German astronomer Gustav Friedrich Wilhelm Spörer who called attention to the Maunder minimum as early as 1887. *See* Little Ice Age, and Maunder Minimum.

Spörer's law The appearance of sunspots at solar lower latitudes over the course of the 11-year solar activity cycle, drifting from mid-latitudes towards the equator as the cycle progresses. It is named after the German astronomer Gustav Friedrich Wilhelm Spörer who first studied it in detail. *See* butterfly diagram, solar activity cycle, and sunspot cycle.

Standard Solar Model A theoretical model of the evolution and internal properties of the Sun, based on physical laws and constrained by an assumed initial composition and age of the Sun as well as its observed mass, radius and luminosity. A mathematical description of the solar interior that specifies the variation with radius of density, temperature, luminosity and pressure. Theoretical neutrino fluxes are also calculated from the model. The Standard Solar Model is consistent with helioseismology measurements of the temperatures inside the Sun. Non-standard models incorporate unlikely features to reduce the predicted solar neutrino output. *See* helioseismology, and solar neutrino problem.

star A self-luminous ball of gas whose radiant energy is produced by nuclear reactions in the stellar core. Most stars are found on the main sequence of the Hertzsprung–Russell diagram; they shine by the conversion of hydrogen nuclei into helium nuclei by either the carbon-–nitrogen–oxygen cycle or the proton–proton chain. When the core hydrogen is depleted, a main-sequence star evolves into a giant star, obtaining its radiant energy by the nuclear burning of helium or heavier elements. The temperature and luminosity of a star are determined by its mass. The most-massive stars are about 120 solar masses, or one hundred times heavier than the Sun. The minimum stellar mass is about 0.08 solar masses, or about 85 times the mass of Jupiter; less-massive stars do not have enough mass and sufficient pressure and temperature to initiate nuclear reactions in their cores. *See* carbon–nitrogen–oxygen cycle, giant star, Hertzsprung–Russell diagram, main sequence, mass-luminosity relation, and proton-proton chain.

Stefan–Boltzmann constant The constant of proportionality, denoted by the symbol σ, relating the radiant flux per unit area from a black-body to the fourth power of its effective temperature. The constant $\sigma = 5.670\ 51 \times 10^{-8}$ J m^{-2} K^{-4} s^{-1}. *See* Stefan–Boltzmann law.

Stefan–Boltzmann law The absolute luminosity of the thermal emission from a black-body is given by the Stefan–Boltzmann law in which $L = 4\pi\sigma R^2 T_e^4$, where $\pi = 3.141\ 59$, the Stefan–Boltzmann constant, $\sigma = 5.670\ 51 \times 10^{-8}$ J m^{-2} K^{-4} s^{-1}, the radius is R and the effective temperature is T_e. *See* absolute luminosity, and luminosity.

stellar activity Emission from a star in excess of that expected from a purely radiative atmosphere, suggesting additional heating sources related to magnetic fields. *See* flare star.

stellar evolution *See* giant star, Hertzsprung–Russell diagram, main sequence and star.

stellar populations A classification of stars according to their age, composition, and locations in the Galaxy. Young Population I stars are found in the spiral arms; older Population II stars are predominantly located in the halo and central nucleus of our Galaxy. Population I stars contain more heavy elements than Population II stars. *See* Population I star and Population II star.

stellar wind The outward flow of charged particles, mostly protons and electrons, from a star. Young stars evolving towards the main sequence have powerful stellar winds, up to a thousand times stronger than the Sun's solar wind. Old stars evolving into red giants also have strong stellar winds. The velocity of stellar winds range from 3×10^5 to 5×10^6 m s^{-1}. *See* solar wind.

steradian A unit of solid angle equal to $32\ 400/\pi^2$ square radian and 32 828 square degrees.

STEREO Acronym for the *Solar TErrestrial RElations Observatory* that is tentatively scheduled for launch by NASA in 2004. STEREO will use two spacecraft, preceding and following the Earth in its orbit, to simultaneously observe coronal mass ejections, providing three-dimensional observations from their onset at the Sun to the Earth's orbit.

stratosphere The layer of a planet's atmosphere in which the temperature remains roughly constant with altitude. The region of the Earth's atmosphere immediately above the troposphere and below the ionosphere. The stratosphere lies between heights of about $(1.5$ to $50) \times 10^4$ m (15 to 50 km).

streamer *See* coronal streamer and helmet streamer.

subsonic Moving at a speed less than that of sound, or with a velocity that is slower than the velocity of sound. In air on Earth, the velocity of sound is about 340 m s^{-1}, but the velocity of sound depends on both the temperature and composition of the gas. *See* velocity of sound.

substorm A frequent disturbance of the magnetosphere that produces geomagnetic activity. *See* geomagnetic activity, and geomagnetic storm.

Sudbury Neutrino Observatory Abbreviated SNO, a massive underground neutrino detector in Sudbury, Ontario, Canada filled with heavy water.

sudden commencement Abbreviated SC, the beginning of a geomagnetic storm, signaled by an abrupt increase or decrease in the northward component of the Earth's magnetic field. *See* geomagnetic storm.

sudden ionospheric disturbance Change in the Earth's ionosphere as a result of a burst of X-rays or ultraviolet radiation during solar flares, producing extra ionization.

Sun The central star of the solar system, around which all the planets, asteroids and comets revolve in their orbits. The Sun is a dwarf star on the main sequence, of spectral type G2 with an effective temperature of 5780 K. It is a ball of hot gas, compacted at the center and becoming more tenuous further out. The Sun's energy source is the nuclear fusion of hydrogen into helium by the proton–proton chain, taking place in the center. Overlying this core is the radiative zone where the high-energy photons produced in the fusion reactions collide with electrons and ions to be re-radiated in the form of light and heat. Beyond the radiative zone is a convective zone in which currents of gas flow upwards to release energy at the photosphere before flowing downwards to be reheated. The visible disk, or photosphere, from which the light we

see comes, is less than 3.5×10^5 m thick. The layer immediately above the photosphere is the chromosphere. The tenuous, million-degree kelvin outermost layers, forming the solar corona, expand outward to form the solar wind. Magnetic fields, generated by dynamo action inside the Sun, pervade the solar atmosphere, accounting for sunspots and powerful explosions called coronal mass ejections and flares. *See* chromosphere, convective zone, corona, coronal mass ejection, flare, granulation, Hertzsprung-Russell diagram, main sequence, photosphere, proton–proton chain, solar activity cycle, solar wind, sunspot, and sunspot cycle.

sunspot A dark, temporary concentration of strong magnetic fields in the Sun's photosphere. A sunspot is cooler than its surroundings and therefore appears darker. A typical spot has a central umbra surrounded by a penumbra, although either feature can exist without the other. In the umbra, the effective temperature can be about 4000 K compared with 5780 K in the surrounding photosphere. Sunspots are associated with strong magnetic fields of 0.2 to 0.4 T, and vary in size from $(1$ to $50) \times 10^6$ m. They occasionally grow to about 2×10^8 m in size, becoming visible to the unaided eye. Their duration varies from a few hours to a few weeks, or months for the very biggest. The number and location of sunspots depend on the 11-year solar activity cycle. They usually occur in pairs or groups of opposite magnetic polarity that move in unison across the face of the Sun as it rotates. The leading (or preceding) spot is called the P spot; the following one is termed the F spot. *See* Evershed effect, penumbra, solar activity cycle, sunspot cycle, sunspot group classification, umbra, Wilson effect, and Zeeman effect.

sunspot belts The heliographic latitude zones where sunspots are found, moving from mid-latitudes to the solar equator during the 11-year sunspot cycle. *See* solar activity cycles and sunspot cycle.

sunspot cycle The recurring, 11-year rise and fall in the number and position of sunspots. The conventional onset for the start of a sunspot cycle is the time when the smoothed number of sunspots (the 12-month moving average) has decreased to its minimum value. At the commencement of a new cycle sunspots erupt around latitudes of 35 to 45 degrees north and south. Over the course of the cycle, subsequent spots emerge closer to the equator, continuing to appear in belts on each side of the equator and finishing at around 7 degrees north and south. This pattern can be demonstrated graphically as a butterfly diagram. See butterfly diagram, solar activity cycle, and Spörer's law.

sunspot group classification
A—A small single unipolar sunspot or very small group of spots without penumbra.
B – Bipolar sunspot group with no penumbra.

C – An elongated bipolar sunspot group having one sunspot with penumbra.

D – An elongated bipolar sunspot group with penumbra on both ends of the group.

E – An elongated sunspot group with penumbra on both ends and a longitudinal extent of the penumbra between 10 and 15 degrees.

F – An elongated bipolar sunspot group with penumbra on both ends and the longitudinal extent of the penumbra exceeding 15 degrees.

H – A unipolar sunspot group with penumbra.

sunspot number A daily index of sunspot activity, R, defined as $R = k (10g + s)$ where k is a factor based on the estimated efficiency of observer and telescope, g is the number of groups of sunspots, irrespective of the number of spots each contains, and s is the total number of individual spots in all the groups. The sunspot number has also been called the Wolf number and the Zurich relative sunspot number, after the pioneering sunspot records begun in 1849 by the Swiss astronomer Rudolf Wolf at the Zurich Federal Observatory.

supergranulation cells Large convective cells seen in the solar photosphere with dimensions of about 3.5×10^7 m that last about 200 hours and have circulatory velocity amplitudes of about 20 to 400 m s^{-1}. The supergranulation pattern covers the entire photosphere except in plages and sunspots. The dominant flow in the observed cells is horizontal and outward from cell center, but there is a weak upward flow at the cell center and downward flow at the cell boundaries. The cells are outlined by the chromospheric network – boundaries containing concentrations of magnetic fields. Spicules are found along the network bounding the supergranulation cells. *See* chromospheric network, and spicules.

supergranule A large convection cell on the Sun that is approximately 35 times as large as a granule, or about 35 million, 3.5×10^7, meters in diameter.

SUPER KAMIOKANDE A massive underground neutrino detector in Japan filled with pure water, replacing the KAMIOKANDE detector.

supernova The explosion of a massive star when its depletes its nuclear fuel. For a week or so, a supernova may outshine the combined light of all the other stars in its galaxy. The inward collapse and immense heat and pressure at the core result in a stellar explosion that ejects most of the mass of the star into interstellar space. Stars more massive than roughly six solar masses may evolve so fast that they become unstable and explode to form a supernova. A neutron star can be formed at the center of a supernova. *See* Crab nebula, neutron star, and supernova remnant.

supernova remnant Abbreviated SNR, the expanding shell of matter thrown off into space during the outburst of a supernova. These remnants are often strong radio sources, emitting synchrotron radiation; the ejected material also collides with the surrounding interstellar gas and heats it up to about 10^6 K, emitting intense X-rays. *See* Crab nebula, supernova, and synchrotron radiation.

supersonic Moving at a speed greater than that of sound, or with a velocity that exceeds the velocity of sound. In air on Earth, the velocity of sound is about 340 m s^{-1}, but the velocity of sound depends on both the temperature and composition of the gas. *See* velocity of sound.

surge Sudden high-velocity upwelling or jet from active regions, seen prominently in the light of the hydrogen-alpha line. The surge originates in the chromosphere and reaches coronal heights. *See* chromosphere and hydrogen-alpha line.

SWAN Acronym for the Solar Wind ANisotropies instrument on *SOHO*.

SWICS Acronym for the Solar Wind Ion Composition Spectrometer on *Ulysses*.

SWOOPS Acronym for the Solar Wind Observations Over the Poles of the Sun plasma experiment on *Ulysses*.

SXT Acronym for the Soft X-ray Telescope on *Yohkoh*.

synchronous orbit *See* geosynchronous orbit.

synchrotron radiation Electromagnetic radiation emitted by an electron traveling almost at the speed of light in the presence of a magnetic field. The name arises because it was first observed in man-made synchrotron particle accelerators. The acceleration of the electrons causes them to emit radiation that is strongly polarized and increases in intensity at longer wavelengths. The wavelength region in which the emission occurs depends on the energy of the electron – 1–MeV electrons radiate mostly in the radio region. *See* Crab nebula, radio galaxy, and supernova remnant.

synodic period The period of apparent rotation or orbital revolution as observed from the Earth.

telescope An instrument for collecting and magnifying electromagnetic radiation from a cosmic object. A refracting telescope uses a lens to collect light, while a reflecting telescope uses a mirror. Different telescopes operate at various electromagnetic wavelengths, including in the X-ray, ultraviolet, visible, infrared and radio regions of the spectrum. *See Hubble Space Telescope*, radio telescope, reflecting telescope, refracting telescope, *SOlar and Heliospheric Observatory*, *Ulysses*, Very Large Array, and *Yohkoh*.

temperature A measure of the heat of an object, namely of the average kinetic energy of the randomly moving particles in an object.

termination shock A discontinuity in the solar-wind flow in the outer heliosphere where the solar wind slows from supersonic to subsonic motion as it interacts with the interstellar plasma. It marks the outer edge of the solar system. *See* heliopause.

terrestrial planets The four rocky planets in the inner part of the solar system: Mercury, Venus, Earth and Mars.

tesla The SI (Systéme International) unit of magnetic flux density, named after Nikola Tesla, a naturalized American pioneer in the field of electrical power generation and distribution. The tesla is a measure of the strength of a magnetic field. The c.g.s. (centimeter–grams-second) unit of magnetic field strength, often used in astrophysics, is the gauss, where 1 tesla = 10 000 gauss = 10^4 gauss. *See* gamma, and gauss.

thermal bremsstrahlung Emission of radiation by energetic electrons in a hot gas moving in the field of a positive ion. *See* bremsstrahlung.

thermal diffusion Heat transport resulting from a temperature gradient in a solid body.

thermal energy Energy associated with the motions of the molecules, atoms, or ions.

thermal equilibrium That equilibrium which is attained by a system that can be characterized by the same, constant temperature at all points. In such an equilibrium, the velocity distribution is characterized by a single temperature. *See* Gaussian distribution, and Maxwellian distribution.

thermal gas A collection of particles that collide with each other and exchange energy frequently, giving a distribution of particle energies that can be characterized by a single temperature. *See* non-thermal particle, and thermal particle.

thermal particle A particle that is part of a thermal gas. *See* non-thermal particle.

thermal radiation Electromagnetic radiation emitted by electrons in a thermal gas, arising by virtue of an object's heat (i.e. temperature). Non-thermal radiation is emitted by energetic electrons that are not necessarily in thermodynamic equilibrium. *See* black-body radiation, and non-thermal radiation.

thermonuclear fusion The combination of atomic nuclei at high temperatures to form more-massive nuclei with the simultaneous release of energy. Thermonuclear fusion is the power source at the core of the Sun. *See* fusion.

thermosphere The atmospheric region where the temperature rises due to heating in the ionosphere.

TIMED Acronym for the *Thermosphere, Ionosphere, Mesosphere Energetics and Dynamics* mission, a planned NASA mission scheduled for launch in 2001 that will sense and image various layers in the Earth's atmosphere. The instruments on *TIMED* will include GUVI, or Global UltraViolet Imaging, SEE, or Solar Extreme-ultraviolet Experiment, TIDI, for TImed Doppler Interferometer, and SABER, for Sounding of the Atmosphere using Broadband Emission Radiometry.

Titius–Bode law A numerical scheme by which a sequence of numbers can be obtained that give the approximate distances of the planets from the Sun in astronomical units, first stated by Johann Elert Titius in 1766 and Johann Daniel Bode in 1772. *See* Bode's law.

torsional oscillations Zones of alternating fast and slow rotation appearing in the photosphere and below, moving from the poles to the equator.

transition region A thin region of the solar atmosphere, less than 10^5 m, between the chromosphere and corona characterized by a large rise of temperature from 10^4 to 10^6 K. The density decreases as the temperature increases in such a way to keep the gas pressure spatially constant in the transition region.

trigonometric parallax *See* annual parallax.

troposphere The lowest layer of the Earth's atmosphere, where most of the weather takes place, from the ground at zero height to about 1.5×10^4 m (15 km) above the surface. The temperature of the troposphere decreases from 290 K at the ground to 240 K at its top. Any region of a planetary atmosphere in which convection normally takes place. *See* convection.

tunnel effect A quantum mechanical effect that permits two colliding protons to overcome the electrical repulsion between them, enabling their nuclear fusion and the release of energy.

turbulence The chaotic mass motions associated with convection.

two-ribbon flare A flare that has developed as a pair of bright, hydrogen-alpha strands (ribbons) on both sides of the main inversion, or neutral, line of the photospheric magnetic field of the active region.

type I radio burst Short (seconds), narrow-band burst detected at meter wavelengths (frequencies 300 to 50 MHz) that may continue for hours, usually superposed upon a noise storm continuum.

type II radio burst Narrow-band emission that drifts slowly (tens of minutes) from high to low frequencies; it begins in the meter-length range (300 MHz in frequency) and sweeps toward decameter wavelengths (10 MHz in frequency). Type II bursts are thought to be caused by shock waves moving outwards at velocities of about 10^6 m s^{-1}, exciting the local plasma frequency in the corona. Type II bursts occur much less frequently than type III bursts and are sometimes accompanied by a large flare.

type III radio burst Narrow-band emission characterized by its brief duration of seconds and rapid drift from decimeter to decameter wavelengths (frequencies of 500 to 5 MHz). They are interpreted in terms of electrons accelerated to energies of 1 to 10 keV in solar active regions, then moving outward through the corona at speeds of 0.05 c to 0.2 c, where c denotes the velocity of light. At each successive height the electron stream excites plasma oscillations at the local plasma frequency.

type IV radio burst A smooth continuum of broad-band bursts primarily in the meter range of wavelengths, or at frequencies of 300 to 30 MHz. These bursts are associated with some major flare events, beginning 10 to 20 minutes after flare maximum, and can last for hours. Type IV radio bursts may be emitted by magnetically-trapped, high-energy electrons.

U burst A radio burst that has a u-shaped appearance in an intensity-frequency plot.

UBV system A system of three-color photometry at ultraviolet (U) wavelengths, peaking at 360 nm, blue (B) peaking at 440 nm, and yellow (V for visual) peaking at 550 nm. See color index.

ultraviolet radiation Electromagnetic radiation with a higher frequency and shorter wavelength than visible blue light. Ultraviolet radiation has wavelengths between about 10^{-8} and 3.5×10^{-7} m, with the extreme ultraviolet lying in the short-wavelength part of this range. Because the Earth's atmosphere absorbs most ultraviolet emission, thorough studies of the Sun's ultraviolet output must be conducted from space. See extreme ultraviolet, or EUV.

Ulysses A joint undertaking of ESA and NASA, *Ulysses* was launched by NASA's *Space Shuttle Discovery* on 6 October 1990 to study the interplanetary medium and the solar wind at different solar latitudes. It provided the first opportunity for measurements to be made over the poles of the Sun, using the gravity-assist technique to take it out of the plane of the solar system. After an encounter with Jupiter in February 1992, the spacecraft moved back towards the Sun to pass over the solar south pole in September 1994 and the north pole in July 1995.

umbra The dark inner core of a sunspot with a penumbra, or a sunspot lacking a penumbra, visible in white light. The magnetic field in an umbra is radial and typically has a strength of 0.2 to 0.4 T. The effective temperature of the sunspot umbra is about 4000 K. Also the inner part of the shadow cast by the Moon during a total solar eclipse. See penumbra.

uncertainty principle Proposed by Werner Heisenberg in 1927, the uncertainty principle states that the momentum and position of a quantum-mechanical particle cannot be simultaneously determined. See quantum mechanics.

unipolar region A large area with weak magnetic fields of a single polarity, often located to the poleward side of the sunspot belt. Such regions could be related to coronal holes and to the source of the high-speed component of the solar wind. See coronal holes.

URAP Acronym for the Unified Radio And Plasma experiment on *Ulysses*.

UV Abbreviation for ultraviolet radiation. See ultraviolet radiation.

UVCS Acronym for the UltraViolet Coronagraph Spectrometer on *SOHO*.

Van Allen belts Two ring-shaped regions of high-energy charged particles that girdle the Earth's equator within the Earth's magnetosphere. They were discovered by the United States' first successful artificial Earth satellite, *Explorer 1*, which was launched on 31 January 1958. The inner belt lies between 1.2 and 4.5 Earth radii (measured from the Earth's center) and the outer belt is located between 4.5 and 6.0 Earth radii. The inner Van Allen belt contains protons with energies greater than 10 MeV and electrons exceeding 0.5 MeV. The outer belt also contains protons and electrons, most of which have energies under 1.5 MeV. The outer belt contains mainly electrons from the solar wind, and the inner belt mainly protons from the solar wind. Another source of radiation-belt particles is neutrons produced when cosmic rays and energetic particles bombard Earth's atmosphere; some of these neutrons decay into protons and electrons that are trapped by the Earth's magnetic field. Within the inner belt is a radiation belt consisting of particles produced by interactions between the solar wind and heavier cosmic-ray particles. Because the Earth's magnetic field is offset from the planet's center by 5×10^5 m, the inner belt dips down towards the surface in the region of the South Atlantic Ocean, off the coast of Brazil. See radiation belt, and South Atlantic Anomaly.

vector magnetic field Magnetic field defined by both the magnitude and direction of the field in the photosphere.

velocity A quantity that measures the rate of movement and the direction of movement of an object.

velocity of light The fastest speed that anything can move, equals $2.997\ 924\ 58 \times 10^8$ m s^{-1}.

velocity of sound Denoted by s, the velocity of sound is proportional to the square root of the gas temperature, T, and inversely proportional to the square root of the mean molecular weight, μ, or $s \propto (T/\mu)^{1/2}$.

Very Large Array Abbreviated VLA, a radio interferometer near Socorro, New Mexico consisting of twenty-seven dishes of 25-m diameter, moveable along the arms of a giant Y with dish separations of up to 34 km. The VLA operates at wavelengths between 0.018 and 0.90 m. See National Radio Astronomy Observatory, and radio telescope.

VHM Acronym for the Vector Helium Magnetometer aboard *Ulysses*.

VIRGO Acronym for the Variability of solar IRradiance and Gravity Oscillations instrument on *SOHO*.

virial theorem For a bound gravitational system, the long-term average of the kinetic energy is one-half of the potential energy.

visible light The form of electromagnetic radiation that can be seen by human eyes. *See* visible radiation.

visible radiation Radiation at the narrow range of wavelengths in the electromagnetic spectrum perceptible to the human eye; namely, light. It extends roughly from violet wavelengths at 385 nm, or 3.85×10^{-7} m, to red wavelengths at 700 nm, or 7.0×10^{-7} m, and lies between the ultraviolet and infrared parts of the electromagnetic spectrum. *See* electromagnetic radiation, and optical radiation.

visual magnitude The absolute, M_v, or apparent, m_v, magnitude of a celestial object in the color region to which the human eye is most sensitive, at wavelengths near 560 nm or 5.6×10^{-7} m. The absolute visual magnitude of the Sun is $+4.83$ and its apparent visual magnitude is -26.74. *See* absolute magnitude, apparent magnitude, and magnitude.

VLA Acronym for the Very Large Array, United States. *See* Very Large Array.

Voyager 1 and 2 Two almost identical planetary probes launched by the United States in 1977.

wavelength The distance between successive crests or troughs of an electromagnetic or other wave. Wavelengths are inversely proportional to frequency. The longer the wavelength, the lower the frequency. The product of the wavelength and the frequency of electromagnetic radiation is equal to the velocity of light.

WBS Acronym for the Wide-Band Spectrometer on *Yohkoh*.

white-dwarf star A very small, hot old star, about the size of the Earth, but with a mass about that of the Sun and an initial effective temperature similar to or hotter than that of the Sun. A white-dwarf star is supported against gravity by degenerate electron pressure. It is the endpoint of the evolution for stars with a mass comparable to that of the Sun. A white-dwarf star cannot have a mass greater than 1.4 solar masses; more-massive stars collapse under their own weight to form a neutron star or black hole at the endpoints of stellar evolution. *See* black hole, Hertzsprung–Russell diagram, and neutron star.

white light The visible portion of sunlight that includes all of its colors. Sunlight integrated over the visible portion of the spectrum (385 to 700 nm) so that all colors are blended to appear white to the eye.

white-light flare An exceptionally intense and rare flare that becomes visible in white light.

Wilson effect The foreshortening of a sunspot umbra that appears displaced towards the Sun's center when the sunspot is near the Sun's limb, accompanied by a widening of the penumbra on the side nearest the limb and a narrowing on the side farthest from the limb. The phenomenon was discovered by the Scottish astronomer Alexander Wilson (1714–86), who attributed it to a depression in the umbral level. However, the cause is now thought to be that sunspots are more transparent than the surrounding photosphere, and the umbra more transparent than the penumbra.

WIND A NASA spacecraft launched on 1 November 1994 to investigate basic plasma processes occurring in the near-Earth solar wind. It provides plasma, energetic particle, and magnetic field information about the solar-wind input to the magnetosphere and ionosphere, as well as data on the magnetospheric output to interplanetary space in the upstream region. *WIND* is a participating mission of the ISTP program.

Wolf number A historic procedure for computing the number of sunspots, originated by Rudolf Wolf of Zurich in 1849. *See* sunspot number.

X The mass fraction of hydrogen. Observations of sunlight and meteorites indicate that $X = 0.705\,83 \pm 0.025$ for the solar material outside its core. Owing to nuclear reactions, there is now less hydrogen in the Sun's core.

X-ray bright point A small X-ray-emitting region in the corona associated with a bipolar magnetic region.

X-ray flare class Rank of a solar flare based on its X-ray energy output. The classification is based on the peak burst intensity, I, at wavelengths of $(1 \text{ to } 8) \times 10^{-10}$ meters, or 0.1 to 0.8 nm, measured just outside the Earth's atmosphere in units of watts per square meter $(W\,m^{-2})$.

B – X-ray intensity is less than $10^{-6}\,W\,m^{-2}$.
C – X-ray intensity is between 10^{-6} and $10^{-5}\,W\,m^{-2}$.
M – X-ray intensity is between 10^{-5} and $10^{-4}\,W\,m^{-2}$.
X – X-ray intensity is greater than $10^{-4}\,W\,m^{-2}$.

X-ray radiation The part of the electromagnetic spectrum whose radiation has somewhat greater frequencies and smaller wavelengths than those of ultraviolet radiation. X-ray radiation covers the wavelength range from about 10^{-8} to 10^{-11} m and the energy range between 0.1 and 100 keV. Soft X-rays have lower energy, between 1 and 10 keV, and hard X-rays have higher energies ranging from 10 to 100 keV. Because X-rays are absorbed by the Earth's atmosphere, X-ray astronomy is performed in space. *See* soft X-rays, and hard X-rays.

Y The mass fraction of helium. Observations of sunlight and meteorites indicate that $Y = 0.2743 \pm 0.026$ for solar material outside the core. Owing to nuclear reactions, there is now more helium in the Sun's core.

Yohkoh A satellite launched by the Japanese Institute of Space and Astronautical Science (ISAS) on 30 August 1991 to study the Sun, particularly solar flares, at X-ray and gamma-ray wavelengths. The primary scientific objective of the Yohkoh, or "sunbeam" mission is the study of high-energy solar phenomena, but its Soft X-ray Telescope (SXT) is sufficiently sensitive for detailed observations of the ever-changing solar corona. Yohkoh is a Japanese–British–American collaboration. *See* hard X-rays, HXT, soft X-rays, *Solar-A*, and SXT.

Z The mass fraction of elements heavier than hydrogen and helium. Observations of sunlight and meteorites indicate that Z = 0.018 86 ± 0.0085 for solar material. A capital Z also denotes the atomic number. *See* atomic number.

Zeeman components The linearly polarized (π) and circularly polarized (σ) components which comprise a line split in the presence of a strong magnetic field. *See* Zeeman effect, and Zeeman splitting.

Zeeman effect A splitting of a spectral line into components by a strong magnetic field. If the components cannot be resolved, there is an apparent broadening or widening of the spectral line. The amount of splitting measures the strength of the magnetic field, and the direction of the magnetic field can be inferred from the polarization of the components. The Zeeman effect is thereby used to determine the direction, distribution and strength of the longitudinal magnetic fields in the photosphere. It has also been used to measure the magnetic fields in other stars and in the interstellar medium. In the simplest case, called normal Zeeman splitting, a line splits into three components. One component (the π component) is not displaced in wavelength or frequency, and is linearly polarized. Two components (the σ components) are shifted by equal amounts to lower and higher wavelengths or frequencies, the magnitude of the shift being proportional to the magnetic field strength. In the general case, the σ components are both circularly and linearly polarized. The phenomenon was predicted by Hendrik Lorentz, and observed in a terrestrial laboratory in 1896 by Pieter Zeeman; the two Dutch physicists shared the 1902 Nobel Prize for Physics. *See* magnetograph, Zeeman components, and Zeeman splitting.

zodiac A belt on the celestial sphere, about 8 degrees on each side of the ecliptic, which forms the background for the apparent motions of the Sun, Moon and planets. *See* ecliptic.

zodiacal cloud Cloud of interplanetary dust in the solar system, lying close to the ecliptic plane. The dust in the zodiacal cloud comes from both comets and asteroids.

zodiacal light A faint conical glow in the night sky caused by sunlight scattering off interplanetary dust near the plane of the ecliptic. A luminous pyramid of light that appears brightest and widest in the direction of the Sun, stretching along the ecliptic or zodiac from the western horizon after evening twilight or from the eastern horizon before morning twilight. It is visible at all seasons in the tropics in the absence of moonlight. The zodiacal light is an extension of the F (dust) component of the solar corona, and results from scattering of sunlight by dust particles in the plane of the ecliptic. The zodiacal dust cloud probably originates from both matter ejected by the Sun and from the decay of comets and asteroids.

Index